Student Solutions Manual

Nicholas Drapela
Oregon State University

to accompany

CHEMISTRY

Matter and Its Changes

Fourth Edition

James E. Brady
St. John's University

Frederick Senese
Frostburg State University

Special Thanks to Michael J. Kenney, of Crabtree and Company, and Cynthia K. Anderson, of Robert E. Lee High School, for their contributions to previous editions of this manual.

JOHN WILEY & SONS, INC.

Cover image: Boron Nitride image courtesy of Zettl Research Group, University of California at Berkeley, and Lawrence Berkeley National Laboratory.

To order books or for customer service, please call 1-800-CALL-WILEY (225-5945).

ISBN 0-471-21518-X

Printed in the United States of America

10 9 8 7 6 5 4 3 2 1

Printed and bound by Courier Kendallville, Inc.

Table of Contents

Chapter 1 1
Chapter 2 5
Chapter 3 9
Chapter 4 19
Chapter 5 41
Chapter 6 61
Chapter 7 77
Chapter 8 87
Chapter 9 95
Chapter 10 107
Chapter 11 117
Chapter 12 131
Chapter 13 139
Chapter 14 145
Chapter 15 157
Chapter 16 171
Chapter 17 187
Chapter 18 197
Chapter 19 239
Chapter 20 267
Chapter 21 285
Chapter 22 303
Chapter 23 309
Chapter 24 313
Chapter 25 315

Chapter 1

Practice Exercises

1.1 a) Fe_2O_3 contains iron (Fe), and oxygen (O)
b) Na_3PO_4 contains sodium (Na), phosphorus (P), and oxygen (O)
c) Al_2O_3 contains aluminum (Al), and oxygen (O)
d) $CaCO_3$ contains calcium (Ca), carbon (C), and oxygen (O)

1.2 The first sample has a ratio of

$$\frac{1.25\,\text{g Cd}}{0.357\,\text{g S}}$$

Therefore, the second sample must have the same ratio of Cd to S:

$$\frac{1.25\,\text{g Cd}}{0.357\,\text{g S}} = \frac{x}{3.50\,\text{g S}}$$

Cross-multiplication gives,

$$(1.25\text{ g Cd})(3.50\text{ g S}) = \text{x}(0.357\text{ g S})$$

$$\frac{(1.25\,\text{g Cd})(3.50\,\text{g S})}{0.357\,\text{g S}} = \text{x}$$

$$\text{x} = 12.3\text{ g Cd}$$

1.3 $2.24845 \times 12\text{ u} = 26.9814\text{ u}$

1.4 Copper is $63.546\text{ u} \div 12\text{ u} = 5.2955$ times as heavy as carbon

1.5 $(0.198 \times 10.0129\text{ u}) + (0.802 \times 11.0093\text{ u}) = 10.8\text{ u}$

1.6 $^{240}_{94}\text{Pu}$

The bottom number is the atomic number, found on the periodic table (number of protons). The top number is the mass number (sum of the number of protons and the number of neutrons).

1.7 $^{35}_{17}\text{Cl}$ contains 17 protons, 17 electrons, and 18 neutrons.

The bottom number is the atomic number (17), found on the periodic table (number of protons). In a neutral atom, the number of electrons equals the number of protons, giving 17 electrons. The top number is the mass number (sum of the number of protons and the number of neutrons), so subtracting the protons gives $35 - 17 = 18$ neutrons.

1.8 a) K, Ar, Al d) Ne
b) Cl e) Li
c) Ba f) Ce

Review Problems

1.66 Compound (c). An authentic sample of laughing gas must have a mass ratio of nitrogen/oxygen of 1.74 to 1.00. The only possibility in this list is item (c), which has the ratio of mass of nitrogen to mass of oxygen of $8.84/5.05 = 1.75$.

1.68 29.3 g nitrogen. From the first ratio we see that there is a ratio of 4.67 g N to 1.00 g H. Multiplying the mass of hydrogen by 4.67 we see that for every 6.28 g hydrogen there will be 29.3 g nitrogen.

$$\frac{4.67\text{ g N}}{1.00\text{ g H}} = \frac{x}{6.28\text{ g H}}$$

$$(4.67\text{ g N})(6.28\text{ g H}) = x(1.00\text{ g H})$$

$$\frac{(4.67\text{ g N})(6.28\text{ g H})}{1.00\text{ g H}} = x$$

$$x = 29.3\text{ g N}$$

1.70 5.54 g ammonia. Using the same method from above we see that for every 4.56 g nitrogen there need be 0.98 g hydrogen. According to the Law of Conservation of Mass then there will be 5.54 g ammonia produced.

1.72 2.286 g oxygen. If there are twice as many oxygen atoms per nitrogen atom, there should be twice the mass of oxygen per mass of nitrogen (2×1.143 g).

1.74 $1.9926482 \times 10^{-23}$ g. The mass of a carbon-12 atom is exactly 12 atomic mass units. If one atomic mass unit has a mass of $1.6605402 \times 10^{-24}$ g, then 12 u = $1.9926482 \times 10^{-23}$ g.

1.76 1.0079 u. From the problem we find the following ratios:

$$\frac{4\text{ atoms H}}{1\text{ atom C}} \text{ and } \frac{0.33597\text{ g H}}{1.0000\text{ g C}}$$

Therefore, 0.33597 represents 4 times as many hydrogen atoms as carbon atoms. To find the relative mass of hydrogen to carbon we can divide this number by 4 to get 0.083993. Since one atom of carbon-12 has a mass of exactly 12 u, the atomic mass of hydrogen must be $12 \times 0.083993 = 1.0079$ u.

1.78 2.0158 u. Regardless of the definition, the ratio of the mass of hydrogen to that of oxygen would be the same. If C-12 were assigned a mass of 24 (twice its accepted value), then hydrogen would also have a mass twice its current value.

1.80 $(0.6916 \times 62.9396 \text{ u}) + (0.3083 \times 64.9278 \text{ u}) = 63.55 \text{ u}$

1.82 a) 138 neutrons, 88 protons, 88 electrons
b) 8 neutrons, 6 protons, 6 electrons
c) 124 neutrons, 82 protons, 82 electrons
d) 12 neutrons, 11 protons, 11 electrons

Additional Exercises

1.84 a) A metal. The element is manganese (Mn), because manganese is defined by having 25 protons in its nucleus (from the periodic table). Manganese is in the region of the periodic table which contains metals.

b) ${}^{55}_{25}Mn$. The average mass for manganese (from the periodic table) is 54.938 u. This is closest to the whole number 25.

c) 30 neutrons. The neutrons may be found by subtracting the protons (25) from the mass number (55).

d) 25 electrons. The number of electrons equals the number of protons in a neutral atom.

e) 4.5782 times heavier than ${}^{12}C$. $(54.938 \div 12.000)$.

1.86 55.85 u
$(0.0580 \times 53.9396 \text{ u}) + (0.9172 \times 55.93949 \text{ u}) + (0.0220 \times 56.9354 \text{ u}) + (0.0028 \times 57.9333 \text{ u}) = 55.85 \text{ u}$

*1.88. Fe_3O_4. From the problem we find the following ratios:

$$\frac{2 \text{ atoms Fe}}{3 \text{ atoms O}} \quad \text{and} \quad \frac{2.325 \text{ g Fe}}{1.000 \text{ g O}}$$

Therefore, 2.325 g represents $\frac{2}{3}$ as many iron atoms as oxygen atoms. To find the relative mass of iron to oxygen we can divide this number by 2 to get 1.1625 for Fe and divide 1.000 by 3 to get 0.3333 for oxygen.

$$\frac{\text{mass of Fe atom}}{\text{mass of O atom}} = \frac{1.1625\text{ g Fe}}{0.33333\text{ g O}} = \frac{3.4878\text{ g Fe}}{1.0000\text{ g O}}$$

In the unknown compound, there are 2.616 g Fe for every 1 g O. This represents 2.616/3.4878 Fe atoms per O atom, or

$$\frac{2.616/3.4878\text{ atoms Fe}}{1\text{ atom O}} = \frac{0.75\text{ atoms Fe}}{1\text{ atom O}} = \frac{3\text{ atoms Fe}}{4\text{ atoms O}}$$

Chapter 2

Practice Exercises

2.1 a) 1 Ni, 2 Cl
b) 1 Fe, 1 S, 4 O
c) 3 Ca, 2 P, 8 O
d) 1 Co, 2 N, 12 O, 12 H

2.2 1 Mg, 2 O, 4 H, and 2 Cl (on each side)

2.3 $Mg(OH)_2\ (s) + 2\ HCl\ (aq) \rightarrow MgCl_2\ (aq) + 2\ H_2O\ (l)$

2.4 a) Colder temperatures mean lower molecular kinetic energies, so the kinetic energies of the molecules in the mixture are reduced.
b) Heat is absorbed into the mixture from the surroundings, as evidenced by a touch with the hand, so energy flows into the mixture and is stored as potential energy. The potential energy of the mixture increases.

2.5 $C_{10}H_{22}$
The formula for an alkane hydrocarbon is C_nH_{2n+2}. Here, n = 10, so 2n + 2 = 22.

2.6 a) C_3H_8O. Propane is C_3H_8, so removing one H and replacing it with OH gives C_3H_8O.

b) $C_4H_{10}O$. Butane is C_4H_{10}, so removing one H and replacing it with OH gives $C_4H_{10}O$.

2.7 a) phosphorus trichloride b) sulfur dioxide c) dichlorine heptaoxide

2.8 a) $AsCl_5$
b) SCl_6
c) S_2Cl_2

2.9 a) 8 protons, 8 electrons
b) 8 protons, 10 electrons
c) 13 protons, 10 electrons
d) 13 protons, 13 electrons

2.10 a) NaF b) Na_2O c) MgF_2 d) Al_4C_3

2.11 a) $CrCl_3$ and $CrCl_2$, Cr_2O_3 and CrO
b) CuCl, $CuCl_2$, Cu_2O and CuO

2.12 a) Na_2CO_3
b) $(NH_4)_2SO_4$
c) $KC_2H_3O_2$
d) $Sr(NO_3)_2$
e) $Fe(C_2H_3O_2)_3$

2.13 a) potassium sulfide b) magnesium phosphide
c) nickel(II) chloride d) iron(III) oxide

2.14 a) Al_2S_3 b) SrF_2 c) TiO_2 d) Au_2O_3

2.15 a) Lithium carbonate b) iron(III) hydroxide

2.16 a) $KClO_3$
b) $Ni_3(PO_4)_2$

2.17 diiodine pentaoxide

2.18 chromium(II) acetate

2.19 *X* is a nonmetal. Nonmetals combine with nonmetals to form molecular substances, which are characterized by low melting points such as this. Also, the inability of the substance to conduct electricity in the liquid state points to a molecular substance rather than an ionic substance, since molten ionic substances conduct electricity readily.

Review Problems

2.71 a) 2 K, 2 C, 4 O
b) 2 H, 1 S, 3 O
c) 12 C, 26 H
d) 4 H, 2 C, 2 O
e) 9 H, 2 N, 1 P, 4 O

2.73 a) Ni, 2 Cl, 8 O
b) 1 Cu, 1 C, 3 O
c) 2 K, 2 Cr, 7 O
d) 2 C, 4 H, 2 O
e) 2 N, 9 H, 1 P, 4 O

2.75 1 Cr, 6 C, 9 H, 6 O

2.77 a) 6 N, 3 O
b) 4 Na, 4 H, 4 C, 12 O
c) 2 Cu, 2 S, 18 O, 20 H

2.79 a) 6 atoms Na
b) 3 atoms C
c) 27 atoms O

2.81 a) K^+
b) Br^-
c) Mg^{2+}
d) S^{2-}
e) Al^{3+}

2.83 a) NaBr
b) KI
c) BaO
d) $MgBr_2$
e) BaF_2

2.85 a) KNO_3
b) $Ca(C_2H_3O_2)_2$
c) NH_4Cl
d) $Fe_2(CO_3)_3$
e) $Mg_3(PO_4)_2$

2.87 a) PbO and PbO_2
b) SnO and SnO_2
c) MnO and MnO_2
d) FeO and Fe_2O_3
e) Cu_2O and CuO

2.89 a) silicon dioxide
b) xenon tetrafluoride
c) tetraphosphorus decaoxide
d) dichlorine heptaoxide

2.91 a) calcium sulfide
b) aluminum bromide
c) sodium phosphide
d) barium arsenide
e) rubidium sulfide

2.93 a) iron(II) sulfide
b) copper(II) oxide
c) tin(IV) oxide
d) cobalt(II) chloride hexahydrate

2.95 a) sodium nitrite
b) potassium permanganate
c) magnesium sulfate heptahydrate
d) potassium thiocyanate

2.97 a) ionic, chromium(II) chloride
b) molecular, disulfur dichloride
c) ionic, ammonium acetate
d) molecule, sulfur trioxide
e) ionic, potassium iodate
f) molecular, tetraphosphorus decaoxide
g) ionic, calcium sulfite
h) ionic, silver cyanide
i) ionic, zinc bromide
j) molecular, hydrogen selenide

2.99 a) Na_2HPO_4
b) Li_2Se
c) $Cr(C_2H_3O_2)_3$
d) S_2F_{10}
e) $Ni(CN)_2$
f) Fe_2O_3
g) SbF_5

2.101 a) $(NH_4)_2S$
b) $Cr_2(SO_4)_3 \cdot 6H_2O$
c) SiF_4
d) MoS_2
e) $SnCl_4$
f) H_2Se
g) P_4S_7

2.103 Se_2S_6 is diselenium hexasulfide
Se_2S_4 is diselenium tetrasulfide

Additional Exercises

2.105 mercury(I) nitrate dihydrate = $Hg_2(NO_3)_2 \cdot 2H_2O$
mercury(II) nitrate monohydrate = $Hg(NO_3)_2 \cdot H_2O$

2.107 a) auric nitrate
b) cupric sulfate
c) plumbic oxide
d) mercurous chloride
e) mercuric chloride
f) cobaltous hydroxide
g) stannous chloride
h) stannic sulfide
i) auric sulfate

2.109 ferric nitrate nonahydrate

Chapter 3

Practice Exercises

3.1 a) $m \cdot v^2$ would have units of $kg \cdot (m/s)^2 = kg \cdot m^2/s^2$
b) mgh would have units of $kg \cdot (m/s^2) \cdot m = kg \cdot m^2/s^2$

3.2 a) μg
b) μm
c) ns

a) 1×10^{-9}
b) 1×10^{-2}
c) 1×10^{-12}

a) cg
b) Mm
c) μs

3.3
$$t_C = (t_F - 32\ ^\circ F)\left(\frac{5\ ^\circ C}{9\ ^\circ F}\right) = (50\ ^\circ F - 32\ ^\circ F)\left(\frac{5\ ^\circ C}{9\ ^\circ F}\right) = 10\ ^\circ C$$

To convert from °F to K we first convert to °C.

$$t_C = (t_F - 32\ ^\circ F)\left(\frac{5\ ^\circ C}{9\ ^\circ F}\right) = (68\ ^\circ F - 32\ ^\circ F)\left(\frac{5\ ^\circ C}{9\ ^\circ F}\right) = 20\ ^\circ C$$

$T_K = 273 + t_C = 273 + 20 = 293$ K

3.4 The data set from Worker C has the best precision.
The data set from Worker A has the best accuracy.

3.5 a) 42.0 g
b) 30.0 mL
c) 54.155 g
d) 11.3 g
e) 0.857 g/mL
f) 3.62 ft (1 and 12 are exact numbers)
g) 8.3 m^3

3.6 a) $$\#\ \text{in.} = (3.00\ \text{yd})\left(\frac{3\ \text{ft}}{1\ \text{yd}}\right)\left(\frac{12\ \text{in.}}{1\ \text{ft}}\right) = 108\ \text{in.}$$

b) $$\#\ \text{cm} = (1.25\ \text{km})\left(\frac{1000\ \text{m}}{1\ \text{km}}\right)\left(\frac{100\ \text{cm}}{1\ \text{m}}\right) = 1.25 \times 10^5\ \text{cm}$$

c) $\# \text{ft} = (3.27\ \text{mm})\left(\frac{1\ \text{m}}{1000\ \text{mm}}\right)\left(\frac{100\ \text{cm}}{1\ \text{m}}\right)\left(\frac{1\ \text{in.}}{2.54\ \text{cm}}\right)\left(\frac{1\ \text{ft}}{12\ \text{in.}}\right) = 0.0107\ \text{ft}$

d) $\frac{\#\ \text{km}}{\text{L}} = \left(\frac{20.2\ \text{mile}}{1\ \text{gal}}\right)\left(\frac{1.609\ \text{km}}{1\ \text{mile}}\right)\left(\frac{1\ \text{gal}}{3.785\ \text{L}}\right) = 8.59\ \text{km/L}$

3.7 density = mass/volume = $(1.24 \times 10^6\ \text{g})/(1.38 \times 10^6\ \text{cm}^3) = 0.899\ \text{g/cm}^3$

3.8 $\text{mass (g)} = (250{,}000\ \text{cm}^3)\left(\frac{19.82\ \text{g}}{1\ \text{cm}^3}\right) = 5.0 \times 10^6\ \text{g (or 5.0 Mg)}$

3.9 $\text{sp. gr.} = \frac{d_{\text{aluminum}}}{d_{\text{water}}} = \frac{2.70\ \text{g/mL}}{1.00\ \text{g/mL}} = 2.70$

$d_{\text{aluminum}} = \text{sp. gr.} \times d_{\text{water}} = 2.70 \times 62.4\ \text{lb/ft}^3 = 168\ \text{lb/ft}^3$

3.10 $d_{\text{ethyl acetate}} = \text{sp. gr.} \times d_{\text{water}} = 0.902 \times 1.00\ \text{g/mL} = 0.902\ \text{g/mL}$

$d_{\text{ethyl acetate}} = \text{sp. gr.} \times d_{\text{water}} = 0.902 \times 8.34\ \text{lb/gal} = 7.52\ \text{lb/gal}$

Review Problems

3.18 a) 0.01 m
b) 1000 m
c) 1×10^{12} pm
d) 0.1 m
e) 0.001 kg
f) 0.01 g

3.20 a) °F = 9/5 (°C) + 32 = (9/5)(50) + 32 = 122 °F
b) °F = 9/5 (°C) + 32 = (9/5)(10) + 32 = 50 °F

c) $t_C = (t_F - 32\ ^\circ\text{F})\left(\frac{5\ ^\circ\text{C}}{9\ ^\circ\text{F}}\right) = (25.5\ ^\circ\text{F} - 32\ ^\circ\text{F})\left(\frac{5\ ^\circ\text{C}}{9\ ^\circ\text{F}}\right) = -3.6\ ^\circ\text{C}$

d) $t_C = (t_F - 32\ ^\circ\text{F})\left(\frac{5\ ^\circ\text{C}}{9\ ^\circ\text{F}}\right) = (49\ ^\circ\text{F} - 32\ ^\circ\text{F})\left(\frac{5\ ^\circ\text{C}}{9\ ^\circ\text{F}}\right) = 9.4\ ^\circ\text{C}$

e) K = °C + 273 = 60 + 273 = 333K
f) K = °C + 273 = (−30) + 273 = 243K

3.22 $t_C = (t_F - 32\ ^\circ\text{F})\left(\frac{5\ ^\circ\text{C}}{9\ ^\circ\text{F}}\right) = (103.5\ ^\circ\text{F} - 32\ ^\circ\text{F})\left(\frac{5\ ^\circ\text{C}}{9\ ^\circ\text{F}}\right) = 39.7\ ^\circ\text{C}$

This dog has a fever; the temperature is out of normal canine range.

3.24 $t_C = (t_F - 32\ ^\circ F)\left(\frac{5\ ^\circ C}{9\ ^\circ F}\right) = (1500\ ^\circ F - 32\ ^\circ F)\left(\frac{5\ ^\circ C}{9\ ^\circ F}\right) = 830\ ^\circ C$

3.26 Range in Kelvin:

$$K = (10\ MK)\left(\frac{1\times10^6\ K}{1\ MK}\right) = 1.0\times10^7\ K$$

$$K = (25\ MK)\left(\frac{1\times10^6\ K}{1\ MK}\right) = 2.5\times10^7\ K$$

Range in degrees Celsius:

$°C = K - 273 = 1.0\times10^7 - 273 \approx 1.0\times10^7\ °C$

$°C = K - 273 = 2.5\times10^7 - 273 \approx 2.5\times10^7\ °C$

Range in degrees Fahrenheit:

$°F = 9/5\ (°C) + 32 = (9/5)(1.0\times10^7) + 32 \approx 1.8\times10^7\ °F$

$°F = 9/5\ (°C) + 32 = (9/5)(2.5\times10^7) + 32 \approx 4.5\times10^7\ °F$

3.28 $°C = K - 273 = 4 - 273 = -269\ °C$

3.30 $°C = K - 273 = 77 - 273 = -196\ °C$

$°F = 9/5\ (°C) + 32 = (9/5)(-196) + 32 = -321\ °F$

This temperature is *lower* than the boiling point of oxygen (–297 °F), so oxygen is a liquid at this temperature.

3.32 a) 4 significant figures
b) 5 significant figures
c) 4 significant figures
d) 2 significant figures
e) 4 significant figures
f) 2 significant figures

3.34 a) 5 significant figures
b) 5 significant figures
c) 2 significant figures
d) 5 significant figures
e) 4 significant figures
f) 1 significant figure

3.36 a) 0.72 m^2
b) 84.24 kg
c) – 0.465 g/cm^3 (dividing a number with 4 sig. figs by one with 3 sig. figs)
d) 19.42 g/mL
e) 857.7 cm^2

3.38 a) 4.34×10^3
b) 3.20×10^7
c) 3.29×10^{-3}
d) 4.20×10^4
e) 8.00×10^{-6}
f) 3.24×10^5

3.40 a) 310,000
b) 0.000,004,35
c) 3,900
d) 0.000,000,000,004,4
e) 0.000,000,356
f) 88,000,000

3.42 a) 4.0×10^7 (think of it as $4.0 \times 10^7 - 0.021 \times 10^7$)
b) 3.0×10^{-2}
c) 4.4×10^{12}
d) 2.5
e) 2.3×10^{18}

3.44 a) $$\#\,\text{km} = (32.0\text{ dm})\left(\frac{1\text{ m}}{10\text{ dm}}\right)\left(\frac{1\text{ km}}{1000\text{ m}}\right) = 3.20\times10^{-3}\text{ km}$$

b) $$\#\,\mu\text{g} = (8.2\text{ mg})\left(\frac{1\text{ g}}{1000\text{ mg}}\right)\left(\frac{1\times10^{6}\ \mu\text{g}}{1\text{ g}}\right) = 8.2\times10^{3}\ \mu\text{g}$$

c) $$\#\,\text{kg} = (75.3\text{ mg})\left(\frac{1\text{ g}}{1000\text{ mg}}\right)\left(\frac{1\text{ kg}}{1000\text{ g}}\right) = 7.53\times10^{-5}\text{ kg}$$

d) $$\#\,\text{L} = (137.5\text{ mL})\left(\frac{1\text{ L}}{1000\text{ mL}}\right) = 0.1375\text{ L}$$

e) $$\#\,\text{mL} = (0.025\text{ L})\left(\frac{1000\text{ mL}}{1\text{ L}}\right) = 25\text{ mL}$$

f) $$\#\,\text{dm} = (342\text{ pm})\left(\frac{1\text{x}10^{-12}\text{ m}}{1\text{ pm}}\right)\left(\frac{10\text{ dm}}{1\text{ m}}\right) = 3.42\times10^{-9}\text{ dm}$$

3.46 a) $$\#\,\text{cm} = (230\text{ km})\left(\frac{1\times10^{3}\text{m}}{1\text{ km}}\right)\left(\frac{1\text{ cm}}{1\times10^{-2}\text{ m}}\right) = 2.3\times10^{7}\text{ cm}$$

b) $$\#\,\text{mg} = (423\text{ kg})\left(\frac{1\times10^{3}\text{g}}{1\text{ kg}}\right)\left(\frac{1\text{ mg}}{1\times10^{-3}\text{ g}}\right) = 4.23\times10^{8}\text{ mg}$$

c) $$\#\,\text{Mg} = (423\text{ kg})\left(\frac{1\times10^{3}\text{g}}{1\text{ kg}}\right)\left(\frac{1\text{ Mg}}{1\times10^{6}\text{ g}}\right) = 4.23\times10^{-1}\text{ Mg}$$

d) $$\#\,\text{mL} = (430\ \mu\text{L})\left(\frac{1\times10^{-6}\text{L}}{1\ \mu\text{L}}\right)\left(\frac{1\text{ mL}}{1\times10^{-3}\text{ L}}\right) = 4.3\times10^{-1}\text{ mL}$$

e) $$\#\,\text{kg} = (27\text{ ng})\left(\frac{1\times10^{-9}\text{g}}{1\text{ ng}}\right)\left(\frac{1\text{ kg}}{1\times10^{3}\text{ g}}\right) = 2.7\times10^{-11}\text{ kg}$$

3.48 a) $$\#\,\text{cm} = (36\text{ in.})\left(\frac{2.54\text{ cm}}{1\text{ in.}}\right) = 91\text{ cm}$$

b) $$\#\,\text{kg} = (5.0\text{ lb})\left(\frac{1\text{ kg}}{2.205\text{ lb}}\right) = 2.3\text{ kg}$$

c) $$\#\,\text{mL} = (3.0\text{ qt})\left(\frac{946.4\text{ mL}}{1\text{ qt}}\right) = 2800\text{ mL}$$

d) $\#\,\text{mL} = (8\text{ oz})\left(\frac{29.6\text{ mL}}{1\text{ oz}}\right) = 200\text{ mL}$

e) $\#\,\text{km/hr} = (55\text{ mi/hr})\left(\frac{1.609\text{ km}}{1\text{ mi}}\right) = 88\text{ km/hr}$

f) $\#\,\text{km} = (50.0\text{ mi})\left(\frac{1.609\text{ km}}{1\text{ mi}}\right) = 80.4\text{ km}$

3.50 $\#\,\text{mL} = (12\text{ oz})\left(\frac{29.6\text{ mL}}{1\text{ oz}}\right) = 360\text{ mL}$

3.52 $\#\,\text{qts} = (4700\text{ mL})\left(\frac{1\text{ qt}}{946.35\text{ mL}}\right) = 5.0\text{ qts}$

(relationship 1 qt = 946.35 is found inside back cover of text)

3.54 $\#\,\text{lb} = (1000\text{ kg})\left(\frac{2.205\text{ lb}}{1\text{ kg}}\right) = 2205\text{ lb}$

3.56 $\#\,\text{mL} = (4.2\text{ qts})\left(\frac{946.35\text{ mL}}{1\text{ qt}}\right) = 4.0 \times 10^3\text{ mL}$ (stomach volume)

4.0×10^3 mL ÷ 0.9 mL = 4,000 pistachios (don't try this at home)

3.58 $\#\,\frac{\text{m}}{\text{s}} = \left(\frac{200\text{ mi}}{1\text{ hr}}\right)\left(\frac{5280\text{ ft}}{1\text{ mi}}\right)\left(\frac{30.48\text{ cm}}{1\text{ ft}}\right)\left(\frac{1\times10^{-2}\text{ m}}{1\text{ cm}}\right)\left(\frac{1\text{ hr}}{60\text{ min}}\right)\left(\frac{1\text{ min}}{60\text{ s}}\right) = 90\,\frac{\text{m}}{\text{s}}$

3.60 $\#\,\text{metric tons} = (2240\text{ lb})\left(\frac{1\text{ kg}}{2.205\text{ lb}}\right)\left(\frac{1\text{ metric ton}}{1000\text{ kg}}\right) = 1.02\text{ metric tons}$

3.62 First, 6'2" = 74" $\#\,\text{cm} = (74\text{ in.})\left(\frac{2.54\text{ cm}}{1\text{ in.}}\right) = 190\text{ cm}$

3.64 a) $\#\,\text{cm}^2 = (8.4\text{ ft}^2)\left(\frac{30.48\text{ cm}}{1\text{ ft}}\right)^2 = 7{,}800\text{ cm}^2$

b) $\#\,\text{km}^2 = (223\text{ mi}^2)\left(\frac{5280\text{ ft}}{1\text{ mi}}\right)^2\left(\frac{30.48\text{ cm}}{1\text{ ft}}\right)^2\left(\frac{1\text{m}}{100\text{ cm}}\right)^2\left(\frac{1\text{ km}}{1\times10^3\text{m}}\right)^2 = 580\text{ km}^2$

c) $\# \text{cm}^3 = (231 \text{ ft}^3)\left(\frac{30.48 \text{ cm}}{1 \text{ ft}}\right)^3 = 6.54\times10^6 \text{ cm}^3$

3.66 a) $\# \text{m}^2 = (1.0 \text{ in}^2)\left(\frac{2.54 \text{ cm}}{1 \text{ in}}\right)^2\left(\frac{1\times10^{-2} \text{ m}}{1 \text{ cm}}\right)^2 = 0.000{,}65 \text{ m}^2$

b) $\# \text{km}^2 = (3.7 \text{ mi}^2)\left(\frac{5280 \text{ ft}}{1 \text{ mi}}\right)^2\left(\frac{30.48 \text{ cm}}{1 \text{ ft}}\right)^2\left(\frac{1\text{m}}{100 \text{ cm}}\right)^2\left(\frac{1 \text{ km}}{1\times10^3 \text{m}}\right)^2 = 9.6 \text{ km}^2$

c) $\# \text{mL} = (144 \text{ in}^3)\left(\frac{2.54 \text{ cm}}{1 \text{ in}}\right)^3\left(\frac{1 \text{ mL}}{1 \text{ cm}^3}\right) = 2{,}360 \text{ mL}$

3.68 $\# \frac{\text{mi}}{\text{hr}} = \left(\frac{3.5 \text{ m}}{1 \text{ s}}\right)\left(\frac{1 \text{ cm}}{1\times10^{-2} \text{m}}\right)\left(\frac{1 \text{ ft}}{30.48 \text{ cm}}\right)\left(\frac{1 \text{ mi}}{5280 \text{ ft}}\right)\left(\frac{60 \text{ s}}{1 \text{ min}}\right)\left(\frac{60 \text{ min}}{1 \text{ hr}}\right) = 7.8 \text{ mi/hr}$

3.70 density = mass/ volume = 36.4 g/45.6 mL = 0.798 g/mL

3.72 $\text{radius} = \frac{\text{diameter}}{2} = \left(\frac{1\times10^{-15} \text{ m}}{2}\right)\left(\frac{100 \text{ cm}}{1 \text{ m}}\right) = 5\times10^{-14} \text{ cm}$

$\text{volume of sphere} = \frac{4}{3}\pi r^3 = \frac{4}{3}(3.1415)(5\times10^{-14})^3 = 5.2\times10^{-40} \text{ cm}^3$

density = mass/volume = 1.66×10^{-24} g/5.2×10^{-40} cm^3 = 3×10^{15} g/m^3

3.74 $\# \text{mL} = 25.0\text{g}\left(\frac{1 \text{ mL}}{0.791 \text{ g}}\right) = 31.6 \text{ mL}$

3.76 $\# \text{g} = 185 \text{ mL}\left(\frac{1.492 \text{ g}}{1 \text{ mL}}\right) = 276 \text{ g}$

3.78 mass of silver = 62.00g – 27.35g = 34.65 g
volume of silver = 18.3 mL –15 mL = 3.3 mL
density of silver = (mass of siver)/(volume of silver) = (34.65 g)/(3.3 mL)
= 11 g/mL

3.80 $\text{sp.gr.} = \left(\frac{d_{\text{substance}}}{d_{\text{water}}}\right) = \left(\frac{0.715 \text{ g/mL}}{1.00 \text{ g/mL}}\right) = 0.715$

3.82 $d_{\text{substance}} = (\text{sp. gr.})_{\text{substance}} \times d_{\text{water}} = 1.47 \times 0.998\text{g/mL} = 1.47 \text{ g/mL}$

$$\text{mass} = 1000\text{ mL}\left(\frac{1.47\text{g}}{1\text{ mL}}\right) = 1470\text{ g}$$

3.84 The density of gold is 1.20×10^3 lb/ft^3

$$\#\text{ lb} = (1\text{ft}^3)\left(\frac{1.20\times10^3\text{lb}}{1\text{ ft}^3}\right) = 1.20 \times 10^3\text{ lb}$$

Additional Exercises

3.86 Several possible answers for this one. Qualitatively, paint chips found on the fenders or in the dents of the cars could be held against the paint on the cars from which they presumably came for comparison. This will be less persuasive in court because many cars could have the same color paint. Quantitatively, the surface areas of the paint chips can be compared to the surface areas of missing paint on the cars. Also, the height of the fender of the red car may be compared to the height of the dent of the white car. These are much more persuasive arguments because it is unlikely that the same data for a randomly-chosen car would match that from a car at the accident site.

3.88 If the density is in metric tons…

$$\#\text{ g} = 1\text{ teaspoon}\left(\frac{4.93\text{ mL}}{1\text{ tsp}}\right)\left(\frac{1\text{ cm}^3}{1\text{ mL}}\right)\left(\frac{100{,}000{,}000\text{ tons}}{1\text{ cm}^3}\right)\left(\frac{1000\text{ kg}}{1\text{ ton}}\right)\left(\frac{1\times10^3\text{g}}{1\text{ kg}}\right)$$
$$= 4.93 \times 10^{14}\text{ g}$$

If the density is in English tons…

$$\#\text{ g} = 1\text{ teaspoon}\left(\frac{4.93\text{ mL}}{1\text{ tsp}}\right)\left(\frac{1\text{ cm}^3}{1\text{ mL}}\right)\left(\frac{100{,}000{,}000\text{ tons}}{1\text{ cm}^3}\right)\left(\frac{2000\text{ lbs}}{1\text{ ton}}\right)\left(\frac{453.59\text{g}}{1\text{ lb}}\right)$$
$$= 4.47 \times 10^{14}\text{ g}$$

3.90 a) In order to determine the volume of the pycnometer, we need to determine the volume of the water that fills it. We will do this using the mass of the water and its density.

mass of water = mass of filled pycnometer – mass of empty pycnometer
= 41.428g – 27.314g = 9.528g

$$\text{volume} = (9.528\text{ g})\left(\frac{1\text{ mL}}{0.99704\text{ g}}\right) = 9.556\text{ mL}$$

b) We know the volume of chloroform from part (a). The mass of chloroform is determined in the same way that we determined the mass of water.

mass of chloroform = mass of filled pycnometer – mass of empty pycnometer

$= 41.428\text{g} - 27.314\text{g} = 14.114\text{g}$

$$\text{Density of chloroform} = \left(\frac{14.114\text{g}}{9.556\text{mL}}\right) = 1.477\text{g/mL}$$

3.92 a) $\$4.50 = 30\ \text{min}$

b) $$\#\$ = \left(\left(1\ \text{hr x} \frac{60\ \text{min}}{\text{hr}}\right) + 45\ \text{min}\right)\left(\frac{\$0.15}{\text{min}}\right) = \$15.75$$

c) $$\#\ \text{min} = (\$17.35)\left(\frac{1\ \text{min}}{\$0.15}\right) = 116\ \text{min}$$

3.94 $$\#\,\text{g} = 2510\ \text{cm}^3\left(\frac{1\ \text{in}}{2.54\ \text{cm}}\right)^3\left(\frac{0.000{,}11\ \text{lbs}}{1\ \text{in}^3}\right)\left(\frac{453.59\ \text{g}}{1\ \text{lb}}\right) = 0.76\ \text{g}$$

3.96 Given both a mass and a volume we can determine density. Then, by experimentation, we could differentiate between two liquids.

3.98 Since the density closely matches the known value, we conclude that this is an authentic sample of ethylene glycol.

3.100 a) $$\frac{\#\ \text{lb}}{\text{gal}} = \left(\frac{785\ \text{kg}}{1\ \text{m}^3}\right)\left(\frac{2.205\ \text{lb}}{1\ \text{kg}}\right)\left(\frac{1\ \text{m}}{100\ \text{cm}}\right)^3\left(\frac{1000\ \text{cm}^3}{1\ \text{L}}\right)\left(\frac{3.785\ \text{L}}{1\ \text{gal}}\right)$$

$= 6.55\ \text{lb/gal}$

b) $$\frac{\#\ \text{lb}}{\text{gal}} = \left(\frac{785\ \text{kg}}{1\ \text{m}^3}\right)\left(\frac{1000\ \text{g}}{1\ \text{kg}}\right)\left(\frac{1\ \text{m}}{100\ \text{cm}}\right)^3\left(\frac{1000\ \text{cm}^3}{1\ \text{L}}\right)$$

$= 785\ \text{g/L}$

3.102 We solve by combining two equations:

$$F = \frac{9}{5}C + 32$$

$$F = C$$

If F = C, we can use the same variable for both temperatures:

$$C = \frac{9}{5}C + 32$$

$$\frac{5}{5}\mathrm{C} = \frac{9}{5}\mathrm{C} + 32$$

$$\frac{-4}{5}\mathrm{C} = 32$$

$C = 32\frac{-5}{4} = -40$, therefore the answer is $-40°$.

3.104 $\#\mathrm{lb} = 1\,\mathrm{pt}\left(\frac{1\,\mathrm{qt}}{2\,\mathrm{pts}}\right)\left(\frac{946.56\,\mathrm{mL}}{1\,\mathrm{qt}}\right)\left(\frac{1.0\,\mathrm{g}}{1\,\mathrm{mL}}\right)\left(\frac{1\,\mathrm{lb}}{453.59\,\mathrm{g}}\right) = 1.0\,\mathrm{pound}$

Chapter 4

Practice Exercises

4.1 $$\text{\# atoms Au} = (15.0 \text{ g Au})\left(\frac{1 \text{ mol Au}}{197.0 \text{ g Au}}\right)\left(\frac{6.022\times10^{23} \text{ atoms Au}}{1 \text{ mol Au}}\right)$$

$= 4.59 \times 10^{22}$ atoms Au

4.2 $CH_3(CH_2)_6CH_3 = 8\,C + 18\,H$ Formula mass = 8(12.01) + 18(1.01)
= 114.26

4.3 $C_{21}H_{23}NO_5$ Formula mass = 21(12.01) + 23(1.01) + 5(16.00)
= 369.41

4.4 Formula mass of heroin = 369.41 g/mol (see above exercise)

$$\text{\# molecules heroin} = (0.010 \text{ g heroin})\left(\frac{1 \text{ mol heroin}}{369.41 \text{ g heroin}}\right)\left(\frac{6.022\times10^{23} \text{ molecules heroin}}{1 \text{ mol heroin}}\right)$$

$= 1.6 \times 10^{19}$ molcules heroin

4.5 $$\text{\# mol S} = (35.6 \text{ g S})\left(\frac{1 \text{ mol S}}{32.07 \text{ g S}}\right) = 1.11 \text{ mol S}$$

4.6 $$\text{\# g Ag} = (0.263 \text{ mol Ag})\left(\frac{107.9 \text{ g Ag}}{1 \text{ mol Ag}}\right) = 28.4 \text{ g Ag}$$

4.7 $0.125 \text{ mol} \times 106 \text{ g/mol} = 13.3 \text{ g}$

4.8 $$\text{\# mol } H_2SO_4 = (45.8 \text{ g } H_2SO_4)\left(\frac{1 \text{ mol } H_2SO_4}{98.1 \text{ g } H_2SO_4}\right) = 0.467 \text{ mol } H_2SO_4$$

4.9 $$\text{\# atoms Pb} = (1.00\times10^{-9} \text{ g Pb})\left(\frac{1 \text{ mol Pb}}{207.2 \text{ g Pb}}\right)\left(\frac{6.022\times10^{23} \text{ atoms Pb}}{1 \text{ mol Pb}}\right)$$

$= 2.91 \times 10^{12}$ atoms Pb

4.10 $$\text{\# mol N} = (8.60 \text{ mol O})\left(\frac{2 \text{ mol N}}{5 \text{ mol O}}\right) = 3.44 \text{ mol N atoms}$$

4.11 $$\text{\# g Fe} = (25.6 \text{ g O})\left(\frac{1 \text{ mol O}}{16.0 \text{ g O}}\right)\left(\frac{2 \text{ mol Fe}}{3 \text{ mol O}}\right)\left(\frac{55.8 \text{ g Fe}}{1 \text{ mol Fe}}\right) = 59.5 \text{ g Fe}$$

4.12 $\# \text{g Fe} = (15.0 \text{ g Fe}_2\text{O}_3)\left(\dfrac{111.7 \text{ g Fe}}{159.7 \text{ g Fe}_2\text{O}_3}\right) = 10.5 \text{ g Fe}$

4.13 % N = 0.1417/0.5462 × 100 = 25.94 % N
% O = 0.4045/0.5462 × 100 = 74.06 % O
Since these two values constitute 100 %, there are no other elements present.

4.14 We first determine the number of grams of each element that are present in one mol of sample:

2 mol N × 14.01 g/mol = 28.02 g N
4 mol O × 16.00 g/mol = 64.00 g O

The percentages by mass are then obtained using the formula mass of the compound (92.02 g):

% N = 28.02/92.02 = 30.45 % N
% O = 64.00/92.02 = 69.55 % O

4.15 We first determine the number of mol of each element as follows:

$$\# \text{mol N} = (0.712 \text{ g N})\left(\frac{1 \text{ mol N}}{14.01 \text{ g N}}\right) = 0.0508 \text{ mol N}$$

We need to know the number of grams of O. Since there is a total of 1.525 g of compound and the only other element present is N, the mass of O = 1.525 g – 0.712 g = 0.813 g.

$$\# \text{mol O} = (0.813 \text{ g O})\left(\frac{1 \text{ mol O}}{16.00 \text{ g O}}\right) = 0.0508 \text{ mol O}$$

Since these two mol amounts are the same, the empirical formula is NO.

4.16 We first determine the number of mol of each element as follows:

$$\# \text{mol N} = (0.522 \text{ g N})\left(\frac{1 \text{ mol N}}{14.01 \text{ g N}}\right) = 0.0373 \text{ mol N}$$

We need to know the number of grams of O. Since there is a total of 2.012 g of compound and the only other element present is N, the mass of O = 2.012 g – 0.522 g = 1.490 g.

$$\# \text{mol O} = (1.490 \text{ g O})\left(\frac{1 \text{ mol O}}{16.00 \text{ g O}}\right) = 0.0931 \text{ mol O}$$

Next, we divide each of these mol amounts by the smallest in order to deduce the simplest whole number ratio:

For N: 0.0373 mol/0.0373 mol = 1.00
For O: 0.0931 mol/0.0373 mol = 2.50

There are 2.5 mol of oxygen atoms for every 1 mol of nitrogen atoms. Therefore, there are 2.5 O atoms for every 1 N atom. Empirical formulas must be whole number ratios: This corresponds to a whole number ratio of…5 O atoms for every 2 N atoms.

The empirical formula is therefore N_2O_5.

4.17 It is convenient to assume that we have 100 g of the sample, so that the % by mass values may be taken directly to represent masses. Thus there is 32.37 g of Na, 22.57 g of S and (100.00 – 32.37 – 22.57) = 45.06 g of O. Now, convert these masses to a number of mol:

$$\#\,\text{mol Na} = (32.37\text{ g Na})\left(\frac{1\text{ mol Na}}{23.00\text{ g Na}}\right) = 1.407\text{ mol Na}$$

$$\#\,\text{mol S} = (22.57\text{ g S})\left(\frac{1\text{ mol S}}{32.06\text{ g S}}\right) = 0.7040\text{ mol S}$$

$$\#\,\text{mol O} = (45.06\text{ g O})\left(\frac{1\text{ mol O}}{16.00\text{ g O}}\right) = 2.816\text{ mol O}$$

Next, we divide each of these mol amounts by the smallest in order to deduce the simplest whole number ratio:

For Na: 1.407 mol/0.7040 mol = 1.999

For S: 0.7040 mol/0.7040 mol = 1.000

For O: 2.816 mol/0.7040 mol = 4.000

The empirical formula is Na_2SO_4.

4.18 Since the entire amount of carbon that was present in the original sample appears among the products only as CO_2, we calculate the amount of carbon in the sample as follows:

$$\#\,\text{g C} = (7.406\text{ g CO}_2)\left(\frac{1\text{ mol CO}_2}{44.01\text{ g CO}_2}\right)\left(\frac{1\text{ mol C}}{1\text{ mol CO}_2}\right)\left(\frac{12.01\text{ g C}}{1\text{ mol C}}\right) = 2.021\text{ g C}$$

Similarly, the entire mass of hydrogen that was present in the original sample appears among the products only as H_2O. Thus the mass of hydrogen in the sample is:

$$\#\,\text{g H} = (3.027\text{ g H}_2\text{O})\left(\frac{1\text{ mol H}_2\text{O}}{18.02\text{ g H}_2\text{O}}\right)\left(\frac{2\text{ mol H}}{1\text{ mol H}_2\text{O}}\right)\left(\frac{1.008\text{ g H}}{1\text{ mol H}}\right) = 0.3386\text{ g H}$$

The mass of oxygen in the original sample is determined by difference:

5.048 g – 2.021 g– 0.3386 g = 2.688 g O

Next, these mass amounts are converted to the corresponding mol amounts:

$$\# \text{mol C} = (2.021 \text{ g C})\left(\frac{1 \text{ mol C}}{12.01 \text{ g C}}\right) = 0.1683 \text{ mol C}$$

$$\# \text{mol H} = (0.3386 \text{ g H})\left(\frac{1 \text{ mol H}}{1.008 \text{ g H}}\right) = 0.3359 \text{ mol H}$$

$$\# \text{mol O} = (2.688 \text{ g O})\left(\frac{1 \text{ mol O}}{16.00 \text{ g O}}\right) = 0.1680 \text{ mol O}$$

The simplest formula is obtained by dividing each of these mol amounts by the smallest:

For C: 0.1683 mol/0.1680 mol= 1.002
for H: 0.3359 mol/0.1680 mol= 1.999
For O: 0.1680 mol/0.1680 mol = 1.000

These values give us the simplest formula directly, namely CH_2O.

4.19 The formula mass of the empirical unit is 1 N + 2 H = 16.03. Since this is half of the molcular mass, the molcular formula is N_2H_4.

4.20 $$\# \text{mol } H_2SO_4 = (0.366 \text{ mol NaOH})\left(\frac{1 \text{ mol } H_2SO_4}{2 \text{ mol NaOH}}\right) = 0.183 \text{ mol } H_2SO_4$$

4.21 $$\# \text{mol } O_2 = (0.575 \text{ mol } CO_2)\left(\frac{5 \text{ mol } O_2}{3 \text{ mol } CO_2}\right) = 0.958 \text{ mol } O_2$$

4.22 $$\# \text{g } Al_2O_3 = (86.0 \text{ g Fe})\left(\frac{1 \text{ mol Fe}}{55.85 \text{ g Fe}}\right)\left(\frac{1 \text{ mol } Al_2O_3}{2 \text{ mol Fe}}\right)\left(\frac{102.0 \text{ g } Al_2O_3}{1 \text{ mol } Al_2O_3 \text{ mol } Al_2O_3}\right)$$

$= 78.5 \text{ g } Al_2O_3$

4.23 $3CaCl_2(aq) + 2K_3PO_4(aq) \rightarrow Ca_3(PO_4)_2(s) + 6KCl(aq)$

4.24 First determine the number of grams of O_2 that would be required to react completely with the given amount of ammonia:

$$\# \text{g } O_2 = (30.00 \text{ g } NH_3)\left(\frac{1 \text{ mol } NH_3}{17.03 \text{ g } NH_3}\right)\left(\frac{5 \text{ mol } O_2}{4 \text{ mol } NH_3}\right)\left(\frac{32.00 \text{ g } O_2}{1 \text{ mol } O_2}\right)$$

$= 70.46 \text{ g } O_2$

Since this is more than the amount that is available, we conclude that oxygen is the limiting reactant. The rest of the calculation is therefore based on the available amount of oxygen:

$$\# \text{g NO} = (40.00 \text{ g } O_2)\left(\frac{1 \text{ mol } O_2}{32.00 \text{ g } O_2}\right)\left(\frac{4 \text{ mol NO}}{5 \text{ mol } O_2}\right)\left(\frac{30.01 \text{ g NO}}{1 \text{ mol NO}}\right)$$

$$= 30.01 \text{ g NO}$$

4.25 First determine the number of grams of C_2H_5OH that would be required to react completely with the given amount of sodium dichromate:

$$\# \text{g } C_2H_5OH = (90.0 \text{ g } Na_2Cr_2O_7)\left(\frac{1 \text{ mol } Na_2Cr_2O_7}{262.0 \text{ g } Na_2Cr_2O_7}\right)\left(\frac{3 \text{ mol } C_2H_5OH}{2 \text{ mol } Na_2Cr_2O_7}\right)\left(\frac{46.08 \text{ g } C_2H_5OH}{1 \text{ mol } C_2H_5OH}\right)$$

$$= 23.7 \text{ g } C_2H_5OH$$

Once this amount of C_2H_5OH is reacted the reaction will stop, even though there are 24.0 g C_2H_5OH present, because the $Na_2Cr_2O_7$ will be used up. Therefore $Na_2Cr_2O_7$ is the limiting reactant. The theoretical yield of acetic acid ($HC_2H_3O_2$) is therefore based on the amount of $Na_2Cr_2O_7$ added. This is calculated below:

$$\# \text{g } HC_2H_3O_2 = (90.0 \text{ g } Na_2Cr_2O_7)\left(\frac{1 \text{ mol } Na_2Cr_2O_7}{262.0 \text{ g } Na_2Cr_2O_7}\right)\left(\frac{3 \text{ mol } HC_2H_3O_2}{2 \text{ mol } Na_2Cr_2O_7}\right)\left(\frac{60.06 \text{ g } HC_2H_3O_2}{1 \text{ mol } HC_2H_3O_2}\right)$$

$$= 30.9 \text{ g } HC_2H_3O_2$$

Now the percentage yield can be calculated from the amount of acetic acid actually produced, 26.6 g:

$$\text{percent yield} = \left(\frac{\text{actual yield}}{\text{theoretical yield}}\right) \times 100 = \left(\frac{26.6 \text{ g } HC_2H_3O_2}{30.9 \text{ g } HC_2H_3O_2}\right) \times 100 = 86.1\%$$

Review Problems

4.22 1:2

4.24 2.59×10^{-3} mol Na

4.26 a) 6:11
b) 12:11
c) 2:1
d) 2:1

4.28 $\# \text{mol Al} = (1.58 \text{ mol O})\left(\frac{2 \text{ mol Al}}{3 \text{ mol O}}\right) = 1.05 \text{ mol Al}$

4.30 $$\#\text{ mol Al} = (2.16\text{ mol }Al_2O_3)\left(\frac{2\text{ mol Al}}{1\text{ mol }Al_2O_3}\right) = 4.32\text{ mol Al}$$

4.32 a) $$\left(\frac{2\text{ mol Al}}{3\text{ mol S}}\right)\text{ or }\left(\frac{3\text{ mol S}}{2\text{ mol Al}}\right)$$

b) $$\left(\frac{3\text{ mol S}}{1\text{ mol }Al_2(SO_4)_3}\right)\text{ or }\left(\frac{1\text{ mol }Al_2(SO_4)_3}{3\text{ mol S}}\right)$$

c) $$\#\text{ mol Al} = (0.900\text{ mol S})\left(\frac{2\text{ mol Al}}{3\text{ mol S}}\right) = 0.600\text{ mol Al}$$

d) $$\#\text{ mol S} = (1.16\text{ mol }Al_2(SO_4)_3)\left(\frac{3\text{ mol S}}{1\text{ mol }Al_2(SO_4)_3}\right) = 3.48\text{ mol S}$$

4.34 Based on the balanced equation:

$$2NH_{3(g)} \rightarrow N_{2(g)} + 3H_{2(g)}$$

From this equation the conversion factors can be written:

$$\left(\frac{1\text{ mol }N_2}{2\text{ mol }NH_3}\right)\text{ and }\left(\frac{3\text{ mol }H_2}{2\text{ mol }NH_3}\right)$$

To determine the mol produced, simply convert from starting mol to end mol:

$$0.145\text{ mol }NH_3\left(\frac{1\text{ mol }N_2}{2\text{ mol }NH_3}\right) = 0.0725\text{ mol }N_2$$

The mol of hydrogen are calculated similarly:

$$0.145\text{ mol }NH_3\left(\frac{3\text{ mol }H_2}{2\text{ mol }NH_3}\right) = 0.218\text{ mol }H_2$$

4.36 $$\#\text{ moles }UF_6 = (1.25\text{ mol }CF_4)\left(\frac{4\text{ mol F}}{1\text{ mol }CF_4}\right)\left(\frac{1\text{ mol }UF_6}{6\text{ mol F}}\right) = 0.833\text{ moles }UF_6$$

4.38 $$\#\text{ atoms C} = (4.13\text{ mol H})\left(\frac{1\text{ mol }C_3H_8}{8\text{ mol H}}\right)\left(\frac{3\text{ mol C}}{1\text{ mol }C_3H_8}\right)\left(\frac{6.022\times10^{23}\text{ atoms C}}{1\text{ mol C}}\right)$$

$$= 9.33 \times 10^{23}\text{ atoms C}$$

4.40 $$\#\text{ atoms C} = (0.260\text{ mol }C_6H_{12}O_6)\left(\frac{6\text{ mol C}}{1\text{ mol }C_6H_{12}O_6}\right)\left(\frac{6.022\times10^{23}\text{ atoms C}}{1\text{ mol C}}\right)$$

$$= 9.39 \times 10^{23}\text{ atoms C}$$

$$\#\,\text{atoms H} = (0.260\ \text{mol C}_6\text{H}_{12}\text{O}_6)\left(\frac{12\ \text{mol H}}{1\ \text{mol C}_6\text{H}_{12}\text{O}_6}\right)\left(\frac{6.022\times10^{23}\,\text{atoms H}}{1\ \text{mol H}}\right)$$

$= 1.88 \times 10^{24}$ atoms H

$$\#\,\text{atoms O} = (0.260\ \text{mol C}_6\text{H}_{12}\text{O}_6)\left(\frac{6\ \text{mol O}}{1\ \text{mol C}_6\text{H}_{12}\text{O}_6}\right)\left(\frac{6.022\times10^{23}\,\text{atoms O}}{1\ \text{mol O}}\right)$$

$= 9.39 \times 10^{23}$ atoms O

$(9.39 \times 10^{23}$ atoms C$) + (1.88 \times 10^{24}$ atoms H$) + (9.39 \times 10^{23}$ atoms O$)$

$= 3.76 \times 10^{24}$ atoms

4.42 6 g ÷ 12 g/mol = 0.5 mol, or 3.01×10^{23} atoms

4.44 a) 23.0 g Na
b) 32.1 g S
c) 35.5 g Cl

4.46 a) $\#\,\text{g Fe} = (1.35\ \text{mol Fe})\left(\frac{55.85\ \text{g Fe}}{1\ \text{mole Fe}}\right) = 75.4\ \text{g Fe}$

b) $\#\,\text{g O} = (24.5\ \text{mol O})\left(\frac{16.0\ \text{g O}}{1\ \text{mole O}}\right) = 392\ \text{g O}$

c) $\#\,\text{g K} = (0.876\ \text{mol Ca})\left(\frac{40.1\ \text{g Ca}}{1\ \text{mole Ca}}\right) = 35.1\ \text{g Ca}$

4.48 $\#\,\text{g K} = (2.00\times10^{12}\ \text{atoms K})\left(\frac{1\ \text{mol K}}{6.022\times10^{23}\,\text{atoms K}}\right)\left(\frac{39.1\ \text{g K}}{1\ \text{mol K}}\right) = 1.30\times10^{-10}\ \text{g K}$

4.50 $\#\,\text{mol Ni} = (17.7\ \text{g Ni})\left(\frac{1\ \text{mol Ni}}{58.69\ \text{g Ni}}\right) = 0.302\ \text{mol Ni}$

4.52 Note: all masses are in g/mol

a) $NaHCO_3$	=	1Na + 1H + 1C + 3O
	=	(22.99) + (1.01) + (12.01) + (3 × 16.00)
	=	84.0 g/mol
b) $K_2Cr_2O_7$	=	2K + 2Cr + 7O
	=	(2 × 39.10) + (2 × 52.00) + (7 × 16.00)
	=	294.2 g/mol
c) $(NH_4)_2CO_3$	=	2N + 8H + C + 3O
	=	(2 × 14.01) + (8 × 1.01) + (12.01) + (3 × 16.00)
	=	96.1 g/mol

d) $Al_2(SO_4)_3$ = $2Al + 3S + 12O$
= $(2 \times 26.98) + (3 \times 32.07) + (12 \times 16.00)$
= 342.2 g/mol

e) $CuSO_4 \cdot 5H_2O$ = $1Cu + 1S + 9O + 10H$
= $63.55 + 32.07 + (9 \times 16.00) + (10 \times 1.01)$
= 249.7 g/mol

4.54 a) # g =(1.25 mol $Ca_3(PO_4)_2$)(310.18 g $Ca_3(PO_4)_2$/1 mol $Ca_3(PO_4)_2$)
= 388 g $Ca_3(PO_4)_2$

b) # g =(0.625 mol $Fe(NO_3)_3$)(241.87 g $Fe(NO_3)_3$/1 mol $Fe(NO_3)_3$
= 151 g $Fe(NO_3)_3$

c) # g = (0.600 mol C_4H_{10})(58.12 g C_4H_{10}/1 mol C_4H_{10}) = 34.9 g C_4H_{10}

d) # g = (1.45 mol $(NH_4)_2CO_3$)(96.11 g/mol) = 139 g $(NH_4)_2CO_3$

4.56 a) $\#\text{moles CaCO}_3 = (21.5\text{ g CaCO}_3)\left(\frac{1\text{ mole CaCO}_3}{100.09\text{ g CaCO}_3}\right) = 0.215\text{ mol CaCO}_3$

b) $\#\text{moles NH}_3 = (1.56\text{ g NH}_3)\left(\frac{1\text{ mole NH}_3}{17.03\text{ g NH}_3}\right) = 9.16\times10^{-2}\text{ mol NH}_3$

c) $\#\text{moles Sr(NO}_3)_2 = (16.8\text{ g Sr(NO}_3)_2)\left(\frac{1\text{ mol Sr(NO}_3)_2}{211.6\text{ g Sr(NO}_3)_2}\right)$
$= 7.94 \times 10^{-2}$ mol $Sr(NO_3)_2$

d) $\#\text{moles Na}_2\text{CrO}_4 = (6.98\text{ X }10^{-6}\text{ g Na}_2\text{CrO}_4)\left(\frac{1\text{ mol Na}_2\text{CrO}_4}{162.0\text{ g Na}_2\text{CrO}_4}\right)$
$= 4.31 \times 10^{-8}$ mol Na_2CrO_4

4.58 The formula CaC_2 indicates that there is 1 mol of Ca for every 2 mol of C. Therefore, if there are 0.150 mol of C there must be 0.0750 mol of Ca.

$$\#\text{g Ca} = (0.075\text{ mol Ca})\left(\frac{40.078\text{ g Ca}}{1\text{ mol Ca}}\right) = 3.01\text{ g Ca}$$

4.60 $\#\text{mol N} = (0.650\text{ mol (NH}_4)_2\text{CO}_3)\left(\frac{2\text{ mol N}}{1\text{ mol (NH}_4)_2\text{CO}_3}\right) = 1.30\text{ mol N}$

$\#\text{g (NH}_4)_2\text{CO}_3 = (0.650\text{ mol (NH}_4)_2\text{CO}_3)\left(\frac{96.09\text{ g (NH}_4)_2\text{CO}_3}{1\text{ mol (NH}_4)_2\text{CO}_3}\right)$
$= 62.5\text{ g (NH}_4)_2\text{CO}_3$

4.62 $\#\text{kg fertilizer} = (1\text{ kg N})\left(\frac{1000\text{ g N}}{1\text{ kg N}}\right)\left(\frac{1\text{ mol N}}{14.01\text{ g N}}\right)\left(\frac{1\text{ mol (NH}_4)_2\text{CO}_3}{2\text{ mol N}}\right)$
$\left(\frac{96.11\text{ g (NH}_4)_2\text{CO}_3}{1\text{ mol (NH}_4)_2\text{CO}_3}\right)\left(\frac{1\text{ kg (NH}_4)_2\text{CO}_3}{1000\text{ g (NH}_4)_2\text{CO}_3}\right) = 3.43\text{ kg fertilizer}$

4.64 $\% \text{ O in morphine} = \frac{48.00 \text{ g O}}{285.36 \text{ g } C_{17}H_{19}NO_3} \times 100\% = 16.82\% \text{ O}$

$\% \text{ O in heroin} = \frac{80.00 \text{ g O}}{369.44 \text{ g } C_{21}H_{23}NO_5} \times 100\% = 21.65\% \text{ O}$

Therefore heroin has a higher percentage oxygen.

4.66 $\% \text{ Cl in Freon - 12} = \frac{70.90 \text{ g Cl}}{285.36 \text{ g } CCl_2F_2} \times 100\% = 24.85\% \text{ Cl}$

$\% \text{ Cl in Freon 141b} = \frac{106.35 \text{ g Cl}}{187.37 \text{ g } C_2Cl_3F_3} \times 100\% = 56.76\% \text{ Cl}$

Therefore Freon 141b has a higher percentage chlorine.

4.68 Assume one mol total for each of the following.

a)

The molar mass of NaH_2PO_4 is 119.98 g/mole.

$\% \text{ Na} = \frac{23.0 \text{ g Na}}{119.98 \text{ g } NaH_2PO_4} \times 100\% = 19.2\%$

$\% \text{ H} = \frac{2.02 \text{ g H}}{119.98 \text{ g } NaH_2PO_4} \times 100\% = 1.68\%$

$\% \text{ P} = \frac{31.0 \text{ g P}}{119.98 \text{ g } NaH_2PO_4} \times 100\% = 25.8\%$

$\% \text{ O} = \frac{64.0 \text{ g O}}{119.98 \text{ g } NaH_2PO_4} \times 100\% = 53.3\%$

b)

The molar mass of $NH_4H_2PO_4$ is 115.05 g/mole.

$\% \text{ N} = \frac{14.0 \text{ g N}}{115.05 \text{ g } NH_4H_2PO_4} \times 100\% = 12.2\%$

$\% \text{ H} = \frac{6.05 \text{ g H}}{115.05 \text{ g } NH_4H_2PO_4} \times 100\% = 5.26\%$

$\% \text{ P} = \frac{31.0 \text{ g P}}{115.05 \text{ g } NH_4H_2PO_4} \times 100\% = 26.9\%$

$\% \text{ O} = \frac{64.0 \text{ g O}}{115.05 \text{ g } NH_4H_2PO_4} \times 100\% = 55.6\%$

c)

The molar mass of $(CH_3)_2CO$ is 58.05 g/mole.

$$\%\,C = \frac{36.0\,g\,C}{58.05\,g\,(CH_3)_2CO} \times 100\,\% = 62.0\,\%$$

$$\%\,H = \frac{6.05\,g\,H}{58.05\,g\,(CH_3)_2CO} \times 100\,\% = 10.4\,\%$$

$$\%\,O = \frac{16.0\,g\,O}{58.05\,g\,(CH_3)_2CO} \times 100\,\% = 27.6\,\%$$

d)

The molar mass of $CaSO_4$ is 136.2 g/mole.

$$\%\,Ca = \frac{40.1\,g\,Ca}{136.2\,g\,CaSO_4} \times 100\,\% = 29.4\,\%$$

$$\%\,S = \frac{32.1\,g\,S}{136.2\,g\,CaSO_4} \times 100\,\% = 23.6\,\%$$

$$\%\,O = \frac{64.0\,g\,O}{136.2\,g\,CaSO_4} \times 100\,\% = 47.0\,\%$$

e)

The molar mass of $CaSO_4 \bullet 2H_2O$ is 172.2 g/mole.

$$\%\,Ca = \frac{40.1\,g\,Ca}{172.2\,g\,CaSO_4 \bullet 2H_2O} \times 100\,\% = 23.3\,\%$$

$$\%\,S = \frac{32.1\,g\,S}{172.2\,g\,CaSO_4 \bullet 2H_2O} \times 100\,\% = 18.6\,\%$$

$$\%\,O = \frac{96.0\,g\,O}{172.2\,g\,CaSO_4 \bullet 2H_2O} \times 100\,\% = 55.7\,\%$$

$$\%\,H = \frac{4.03\,g\,H}{172.2\,g\,CaSO_4 \bullet 2H_2O} \times 100\,\% = 2.34\,\%$$

4.70 $$\%\,P = \frac{0.539\,g\,P}{2.35\,g\,\text{compound}} \times 100\% = 22.9\%$$

$\%\,Cl = 100\% - 22.9\% = 77.1\%$

4.72 For $C_{17}H_{25}N$, the molar mass (17C + 25H + 1N) equals 243.39 g/mol, and the three theoretical values for % by weight are calculated as follows:

$$\% C = \frac{204.2 \text{ g C}}{243.4 \text{ g } C_{17}H_{25}N} \times 100\,\% = 83.89\,\%$$

$$\% H = \frac{25.20 \text{ g H}}{243.4 \text{ g } C_{17}H_{25}N} \times 100\,\% = 10.35\,\%$$

$$\% N = \frac{14.01 \text{ g N}}{243.4 \text{ g } C_{17}H_{25}N} \times 100\,\% = 5.76\,\%$$

These data are consistent with the experimental values cited in the problem.

4.74 $\# \text{ g O} = (7.14 \text{ x } 10^{21} \text{ atoms N})\left(\frac{1 \text{ mol N}}{6.02 \text{ x } 10^{23} \text{ atoms N}}\right)\left(\frac{5 \text{ mol O}}{2 \text{ mol N}}\right)\left(\frac{16.0 \text{ g O}}{1 \text{ mol O}}\right)$

$= 0.474 \text{ g O}$

4.76 The molcular formula is some integer multiple of the empirical formula. This means that we can divide the molcular formula by the largest possible whole number that gives an integer ratio among the atoms in the empirical formula.
a) SCl b) CH_2O c) NH_3 d) AsO_3 e) OH

4.78 We begin by realizing that the mass of oxygen in the compound may be determined by difference: 0.8961 g total – (0.1114 g Na + 0.4748 g Tc) = 0.3099 g O. Next we can convert each mass of an element into the corresponding number of mol of that element as follows:

$$\# \text{ moles Na} = (0.111 \text{ g Na})\left(\frac{1 \text{ mole Na}}{23.00 \text{ g Na}}\right) = 4.83 \text{ X } 10^{-3} \text{ moles Na}$$

$$\# \text{ moles Tc} = (0.477 \text{ g Tc})\left(\frac{1 \text{ mole Tc}}{98.9 \text{ g Tc}}\right) = 4.82 \text{ X } 10^{-3} \text{ moles Tc}$$

$$\# \text{ moles O} = (0.308 \text{ g O})\left(\frac{1 \text{ mole O}}{16.0 \text{ g O}}\right) = 1.93 \text{ X } 10^{-2} \text{ moles O}$$

Now we divide each of these numbers of mol by the smallest of the three numbers, in order to obtain the simplest mol ratio among the three elements in the compound:

for Na, 4.83×10^{-3} mol / 4.82×10^{-3} mol = 1.00
for Tc, 4.82×10^{-3} mol / 4.82×10^{-3} mol = 1.00
for O, 1.93×10^{-2} mol / 4.82×10^{-3} mol = 4.00

These relative mol amounts give us the empirical formula: $NaTcO_4$.

4.80 Assume a 100g sample:

$$\#\,\text{mol C} = (14.5\text{ g C})\left(\frac{1\text{ mol C}}{12.01\text{ g C}}\right) = 1.21\text{ mol C}$$

$$\#\,\text{mol Cl} = (85.5\text{ g Cl})\left(\frac{1\text{ mol Cl}}{35.45\text{ g Cl}}\right) = 2.41\text{ mol Cl}$$

Now we divide each of these numbers of mol by the smallest of the three numbers, in order to obtain the simplest mol ratio among the three elements in the compound:

for C, 1.21 mol / 1.21 mol = 1.00
for Cl, 2.41 mol / 1.21mol = 2.000

These relative mol amounts give us the empirical formula CCl_2

4.82 Assume a 100g sample:

$$\#\,\text{mol Na} = (22.9\text{ g Na})\left(\frac{1\text{ mol Na}}{22.99\text{ g Na}}\right) = 0.996\text{ mol Na}$$

$$\#\,\text{mol B} = (21.5\text{ g B})\left(\frac{1\text{ mol B}}{10.81\text{ g B}}\right) = 1.99\text{ mol B}$$

$$\#\,\text{mol O} = (55.7\text{ g O})\left(\frac{1\text{ mol O}}{16.00\text{ g O}}\right) = 3.48\text{ mol O}$$

Now we divide each of these numbers of mol by the smallest of the three numbers, in order to obtain the simplest mol ratio among the three elements in the compound:

for Na, 0.996 mol ÷ 0.996 mol = 1.00
for B, 1.99 mol ÷ 0.996 mol = 2.00
for O, 3.48 mol 0.996 mol = 3.49

These relative mol amounts give us the empirical formula $Na_2B_4O_7$

4.84 All of the carbon is converted to carbon dioxide so,

$$\#\,\text{g C} = (1.312\text{ g CO}_2)\left(\frac{1\text{ mol CO}_2}{44.01\text{ g CO}_2}\right)\left(\frac{1\text{ mol C}}{1\text{ mol CO}_2}\right)\left(\frac{12.01\text{ g C}}{1\text{ mol C}}\right) = 0.358\text{ g C}$$

$$\#\,\text{moles C} = (0.358\text{ g C})\left(\frac{1\text{ mol C}}{12.01\text{ g C}}\right) = 2.98\text{ x }10^{-2}\text{ mol C}$$

All of the hydrogen is converted to H_2O so,

$$\#\,\text{g H} = (0.805\text{ g H}_2\text{O})\left(\frac{1\text{ mol H}_2\text{O}}{18.02\text{ g H}_2\text{O}}\right)\left(\frac{2\text{ mol H}}{1\text{ mol H}_2\text{O}}\right)\left(\frac{1.008\text{ g H}}{1\text{ mol H}}\right) = 0.0901\text{ g H}$$

$$\#\,\text{mol H} = (0.0901\text{ g H})\left(\frac{1\text{ mol H}}{1.008\text{ g H}}\right) = 8.93\text{ x }10^{-2}\text{ moles H}$$

The amount of O in the compound is determined by subtracting the mass of C and the mass of H from the sample.

$$\# \text{ g O} = 0.684\text{g} - 0.358 \text{ g} - 0.0901 \text{ g} = 0.236 \text{ g}$$

$$\# \text{ mol O} = (0.236 \text{ g O})\left(\frac{1 \text{ mol O}}{16 \text{ g O}}\right) = 1.48 \text{ x } 10^{-2} \text{ moles}$$

The relative mol ratios are:
for C, 0.0298 mol / 0.0148 mol = 2.01
for H, 0.0893 mol/ 0.0148 mol = 6.03
for O, 0.0148 mol / 0.0148 mol = 1.00

The relative mol amounts give the empirical formula C_2H_6O

4.86 This type of combustion analysis takes advantage of the fact that the entire amount of carbon in the original sample appears as CO_2 among the products. Hence the mass of carbon in the original sample must be equal to the mass of carbon that is found in the CO_2.

$$\# \text{g C} = (19.73 \times 10^{-3} \text{ g CO}_2)\left(\frac{1 \text{ mole CO}_2}{44.01 \text{ g CO}_2}\right)\left(\frac{1 \text{ mole C}}{1 \text{ mole CO}_2}\right)\left(\frac{12.011 \text{ g C}}{1 \text{ mole C}}\right)$$
$$= 5.385 \times 10^{-3} \text{ g C}$$

Similarly, the entire mass of hydrogen that was present in the original sample ends up in the products as H_2O:

$$\# \text{g H} = (6.391 \times 10^{-3} \text{ g H}_2\text{O})\left(\frac{1 \text{ mole H}_2\text{O}}{18.02 \text{ g H}_2\text{O}}\right)\left(\frac{2 \text{ mole H}}{1 \text{ mole H}_2\text{O}}\right)\left(\frac{1.008 \text{ g H}}{1 \text{ mole H}}\right)$$
$$= 7.150 \times 10^{-4} \text{ g H}$$

The mass of oxygen is determined by subtracting the mass due to C and H from the total mass: 6.853 mg total – (5.385 mg C + 0.7150 mg H) = 0.753 mg O. Now, convert these masses to a number of mol:

$$\# \text{moles C} = (5.385 \times 10^{-3} \text{ g C})\left(\frac{1 \text{ mole C}}{12.011 \text{ g C}}\right) = 4.483 \times 10^{-4} \text{ moles C}$$

$$\# \text{moles H} = (7.150 \times 10^{-4} \text{ g H})\left(\frac{1 \text{ mole H}}{1.0079 \text{ g H}}\right) = 7.094 \times 10^{-4} \text{ moles H}$$

$$\# \text{moles O} = (7.53 \times 10^{-4} \text{ g O})\left(\frac{1 \text{ mole O}}{15.999 \text{ g O}}\right) = 4.71 \times 10^{-5} \text{ moles O}$$

The relative mol amounts are:

for C, 4.483×10^{-4} mol ÷ 4.71×10^{-5} mol = 9.52

for H, 7.094×10^{-4} mol ÷ 4.71×10^{-5} mol = 15.1

for O, 4.71×10^{-5} mol ÷ 4.71×10^{-5} mol = 1.00

The relative mol amounts are not nice whole numbers as we would like. However, we see that if we double the relative number of mol of each compound, there are approximately 19 mol of C, 30 mol of H and 2 mol of O. If we assume these numbers are correct, the empirical formula is $C_{19}H_{30}O_2$, for which the formula weight is 290 g/mol.

In most problems where we attempt to determine an empirical formula, the relative mol amounts should work out to give a "nice" set of values for the formula. Rarely will a problem be designed that gives very odd coefficients. With experience and practice, you will recognize when a set of values is reasonable.

4.88 a) Formula mass = 135.1 g

$$\frac{270.4\ \text{g/mol}}{135.1\ \text{g/mol}} = 2.001$$

The molcular formula is $Na_2S_4O_6$

b) Formula mass = 73.50 g

$$\frac{147.0\ \text{g/mol}}{73.50\ \text{g/mol}} = 2.000$$

The molcular formula is $C_6H_4Cl_2$

c) Formula mass = 60.48 g

$$\frac{181.4\ \text{g/mol}}{60.48\ \text{g/mol}} = 2.999$$

The molcular formula is $C_6H_3Cl_3$

4.90 The molcular mass for the compound $C_{19}H_{30}O_2$ is 290. Since the molcular mass is equal to the mass of this empirical formula, the molcular formula is the same as this empirical formula, $C_{19}H_{30}O_2$.

4.92 From the information provided, we can determine the mass of mercury as the difference between the total mass and the mass of bromine:

$$\#\ \text{g Hg} = 0.389\ \text{g compound} - 0.111\ \text{g Br} = 0.278\ \text{g Hg}$$

To determine the empirical formula first convert the two masses to a number of mol.

$$\# \text{moles Hg} = (0.278 \text{ g Hg})\left(\frac{1 \text{ mole Hg}}{200.59 \text{ g Hg}}\right) = 1.39 \times 10^{-3} \text{ moles Hg}$$

$$\# \text{moles Br} = (0.111 \text{ g Br})\left(\frac{1 \text{ mole Br}}{79.904 \text{ g Br}}\right) = 1.389 \times 10^{-3} \text{ moles Br}$$

Now, we would divide each of these values by the smaller quantity to determine the simplest mol ratio between the two elements. By inspection, though, we can see there are the same number of mol of Hg and Br. Consequently, the simplest mol ratio is 1:1 and the empirical formula is HgBr.

To determine the molecular formula, recall that the ratio of the molecular mass to the empirical mass is equivalent to the ratio of the molecular formula to the empirical formula. Thus, we need to calculate an empirical mass: (1 mol Hg)(200.59 g Hg/mol Hg) + (1 mol Br)(79.904 g Br/mol Br) = 280.49 g/mol HgBr. The molecular mass, as reported in the problem is 561 g/mol. The ratio of these is:

$$\frac{561 \text{ g/mole}}{280.49 \text{ g/mole}} = 2.00$$

So, the molecular formula is two times the empirical formula or Hg_2Br_2.

4.94 First, determine the amount of oxygen in the sample by subtracting the masses of the other elements from the total mass: 0.6216 g – (0.1735 g C + 0.01455 g H + 0.2024 g N) = 0.2312 g O. Now, convert these masses into a number of mol for each element:

$$\# \text{moles C} = (0.1735 \text{ g C})\left(\frac{1 \text{ mole C}}{12.011 \text{ g C}}\right) = 1.445 \times 10^{-2} \text{ moles C}$$

$$\# \text{moles H} = (0.01455 \text{ g H})\left(\frac{1 \text{ mole H}}{1.0079 \text{ g H}}\right) = 1.444 \times 10^{-2} \text{ moles H}$$

$$\# \text{moles N} = (0.2024 \text{ g N})\left(\frac{1 \text{ mole N}}{14.007 \text{ g N}}\right) = 1.445 \times 10^{-2} \text{ moles N}$$

$$\# \text{moles O} = (0.2312 \text{ g O})\left(\frac{1 \text{ mole O}}{15.999 \text{ g O}}\right) = 1.445 \times 10^{-2} \text{ moles O}$$

These are clearly all the same mol amounts, and we deduce that the empirical formula is CHNO, which has a formula weight of 43. It can be seen that the number 43 must be multiplied by the integer 3 in order to obtain the molar mass ($3 \times 43 = 129$), and this means that the empirical formula should similarly be multiplied by 3 in order to arrive at the molecular formula, $C_3H_3N_3O_3$.

4.96 36 mol of hydrogen

4.98 $4Fe(s) + 3O_2(g) \rightarrow 2Fe_2O_3(s)$

4.100 a) $Ca(OH)_2 + 2HCl \rightarrow CaCl_2 + 2H_2O$

b) $2AgNO_3 + CaCl_2 \rightarrow Ca(NO_3)_2 + 2AgCl$

c) $2Fe_2O_3 + 3C \rightarrow 4Fe + 3CO_2$

d) $2NaHCO_3 + H_2SO_4 \rightarrow Na_2SO_4 + 2H_2O + 2CO_2$

e) $2C_4H_{10} + 13O_2 \rightarrow 8CO_2 + 10H_2O$

4.102 a) $Mg(OH)_2 + 2HBr \rightarrow MgBr_2 + 2H_2O$

b) $2HCl + Ca(OH)_2 \rightarrow CaCl_2 + 2H_2O$

c) $Al_2O_3 + 3H_2SO_4 \rightarrow Al_2(SO_4)_3 + 3H_2O$

d) $2KHCO_3 + H_3PO_4 \rightarrow K_2HPO_4 + 2H_2O + 2CO_2$

e) $C_9H_{20} + 14O_2 \rightarrow 9CO_2 + 10H_2O$

4.104 a) $\#\,\text{mol Na}_2\text{S}_2\text{O}_3 = (0.12\text{ mol Cl}_2)\left(\frac{1\text{ mol Na}_2\text{S}_2\text{O}_3}{4\text{ mol Cl}_2}\right)$

$= 0.030\text{ mol Na}_2\text{S}_2\text{O}_3$

b) $\#\,\text{mol HCl} = (0.12\text{ mol Cl}_2)\left(\frac{8\text{ mol HCl}}{4\text{ mol Cl}_2}\right) = 0.24\text{ mol HCl}$

c) $\#\,\text{mol H}_2\text{O} = (0.12\text{ mol Cl}_2)\left(\frac{5\text{ mol H}_2\text{O}}{4\text{ mol Cl}_2}\right) = 0.15\text{ mol H}_2\text{O}$

d) $\#\,\text{mol H}_2\text{O} = (0.24\text{ mol HCl})\left(\frac{5\text{ mol H}_2\text{O}}{8\text{ mol HCl}}\right) = 0.15\text{ mol H}_2\text{O}$

4.106 a) $\#\,\text{g Zn} = 0.11\text{ moles Au(CN)}_2^-\left(\frac{1\text{ mol Zn}}{2\text{ mol Au(CN)}_2^-}\right)\left(\frac{65.39\text{ g Zn}}{1\text{ mol Zn}}\right) = 3.6\text{ g Zn}$

b) $\#\,\text{g Au} = 0.11\text{ mol Au(CN)}_2^-\left(\frac{1\text{ mol Au}}{1\text{ mol Au(CN)}_2^-}\right)\left(\frac{197.00\text{ g Au}}{1\text{ mol Au}}\right) = 22\text{ g Au}$

c) $\#\,\text{g Au(CN)}_2^- = 0.11\text{ mol Zn}\left(\frac{2\text{ mol Au(CN)}_2^-}{1\text{ mol Zn}}\right)\left(\frac{249.04\text{ g Au(CN)}_2^-}{1\text{ mol Au(CN)}_2^-}\right)$

$= 55\text{ g Au(CN)}_2^-$

4.108 a) $4P + 5O_2 \rightarrow P_4O_{10}$

b) $\#\,\text{g O}_2 = (6.85\text{ g P})\left(\frac{1\text{ mol P}}{30.97\text{ g P}}\right)\left(\frac{5\text{ mol O}_2}{4\text{ mol P}}\right)\left(\frac{32.0\text{ g O}_2}{1\text{ mol O}_2}\right) = 8.85\text{ g O}_2$

c) $\#\,\text{g P}_4\text{O}_{10} = (8.00\text{ g O}_2)\left(\frac{1\text{ mol O}_2}{32.00\text{ g O}_2}\right)\left(\frac{1\text{ mol P}_4\text{O}_{10}}{5\text{ mol O}_2}\right)\left(\frac{283.9\text{ g P}_4\text{O}_{10}}{1\text{ mol P}_4\text{O}_{10}}\right)$

$= 14.2\text{ g P}_4\text{O}_{10}$

d) $\#\text{g P} = (7.46\text{ g P}_4\text{O}_{10})\left(\frac{1\text{ mol P}_4\text{O}_{10}}{283.9\text{ g P}_4\text{O}_{10}}\right)\left(\frac{4\text{ mol P}}{1\text{ mol P}_4\text{O}_{10}}\right)\left(\frac{30.97\text{ g P}}{1\text{ mol P}}\right) = 3.26\text{ g P}$

4.110 $\#\text{g of HNO}_3 = (11.45\text{ g Cu})\left(\frac{1\text{ mol Cu}}{63.546\text{ g Cu}}\right)\left(\frac{8\text{ mol HNO}_3}{3\text{ mol Cu}}\right)\left(\frac{63.08\text{ g HNO}_3}{1\text{ mol HNO}_3}\right)$

$= 30.31\text{ g HNO}_3$

4.112 $\#\text{kg O}_2 = (1.0\text{ kg H}_2\text{O}_2)\left(\frac{1000\text{ g H}_2\text{O}_2}{1\text{ kg H}_2\text{O}_2}\right)\left(\frac{1\text{ mol H}_2\text{O}_2}{34.02\text{ g H}_2\text{O}_2}\right)\left(\frac{1\text{ mol O}_2}{2\text{ mol H}_2\text{O}_2}\right)$

$\left(\frac{32.00\text{ g O}_2}{1\text{ mol O}_2}\right)\left(\frac{1\text{ kg O}_2}{1000\text{ g O}_2}\right) = 0.47\text{ kg O}_2$

4.114 a) First determine the amount of Fe_2O_3 that would be required to react completely with the given amount of Al:

$$\#\text{mol Fe}_2\text{O}_3 = (4.20\text{ mol Al})\left(\frac{1\text{ mol Fe}_2\text{O}_3}{2\text{ mol Al}}\right) = 2.10\text{ mol Fe}_2\text{O}_3$$

Since only 1.75 mol of Fe_2O_3 are supplied, it is the limiting reactant. This can be confirmed by calculating the amount of Al that would be required to react completely with all of the available Fe_2O_3:

$$\#\text{moles Al} = (1.75\text{ mol Fe}_2\text{O}_3)\left(\frac{2\text{ mol Al}}{1\text{ mol Fe}_2\text{O}_3}\right) = 3.50\text{ mol Al}$$

Since an excess (4.20 mol – 3.50 mol = 0.70 mol) of Al is present, Fe_2O_3 must be the limiting reactant, as determined above.

b) $\#\text{g Fe} = (1.75\text{ mol Fe}_2\text{O}_3)\left(\frac{2\text{ mol Fe}}{1\text{ mol Fe}_2\text{O}_3}\right)\left(\frac{55.847\text{ g Fe}}{1\text{ mol Fe}}\right) = 195\text{ g Fe}$

4.116 $3AgNO_3 + FeCl_3 \rightarrow 3AgCl + Fe(NO_3)_3$
Calculate the amount of $FeCl_3$ that are required to react completely with all of the available silver nitrate:

$$\#\text{g FeCl}_3 = (18.0\text{ g AgNO}_3)\left(\frac{1\text{ mol AgNO}_3}{169.87\text{ g AgNO}_3}\right)\left(\frac{1\text{ mol FeCl}_3}{3\text{ mol AgNO}_3}\right)\left(\frac{162.21\text{ g FeCl}_3}{1\text{ mol FeCl}_3}\right)$$

$= 5.73\text{ g FeCl}_3$

Since more than this minimum amount is available, $FeCl_3$ is present in excess, and $AgNO_3$ must be the limiting reactant.

We know that only 5.73 g $FeCl_3$ will be used. Therefore, the amount left unused is: 32.4 g total – 5.73 g used = 26.7 g $FeCl_3$

4.118 First, calculate the amount of H_2O needed to completely react with the available NO_2;

$$\# \text{ g water} = (1.0 \text{ mg NO}_2)\left(\frac{1 \text{ g NO}_2}{1000 \text{ mg NO}_2}\right)\left(\frac{1 \text{ mol NO}_2}{46.01 \text{ g NO}_2}\right)\left(\frac{1 \text{ mole H}_2\text{O}}{3 \text{ mol NO}_2}\right)$$

$$\times\left(\frac{18.02 \text{ H}_2\text{O}}{1 \text{ mol H}_2\text{O}}\right) = 0.00013 \text{ g H}_2\text{O}$$

There are 0.050 g water available, which is *more* than the required amount above, so NO_2 is the limiting reactant, and our yield will be based upon this:

$$\# \text{ g HNO}_3 = (1.0 \text{ mg NO}_2)\left(\frac{1 \text{ g NO}_2}{1000 \text{ mg NO}_2}\right)\left(\frac{1 \text{ mol NO}_2}{46.01 \text{ g NO}_2}\right)\left(\frac{2 \text{ mol HNO}_3}{3 \text{ mol NO}_2}\right)$$

$$\times\left(\frac{63.02 \text{ g HNO}_3}{1 \text{ mole HNO}_3}\right) = 0.00091 \text{ g HNO}_3$$

4.120 First determine the theoretical yield:

$$\# \text{ g BaSO}_4 = (75.00 \text{ g Ba(NO}_3)_2)\left(\frac{1 \text{ mole Ba(NO}_3)_2}{261.34 \text{ g Ba(NO}_3)_2}\right)$$

$$\times\left(\frac{1 \text{ mole BaSO}_4}{1 \text{ mole Ba(NO}_3)_2}\right)\left(\frac{233.39 \text{ g BaSO}_4}{1 \text{ mole BaSO}_4}\right)$$

$$= 66.98 \text{ g BaSO}_4$$

Then calculate a % yield:

$$\% \text{ yield} = \frac{\text{actual yield}}{\text{theoretical yield}} \times 100 = \frac{64.45 \text{ g}}{66.98 \text{ g}} \times 100 = 96.22\,\%$$

4.122 First, determine how much H_2SO_4 is needed to completely react with the $AlCl_3$

$$\# \text{ g H}_2\text{SO}_4 = (25.00 \text{ g AlCl}_3)\left(\frac{1 \text{ mole AlCl}_3}{133.34 \text{ g AlCl}_3}\right)$$

$$\times\left(\frac{3 \text{ mole H}_2\text{SO}_4}{2 \text{ mole AlCl}_3}\right)\left(\frac{98.08 \text{ g H}_2\text{SO}_4}{1 \text{ mole H}_2\text{SO}_4}\right)$$

$$= 27.58 \text{ g H}_2\text{SO}_4$$

There is an excess of H_2SO_4 present.

Determine the theoretical yield:

$$\# \text{g Al}_2(\text{SO}_4)_3 = (25.00 \text{ g AlCl}_3)\left(\frac{1 \text{ mole AlCl}_3}{133.34 \text{ g AlCl}_3}\right)$$

$$\times \left(\frac{1 \text{ mole Al}_2(\text{SO}_4)_3}{2 \text{ mole AlCl}_3}\right)\left(\frac{342.15 \text{ g Al}_2(\text{SO}_4)_3}{1 \text{ mole Al}_2(\text{SO}_4)_3}\right)$$

$$= 32.07 \text{ g Al}_2(\text{SO}_4)_3$$

$$\% \text{ yield} = \frac{\text{actual yield}}{\text{theoretical yield}} \times 100\,\% = \frac{28.46 \text{ g}}{32.07 \text{ g}} \times 100\,\% = 88.74\,\%$$

*4.124 If the yield for this reaction is only 71 % and we need to have 11.5 g of product, we will attempt to make 16 g of product. This is determined by dividing the actual yield by the percent yield. Recall that

$$\% \text{ yield} = \frac{\text{actual yield}}{\text{theoretical yield}} \times 100.$$

If we rearrange this equation we can see that

$$\text{theoretical yield} = \frac{\text{actual yield}}{\% \text{ yield}} \times 100.$$

Substituting the values from this problem gives the 16 g of product mentioned above.

$$\# \text{g C}_7\text{H}_8 = (16 \text{ g KC}_7\text{H}_5\text{O}_2)\left(\frac{1 \text{ mole KC}_7\text{H}_5\text{O}_2}{160.21 \text{ g KC}_7\text{H}_5\text{O}_2}\right)$$

$$\times \left(\frac{1 \text{ mole C}_7\text{H}_8}{1 \text{ mole KC}_7\text{H}_5\text{O}_2}\right)\left(\frac{92.14 \text{ g C}_7\text{H}_8}{1 \text{ mole C}_7\text{H}_8}\right)$$

$$= 9.2 \text{ g C}_7\text{H}_8$$

Additional Exercises

4.126 1.0 % of 263 tons = 2.6 tons Hg

Since there are 200.59 g Hg in one mol of $(CH_3)_2Hg$ (230.67 g), it follows that:

$$\# \text{lbs (CH}_3)_2\text{Hg} = (2.6 \text{ tons Hg})\left(\frac{230.67 \text{ tons (CH}_3)_2\text{Hg}}{200.59 \text{ tons Hg}}\right)$$

$$\times \left(\frac{2{,}000 \text{ pounds (CH}_3)_2\text{Hg}}{1 \text{ ton (CH}_3)_2\text{Hg}}\right) = 6{,}000 \text{ pounds (or } 6.0\times10^3 \text{ pounds)}$$

4.128 Assume a 100g sample:

$$\# \text{ mol Mg} = (52.9 \text{ g Mg})\left(\frac{1 \text{ mol Mg}}{24.30 \text{ g Mg}}\right) = 2.18 \text{ mol Mg}$$

$$\# \text{ mol B} = (47.1 \text{ g B})\left(\frac{1 \text{ mol B}}{10.81 \text{ g B}}\right) = 4.36 \text{ mol B}$$

Now we divide each of these numbers of mol by the smallest of the two numbers, in order to obtain the simplest mol ratio among the three elements in the compound:

for Mg, 2.18 mol / 2.18 mol = 1.00
for B, 4.36 mol / 2.18 mol = 2.00

These relative mol amounts suggest an empirical formula of MgB_2

4.130

$$\# \text{ yrs} = (6.02 \times 10^{23} \text{ pennies})\left(\frac{1 \text{ dollar}}{100 \text{ pennies}}\right)\left(\frac{1 \text{ second}}{5.00 \times 10^{8} \text{ dollars}}\right)$$
$$\times \left(\frac{1 \text{ min}}{60 \text{ sec}}\right)\left(\frac{1 \text{ hr}}{60 \text{ min}}\right)\left(\frac{1 \text{ day}}{24 \text{ hrs}}\right)\left(\frac{1 \text{ yr}}{365 \text{ days}}\right) = 382{,}000 \text{ yrs}$$

*4.132 First determine the percentage by weight of each element in the respective original samples. This is done by determining the mass of the element in question present in each of the original samples. The percentage by weight of each element in the unknown will be the same as the values we calculate.

$$\# \text{ g Ca} = (0.160 \text{ g } CaCO_3)\left(\frac{1 \text{ mole } CaCO_3}{100.09 \text{ g } CaCO_3}\right)\left(\frac{1 \text{ mole Ca}}{1 \text{ mole } CaCO_3}\right)\left(\frac{40.1 \text{ g Ca}}{1 \text{ mole Ca}}\right)$$
$$= 0.0641 \text{ g Ca}$$

$$\% \text{ Ca} = (0.0641/0.250) \times 100 = 25.6 \text{ \% Ca}$$

$$\# \text{ g S} = (0.344 \text{ g } BaSO_4)\left(\frac{1 \text{ mole } BaSO_4}{233.8 \text{ g } BaSO_4}\right)\left(\frac{1 \text{ mole S}}{1 \text{ mole } BaSO_4}\right)\left(\frac{32.07 \text{ g S}}{1 \text{ mole S}}\right)$$
$$= 0.0472 \text{ g S}$$

$$\% \text{ S} = (0.0472/0.115) \times 100 = 41.0 \text{ \% S}$$

$$\# \text{ g N} = (0.155 \text{ g } NH_3)\left(\frac{1 \text{ mole } NH_3}{17.03 \text{ g } NH_3}\right)\left(\frac{1 \text{ mole N}}{1 \text{ mole } NH_3}\right)\left(\frac{14.01 \text{ g N}}{1 \text{ mole N}}\right)$$
$$= 0.128 \text{ g N}$$

$\% N = (0.128/0.712) \times 100 = 18.0 \% N$

$\% C = 100.0 - (25.6 + 41.0 + 18.0) = 15.4 \% C$. Next, we assume 100 g of the compound, and convert these weight percentages into mol amounts:

$$\#\text{ moles Ca} = (25.6 \text{ g Ca})\left(\frac{1 \text{ mole Ca}}{40.08 \text{ g Ca}}\right) = 0.639 \text{ moles Ca}$$

$$\#\text{ moles S} = (41.0 \text{ g S})\left(\frac{1 \text{ mole S}}{32.07 \text{ g S}}\right) = 1.28 \text{ moles S}$$

$$\#\text{ moles N} = (18.0 \text{ g N})\left(\frac{1 \text{ mole N}}{14.07 \text{ g N}}\right) = 1.28 \text{ moles N}$$

$$\#\text{ moles C} = (15.4 \text{ g C})\left(\frac{1 \text{ mole C}}{12.01 \text{ g C}}\right) = 1.28 \text{ moles C}$$

Dividing each of these mol amounts by the smallest, we have:

For Ca: 0.639 mol / 0.639 mol = 1.00
For S: 1.28 mol / 0.639 mol = 2.00
For N: 1.28 mol / 0.639 mol = 2.00
For C: 1.28 mol / 0.639 mol = 2.00

The empirical formula is therefore $CaC_2S_2N_2$, and the mass of the empirical unit is Ca + 2S + 2N + 2C = 156 g/mol. Since the molcular mass is the same as the empirical mass, the molcular formula is $CaC_2S_2N_2$.

4.134

$$\#\text{ g } (NH_2)_2CO = (6.00 \text{ g N})\left(\frac{1 \text{ mol N}}{14.007 \text{ g N}}\right)\left(\frac{1 \text{ mole } (NH_2)_2CO}{2 \text{ moles N}}\right)\left(\frac{60.06 \text{ g } (NH_2)_2CO}{1 \text{ mol } (NH_2)_2CO}\right)$$
$$= 12.9 \text{ g } (NH_2)_2CO$$

4.136 Assume the hydrogen is the limiting reactant.

$$\text{lb } O_2 = (227{,}641 \text{ lb } H_2)\left(\frac{453.59237 \text{ g}}{1 \text{ lb}}\right)\left(\frac{1 \text{ mol } H_2}{2.01588 \text{ g } H_2}\right)\left(\frac{1 \text{ mol } O_2}{2 \text{ mol } H_2}\right)$$
$$\times \left(\frac{31.9988 \text{ g } O_2}{1 \text{ mol } O_2}\right)\left(\frac{1 \text{ lb } O_2}{453.59237 \text{ g } O_2}\right) = 1{,}806{,}714 \text{ lb } O_2$$

Since this is more than the amount of O_2 that is supplied, the limiting reactant must be O_2.Next calculate the amount of H_2 needed to react completely with all of the available O_2.

$$\text{lb H}_2 = (1{,}361{,}936 \text{ lb O}_2)\left(\frac{453.59237 \text{ g}}{1 \text{ lb}}\right)\left(\frac{1 \text{ mol O}_2}{31.9988 \text{ g O}_2}\right)\left(\frac{2 \text{ mol H}_2}{1 \text{ mol O}_2}\right)$$

$$\times \left(\frac{2.01588 \text{ g H}_2}{1 \text{ mol H}_2}\right)\left(\frac{1 \text{ lb H}_2}{453.59237 \text{ g H}_2}\right) = 171{,}600 \text{ lb H}_2$$

Since only 171,600 lb. of H_2 reacted, there are 227,641 lb. – 171,600 lb. = 56,041 lb. of unreacted H_2.

4.138 $$\text{g F} = (1.0\times10^{-9} \text{ g Cl})\left(\frac{1 \text{ mol Cl}}{35.45 \text{ g Cl}}\right)\left(\frac{1 \text{ mol CF}_2\text{Cl}_2}{2 \text{ mol Cl}}\right)\left(\frac{2 \text{ mol F}}{1 \text{ mol CF}_2\text{Cl}_2}\right)$$

$$\times \left(\frac{19.00 \text{ g F}}{1 \text{ mol F}}\right) = 5.4\times10^{-10} \text{ g F}$$

Chapter 5

Practice Exercises

5.1 a) $MgCl_2(s) \rightarrow Mg^{2+}(aq) + 2Cl^-(aq)$
b) $Al(NO_3)_3(s) \rightarrow Al^{3+}(aq) + 3NO_3^-(aq)$
c) $Na_2CO_3(s) \rightarrow 2Na^+(aq) + CO_3^{2-}(aq)$

5.2 molecular: $CdCl_2(aq) + Na_2S(aq) \rightarrow CdS(s) + 2NaCl(aq)$
ionic: $Cd^{2+}(aq) + 2Cl^-(aq) + 2Na^+(aq) + S^{2-}(aq) \rightarrow CdS(s) + 2Na^+(aq) + 2Cl^-(aq)$
net ionic: $Cd^{2+}(aq) + S^{2-}(aq) \rightarrow CdS(s)$

5.3 a) molecular: $AgNO_3(aq) + NH_4Cl(aq) \rightarrow AgCl(s) + NH_4NO_3(aq)$
ionic: $Ag^+(aq) + NO_3^-(aq) + NH_4^+(aq) + Cl^-(aq) \rightarrow AgCl(s) + NH_4^+(aq) + NO_3^-(aq)$
net ionic: $Ag^+(aq) + Cl^-(aq) \rightarrow AgCl(s)$

b) molecular: $Na_2S(aq) + Pb(C_2H_3O_2)_2(aq) \rightarrow 2NaC_2H_3O_2(aq) + PbS(s)$
ionic: $2Na^+(aq) + S^{2-}(aq) + Pb^{2+}(aq) + 2C_2H_3O_2^-(aq) \rightarrow 2Na^+(aq) + 2C_2H_3O_2^-(aq) + PbS(s)$
net ionic: $S^{2-}(aq) + Pb^{2+}(aq) \rightarrow PbS(s)$

5.4 $HCHO_2(aq) + H_2O \rightarrow H_3O^+(aq) + CHO_2^-(aq)$

5.5 $H_3C_6H_5O_7(s) + H_2O \rightarrow H_3O^+(aq) + H_2C_6H_5O_7^-(aq)$
$H_2C_6H_5O_7^-(aq) + H_2O \rightarrow H_3O^+(aq) + HC_6H_5O_7^{2-}(aq)$
$HC_6H_5O_7^{2-}(aq) + H_2O \rightarrow H_3O^+(aq) + C_6H_5O_7^{3-}(aq)$

5.6 $HONH_2(aq) + H_2O \rightarrow HONH_3^+(aq) + OH^-(aq)$

5.7 HF: Hydrofluoric acid, sodium salt = sodium fluoride (NaF)
HBr: Hydrobromic acid, sodium salt = sodium bromide (NaBr)

5.8 Sodium arsenate

5.9 Calcium formate

5.10 $H_3PO_4(aq) + NaOH(aq) \rightarrow NaH_2PO_4(aq) + H_2O$
$NaH_2PO_4(aq) + NaOH(aq) \rightarrow Na_2HPO_4(aq) + H_2O$
$Na_2HPO4(aq) + NaOH(aq) \rightarrow Na_3PO_4(aq) + H_2O$

5.11 $NaHSO_3$ (sodium hydrogen sulfite)

5.12 $HNO_2(aq) + H_2O \rightleftharpoons H_3O^+(aq) + NO_2^-(aq)$

5.13 $CH_3NH_2(aq) + H_2O \rightleftharpoons CH_3NH_3^+(aq) + OH^-(aq)$

5.14 molecular: $2HCl(aq) + Ca(OH)_2(aq) \rightarrow CaCl_2(aq) + 2H_2O$
ionic: $2H^+(aq) + 2Cl^-(aq) + Ca^{2+}(aq) + 2OH^-(aq) \rightarrow Ca^{2+}(aq) + 2Cl^-(aq) + 2H_2O$
net ionic: $2H^+(aq) + 2OH^-(aq) \rightarrow 2H_2O$

5.15 a) molecular: $HCl(aq) + KOH(aq) \rightarrow H_2O + KCl(aq)$
ionic: $H^+(aq) + Cl^-(aq) + K^+(aq) + OH^-(aq) \rightarrow H_2O + K^+(aq) + Cl^-(aq)$
net ionic: $H^+(aq) + OH^-(aq) \rightarrow H_2O$

b) molecular: $HCHO_2(aq) + LiOH(aq) \rightarrow H_2O + LiCHO_2(aq)$
ionic: $HCHO_2(aq) + Li^+(aq) + OH^-(aq) \rightarrow H_2O + Li^+(aq) + CHO_2^-(aq)$
net ionic: $HCHO_2(aq) + OH^-(aq) \rightarrow H_2O + CHO_2^-(aq)$

c) molecular: $N_2H_4(aq) + HCl(aq) \rightarrow N_2H_5Cl(aq)$
ionic: $N_2H_4(aq) + H^+(aq) + Cl^-(aq) \rightarrow N_2H_5^+(aq) + Cl^-(aq)$
net ionic: $N_2H_4(aq) + H^+(aq) \rightarrow N_2H_5^+(aq)$

5.16 molecular: $CH_3NH_2(aq) + HCHO_2(aq) \rightarrow CH_3NH_3CHO_2(aq)$
ionic: $CH_3NH_2(aq) + HCHO_2(aq) \rightarrow CH_3NH_3^+(aq) + CHO_2^-(aq)$
net ionic: $CH_3NH_2(aq) + HCHO_2(aq) \rightarrow CH_3NH_3^+(aq) + CHO_2^-(aq)$

5.17 molecular: $Al(OH)_3(s) + 3HCl(aq) \rightarrow AlCl_3(aq) + 3H_2O$
ionic: $Al(OH)_3(s) + 3H^+(aq) + 3Cl^-(aq) \rightarrow Al^{3+}(aq) + 3Cl^-(aq) + 3H_2O$
net ionic: $Al(OH)_3(s) + 3H^+(aq) \rightarrow Al^{3+}(aq) + 3H_2O$

5.18 a) Formic acid, a weak acid will form.
Net ionic equation: $H^+(aq) + CHO_2^-(aq) \rightleftharpoons HCHO_2(aq)$
b) Carbonic acid will form and it will further dissociate to water and carbon dioxide:
$CuCO_3(s) + 2H^+(aq) \rightarrow 2CO_2(g) + 2H_2O + Cu^{2+}(aq)$
c) NR
d) Insoluble nickel hydroxide will precipitate.
$Ni^{2+}(aq) + 2OH^-(aq) \rightarrow Ni(OH)_2(s)$

5.19 $$\#\,\text{mol Na}_2\text{SO}_4 = (3.550\text{ g Na}_2\text{SO}_4)\left(\frac{1\text{ mol Na}_2\text{SO}_4}{142.1\text{ g Na}_2\text{SO}_4}\right) = 0.02498\text{ mol Na}_2\text{SO}_4$$

$$\#\,\text{L solution} = (100.0\text{ mL})\left(\frac{1\text{ L}}{1000\text{ mL}}\right) = 0.1000\text{ L solution}$$

$$\text{M} = \left(\frac{\text{moles solute}}{\text{L solution}}\right) = \left(\frac{0.02498\text{ mol Na}_2\text{SO}_4}{0.1000\text{ L solution}}\right) = 0.2498\text{ M}$$

5.20 $\# \text{mL solution} = (0.0500 \text{ mol KCl})\left(\frac{1 \text{ L solution}}{0.150 \text{ mol KCl}}\right)\left(\frac{1000 \text{ mL solution}}{1 \text{ L solution}}\right) = 333 \text{ mL}$

5.21 $\# \text{g AgNO}_3 = (250 \text{ mL solution})\left(\frac{1 \text{ L solution}}{1000 \text{ mL solution}}\right)\left(\frac{0.0125 \text{ mol AgNO}_3}{1 \text{ L solution}}\right)$

$$\times\left(\frac{169.90 \text{ g AgNO}_3}{1 \text{ mol AgNO}_3}\right) = 0.53 \text{ g AgNO}_3$$

5.22 $(V_{dil})(M_{dil}) = (V_{conc})(M_{conc})$

$(100 \text{ mL})(0.125M) = (V_{conc})(0.500M)$

$V_{conc} = (100 \text{ mL})(0.125M)/(0.500M) = 25.0 \text{ mL}$

Therefore, mix 25.0 mL of 0.500 M H_2SO_4 with enough water to make 100 mL of total solution.

5.23 $\# \text{mL NaOH} = (15.4 \text{ mL H}_2\text{SO}_4)\left(\frac{1 \text{ L H}_2\text{SO}_4}{1000 \text{ mL H}_2\text{SO}_4}\right)\left(\frac{0.108 \text{ mol H}_2\text{SO}_4}{1 \text{ L H}_2\text{SO}_4}\right)$

$$\times\left(\frac{2 \text{ mol NaOH}}{1 \text{ mol H}_2\text{SO}_4}\right)\left(\frac{1 \text{ L NaOH}}{0.124 \text{ mol NaOH}}\right)\left(\frac{1000 \text{ mL NaOH}}{1 \text{ L NaOH}}\right) = 26.8 \text{ mL NaOH}$$

5.24 $FeCl_3 \rightarrow Fe^{3+} + 3Cl^-$

$$\text{M Fe}^{3+} = \left(\frac{0.40 \text{ mol FeCl}_3}{1 \text{ L FeCl}_3 \text{ soln}}\right)\left(\frac{1 \text{ mol Fe}^{3+}}{1 \text{ mol FeCl}_3}\right) = 0.40 \text{ M Fe}^{3+}$$

$$\text{M Cl}^- = \left(\frac{0.40 \text{ mol FeCl}_3}{1 \text{ L FeCl}_3 \text{ soln}}\right)\left(\frac{3 \text{ mol Cl}^-}{1 \text{ mol FeCl}_3}\right) = 1.2 \text{ M Cl}^-$$

5.25 $\text{M Na}^+ = \left(\frac{0.250 \text{ mol PO}_4^{\,3-}}{1 \text{ L Na}_3\text{PO}_4 \text{ soln}}\right)\left(\frac{3 \text{ mol Na}^+}{1 \text{ mol PO}_4^{\,3-}}\right) = 0.750 \text{ M Na}^+$

5.26 The balanced net ionic equation is: $Fe^{2+}(aq) + 2OH^-(aq) \rightarrow Fe(OH)_2(s)$.
First determine the number of moles of Fe^{2+} present.

$$\# \text{moles Fe}^{2+} = (60.0 \text{ mL FeCl}_2 \text{ solution})\left(\frac{0.250 \text{ mol FeCl}_2}{1000 \text{ mL solution}}\right)\left(\frac{1 \text{ mol Fe}^{2+}}{1 \text{ mol FeCl}_2}\right)$$

$$= 1.50 \times 10^{-2} \text{ mol Fe}^{2+}$$

Now, determine the amount of KOH needed to react with the Fe^{2+}.

$$\# \text{ mL KOH} = (1.50\times10^{-2} \text{ mol Fe}^{2+})\left(\frac{2 \text{ mol OH}^-}{1 \text{ mol Fe}^{2+}}\right)$$

$$\times\left(\frac{1 \text{ mol KOH}}{1 \text{ mol OH}^-}\right)\left(\frac{1000 \text{ mL solution}}{0.500 \text{ mol KOH}}\right)$$

$$= 60.0 \text{ mL KOH}$$

5.27 The net ionic equation is $Ba^{2+}(aq) + SO_4^{2-}(aq) \rightarrow BaSO_4(s)$
First, determine the initial number of moles of Ba^{2+} ion that are present:

$$\# \text{ mol Ba}^{2+} = (20.0 \text{ mL BaCl}_2 \text{ soln})\left(\frac{0.600 \text{ mol BaCl}_2}{1000 \text{ mL BaCl}_2 \text{ soln}}\right)\left(\frac{1 \text{ mol Ba}^{2+}}{1 \text{ mol BaCl}_2}\right)$$

$$= 1.20\times10^{-2} \text{ mol Ba}^{2+}$$

Next, determine the initial number of moles of sulfate ion that are present:

$$\# \text{ mol SO}_4^{2-} = (30.0 \text{ mL MgSO}_4 \text{ soln})\left(\frac{0.500 \text{ mol MgSO}_4}{1000 \text{ mL MgSO}_4 \text{ soln}}\right)\left(\frac{1 \text{ mol SO}_4^{2-}}{1 \text{ mol MgSO}_4}\right)$$

$$= 1.50\times10^{-2} \text{ mol SO}_4^{2-}$$

Now determine the number of moles of barium ion that are required to react with this much sulphate ion, and compare the result to the amount of barium ion that is available:

$$\# \text{ mol Ba}^{2+} = (1.50\times10^{-2} \text{ mol SO}_4^{2-})\left(\frac{1 \text{ mol Ba}^{2+}}{1 \text{ mol SO}_4^{2-}}\right)$$

$$= 1.50\times10^{-2} \text{ mol Ba}^{2+}$$

Since there is not this much Ba^{2+} available according to the above calculation, then we can conclude that Ba^{2+} must be the limiting reactant, and that subsequent calculations should be based on the number of moles of it that are present:

Since this reaction is 1:1, we know that 1.20×10^{-2} mole of $BaSO_4$ will be formed.

If we assume that the $BaSO_4$ is completely insoluble, then the concentration of barium ion will be essentially zero. The concentrations of the other ions are determined as follows:

$$\# \text{M Cl}^- = \frac{(20.0 \text{ mL BaCl}_2 \text{ soln})\left(\frac{0.600 \text{ mol BaCl}_2}{1000 \text{ mL BaCl}_2 \text{ soln}}\right)\left(\frac{2 \text{ mol Cl}^-}{1 \text{ mol BaCl}_2}\right)}{((20.0+30.0) \text{ mL soln})\left(\frac{1 \text{ L soln}}{1000 \text{ mL soln}}\right)}$$

$$= 0.480 \text{ M Cl}^-$$

$$\# \text{M Mg}^{2+} = \frac{(30.0 \text{ mL MgSO}_4 \text{ soln})\left(\frac{0.500 \text{ mol MgSO}_4}{1000 \text{ mL MgSO}_4 \text{ soln}}\right)\left(\frac{1 \text{ mol Mg}^{2+}}{1 \text{ mol MgSO}_4}\right)}{((30.0+20.0) \text{ mL soln})\left(\frac{1 \text{ L soln}}{1000 \text{ mL soln}}\right)}$$

$$= 0.300 \text{ M Mg}^{2+}$$

For sulfate, we subtract the amount that reacted with the Ba^{2+}:
\# mol SO_4^{2-} = $1.50 \text{ x } 10^{-2}$ mol – 1.20×10^{-2} mol = 3.0×10^{-3} mol

This allows a calculation of the final sulfate concentration:

$$\# \text{M SO}_4^{\ 2-} = \frac{3.0 \text{ X } 10^{-3} \text{ mol SO}_4^{\ 2-}}{((30.0+20.0) \text{ mL soln})\left(\frac{1 \text{ L soln}}{1000 \text{ mL soln}}\right)}$$

$$= 6.0 \times 10^{-2} \text{ M SO}_4^{\ 2-}$$

5.28 a) $\# \text{mol Ca}^{2+} = (0.736 \text{ g CaSO}_4)\left(\frac{1 \text{ mol CaSO}_4}{136.14 \text{ g CaSO}_4}\right)\left(\frac{1 \text{ mol Ca}^{2+}}{1 \text{ mol CaSO4}}\right)$

$$= 5.41 \times 10^{-3} \text{ mol Ca}^{2+}$$

b) Since all of the Ca^{2+} is precipitated as $CaSO_4$, there were originally 5.41×10^{-3} moles of Ca^{2+} in the sample.

c) All of the Ca^{2+} comes from $CaCl_2$, so there were 5.41×10^{-3} moles of $CaCl_2$ in the sample.

d) $\# \text{g CaCl}_2 = (5.41 \times 10^{-3} \text{ mol CaCl}_2)\left(\frac{110.98 \text{ g CaCl}_2}{1 \text{ mol CaCl}_2}\right) = 0.600 \text{ g CaCl}_2$

e) $\% \text{ CaCl}_2 = \frac{0.600 \text{ g CaCl}_2}{2.000 \text{ g sample}} \times 100 = 30.0 \% \text{ CaCl}_2$

5.29

$$M\,H_2SO_4 = \left[\frac{(36.42\text{ mL NaOH soln})\left(\frac{1\text{ L NaOH soln}}{1000\text{ mL NaOH soln}}\right)\left(\frac{0.147\text{ mol NaOH}}{1\text{ L NaOH soln}}\right)\left(\frac{1\text{ mol }H_2SO_4}{2\text{ mol NaOH}}\right)}{(15.00\text{ mL }H_2SO_4\text{ soln})\left(\frac{1\text{ L }H_2SO_4\text{ soln}}{1000\text{ mL }H_2SO_4\text{ soln}}\right)}\right]$$

$$= 0.178\text{ M }H_2SO_4$$

5.30

$$M\,HCl = \left[\frac{(11.00\text{ mL KOH soln})\left(\frac{1\text{ L KOH soln}}{1000\text{ mL NaOH soln}}\right)\left(\frac{0.0100\text{ mol KOH}}{1\text{ L KOH soln}}\right)\left(\frac{1\text{ mol HCl}}{1\text{ mol KOH}}\right)}{(5.00\text{ mL HCl soln})\left(\frac{1\text{ L HCl soln}}{1000\text{ mL HCl soln}}\right)}\right]$$

$$= 0.0220\text{ M HCl}$$

$$\#\text{ g HCl} = (5.00\text{ mL HCl})\left(\frac{0.0220\text{ mol HCl}}{1000\text{ mL HCl}}\right)\left(\frac{36.5\text{ g HCl}}{1\text{ mol HCl}}\right) = 4.02\times10^{-3}\text{ g HCl}$$

$$\text{weight }\% = \frac{4.02\text{ x }10^{-3}\text{ g}}{5.00\text{ g}}\times100\% = 0.0803\%$$

Review Problems

5.60 a) $LiCl(s) \rightarrow Li^{+}(aq) + Cl^{-}(aq)$
b) $BaCl_2(s) \rightarrow Ba^{2+}(aq) + 2Cl^{-}(aq)$
c) $Al(C_2H_3O_2)_3(s) \rightarrow Al^{3+}(aq) + 3C_2H_3O_2^{-}(aq)$
d) $(NH_4)_2CO_3(s) \rightarrow 2NH_4^{+}(aq) + CO_3^{2-}(aq)$
e) $FeCl_3(s) \rightarrow Fe^{3+}(aq) + 3Cl^{-}(aq)$

5.62 a) ionic: $2NH_4^{+}(aq) + CO_3^{2-}(aq) + Mg^{2+}(aq) + 2Cl^{-}(aq) \rightarrow$
$2NH_4^{+}(aq) + 2Cl^{-}(aq) + MgCO_3(s)$
net: $Mg^{2+}(aq) + CO_3^{2-}(aq) \rightarrow MgCO_3(s)$

b) ionic: $Cu^{2+}(aq) + 2Cl^{-}(aq) + 2Na^{+}(aq) + 2OH^{-}(aq) \rightarrow$
$Cu(OH)_2(s) + 2Na^{+}(aq) + 2Cl^{-}(aq)$
net: $Cu^{2+}(aq) + 2OH^{-}(aq) \rightarrow Cu(OH)_2(s)$

c) ionic: $3Fe^{2+}(aq) + 3SO_4^{2-}(aq) + 6Na^{+}(aq) + 2PO_4^{3-}(aq) \rightarrow$
$Fe_3(PO_4)_2(s) + 6Na^{+}(aq) + 3SO_4^{2-}(aq)$
net: $3Fe^{2+}(aq) + 2PO_4^{3-}(aq) \rightarrow Fe_3(PO_4)_2(s)$

d) ionic: $2Ag^{+}(aq) + 2C_2H_3O_2^{-}(aq) + Ni^{2+}(aq) + 2Cl^{-}(aq) \rightarrow$
$2AgCl(s) + Ni^{2+}(aq) + 2C_2H_3O_2^{-}(aq)$
net: $2Ag^{+}(aq) + 2Cl^{-}(aq) \rightarrow 2AgCl(s)$

5.64 molecular: $Na_2S(aq) + Cu(NO_3)_2(aq) \rightarrow 2NaNO_3(aq) + CuS(s)$
ionic: $2Na^+(aq) + S^{2-}(aq) + Cu^{2+}(aq) + 2NO_3^-(aq) \rightarrow$
$2Na^+(aq) + 2NO_3^-(aq) + CuS(s)$
net: $S^{2-}(aq) + Cu^{2+}(aq) \rightarrow CuS(s)$

5.66 molecular: $AgNO_3(aq) + NaBr(aq) \rightarrow AgBr(s) + NaNO_3(aq)$
ionic: $Ag^+(aq) + NO_3^-(aq) + Na^+(aq) + Br^-(aq) \rightarrow AgBr(s) + Na^+(aq) + NO_3^-(aq)$
net: $Ag^+(aq) + Br^-(aq) \rightarrow AgBr(s)$

5.68 This is an ionization reaction: $HClO_4(aq) + H_2O \rightarrow H_3O^+(aq) + ClO_4^-(aq)$

5.70 $N_2H_4(aq) + H_2O \rightleftharpoons N_2H_5^+(aq) + OH^-(aq)$

5.72 $HNO_2(aq) + H_2O \rightleftharpoons H_3O^+(aq) + NO_2^-(aq)$

5.74 $H_2CO_3(aq) + H_2O \rightleftharpoons H_3O^+(aq) + HCO_3^-(aq)$
$HCO_3^-(aq) + H_2O \rightleftharpoons H_3O^+(aq) + CO_3^{2-}(aq)$

5.76 a) molecular: $Ca(OH)_2(aq) + 2HNO_3(aq) \rightarrow Ca(NO_3)_2(aq) + 2H_2O$
ionic: $Ca^{2+}(aq) + 2OH^-(aq) + 2H^+(aq) + 2NO_3^-(aq) \rightarrow$
$Ca^{2+}(aq) + 2NO_3^-(aq) + 2H_2O$
net: $H^+(aq) + OH^-(aq) \rightarrow H_2O$

b) molecular: $Al_2O_3(s) + 6HCl(aq) \rightarrow 2AlCl_3(aq) + 3H_2O$
ionic: $Al_2O_3(s) + 6H^+(aq) + 6Cl^-(aq) \rightarrow 2Al^{3+}(aq) + 6Cl^-(aq) + 3H_2O$
net: $Al_2O_3(s) + 6H^+(aq) \rightarrow 2Al^{3+}(aq) + 3H_2O$

c) molecular: $Zn(OH)_2(s) + H_2SO_4(aq) \rightarrow ZnSO_4(aq) + 2H_2O$
ionic: $Zn(OH)_2(s) + 2H^+(aq) + SO_4^{2-}(aq) \rightarrow Zn^{2+}(aq) + SO_4^{2-}(aq) + 2H_2O$
net: $Zn(OH)_2(s) + 2H^+(aq) \rightarrow Zn^{2+}(aq) + 2H_2O$

5.78 a) $2H^+(aq) + CO_3^{2-}(aq) \rightarrow H_2O + CO_2(g)$
b) $NH_4^+(aq) + OH^-(aq) \rightarrow NH_3(aq) + H_2O$

5.80 These reactions have the following "driving forces":
a) formation of insoluble $Cr(OH)_3$
b) formation of water, a weak electrolyte

5.82 The soluble ones are (a), (b), and (d).

5.84 The insoluble ones are (a), (d), and (f).

5.86 a) $3HNO_3(aq) + Cr(OH)_3(s) \rightarrow Cr(NO_3)_3(aq) + 3H_2O$
ionic: $3H^+(aq) + 3NO_3^-(aq) + Cr(OH)_3(s) \rightarrow$
$Cr^{3+}(aq) + 3NO_3^-(aq) + 3H_2O$

net: $3H^+(aq) + Cr(OH)_3(s) \rightarrow Cr^{3+}(aq) + 3H_2O$

b) $HClO_4(aq) + NaOH(aq) \rightarrow NaClO_4(aq) + H_2O$
ionic: $H^+(aq) + ClO_4^-(aq) + Na^+(aq) + OH^-(aq) \rightarrow$
$Na^+(aq) + ClO_4^-(aq) + H_2O$
net: $H^+(aq) + OH^-(aq) \rightarrow H_2O$

c) $Cu(OH)_2(s) + 2HC_2H_3O_2(aq) \rightarrow Cu(C_2H_3O_2)_2(aq) + 2H_2O$
ionic: $Cu(OH)_2(s) + 2HC_2H_3O_2(aq) \rightarrow$
$Cu^{2+}(aq) + 2C_2H_3O_2^-(aq) + 2H_2O$
net: $Cu(OH)_2(s) + 2HC_2H_3O_2(aq) \rightarrow$
$Cu^{2+}(aq) + 2C_2H_3O_2^-(aq) + 2H_2O$

d) $ZnO(s) + H_2SO_4(aq) \rightarrow ZnSO_4(aq) + H_2O$
ionic: $ZnO(s) + 2H^+(aq) + SO_4^{2-}(aq) \rightarrow Zn^{2+}(aq) + SO_4^{2-}(aq) + H_2O$
net: $ZnO(s) + 2H^+(aq) \rightarrow Zn^{2+}(aq) + H_2O$

5.88 a) $Na_2SO_3(aq) + Ba(NO_3)_2(aq) \rightarrow BaSO_3(s) + 2NaNO_3(aq)$
ionic: $2Na^+(aq) + SO_3^{2-}(aq) + Ba^{2+}(aq) + 2NO_3^-(aq) \rightarrow$
$BaSO_3(s) + 2Na^+(aq) + 2NO_3^-(aq)$
net: $Ba^{2+}(aq) + SO_3^{2-}(aq) \rightarrow BaSO_3(s)$

b) $2HCHO_2(aq) + K_2CO_3(aq) \rightarrow 2KCHO_2(aq) + H_2O + CO_2(g)$
ionic: $2HCHO_2(aq) + 2K^+(aq) + CO_3^{2-}(aq)) \rightarrow 2K^+(aq) + 2CHO_2^-(aq)$
$+ H_2O + CO_2(g)$
net: $2HCHO_2(aq) + CO_3^{2-}(aq)) \rightarrow 2CHO_2^-(aq) + H_2O + CO_2(g)$

c) $2NH_4Br(aq) + Pb(C_2H_3O_2)_2(aq) \rightarrow 2NH_4C_2H_3O_2(aq) + PbBr_2(s)$
ionic: $2NH_4^+(aq) + 2Br^-(aq) + Pb^{2+}(aq) + 2C_2H_3O_2^-(aq) \rightarrow$
$2NH_4^+(aq) + 2C_2H_3O_2^-(aq) + PbBr_2(s)$
net: $Pb^{2+}(aq) + 2Br^-(aq) \rightarrow PbBr_2(s)$

d) $2NH_4ClO_4(aq) + Cu(NO_3)_2(aq) \rightarrow Cu(ClO_4)_2(aq) + 2NH_4NO_3(aq)$
ionic: $2NH_4^+(aq) + 2ClO_4^-(aq) + Cu^{2+}(aq) + 2NO_3^-(aq) \rightarrow$
$Cu^{2+}(aq) + 2ClO_4^-(aq) + 2NO_3^-(aq) + 2NH_4^+(aq)$
net: N.R.

*5.90 There are numerous possible answers. One of many possible sets of answers would be:
a) $NaHCO_3(aq) + HCl(aq) \rightarrow NaCl(aq) + CO_2(g) + H_2O$
b) $FeCl_2(aq) + 2NaOH(aq) \rightarrow Fe(OH)_2(s) + 2NaCl(aq)$
c) $Ba(NO_3)_2(aq) + K_2SO_3(aq) \rightarrow BaSO_3(s) + 2KNO_3(aq)$
d) $2AgNO_3(aq) + Na_2S(aq) \rightarrow Ag_2S(s) + 2NaNO_3(aq)$
e) $ZnO(s) + 2HCl(aq) \rightarrow ZnCl_2(aq) + H_2O$

5.92 $\dfrac{0.25\text{ mol HCl}}{1\text{ L HCl soln}}$ and $\dfrac{1\text{ L HCl soln}}{0.25\text{ mol HCl}}$

5.94 For each of the following recall that molarity is defined as moles of solute divided by liters of solution.

a)

$$\text{M NaOH} = \left(\frac{4.00\text{ g NaOH}}{100.0\text{ mL NaOH soln}}\right)\left(\frac{1\text{ mol NaOH}}{40.00\text{ g NaOH}}\right) \times \left(\frac{1000\text{ mL NaOH soln}}{1\text{ L NaOH soln}}\right)$$

$$= 1.00\text{ M NaOH}$$

b)

$$\text{M CaCl}_2 = \left(\frac{16.0\text{ g CaCl}_2}{250.0\text{ mL CaCl}_2\text{ soln}}\right)\left(\frac{1\text{ mol CaCl}_2}{110.98\text{ g CaCl}_2}\right) \times \left(\frac{1000\text{ mL CaCl}_2\text{ soln}}{1\text{ L CaCl}_2\text{ soln}}\right)$$

$$= 0.577\text{ M CaCl}_2$$

c)

$$\text{M KOH} = \left(\frac{14.0\text{ g KOH}}{75.0\text{ mL KOH soln}}\right)\left(\frac{1\text{ mol KOH}}{56.11\text{ g KOH}}\right) \times \left(\frac{1000\text{ mL KOH soln}}{1\text{ L KOH soln}}\right)$$

$$= 3.33\text{ M KOH}$$

d)

$$\text{M H}_2\text{C}_2\text{O}_4 = \left(\frac{6.75\text{ g H}_2\text{C}_2\text{O}_4}{500\text{ mL H}_2\text{C}_2\text{O}_4\text{ soln}}\right)\left(\frac{1\text{ mol H}_2\text{C}_2\text{O}_4}{90.0\text{ g H}_2\text{C}_2\text{O}_4}\right) \times \left(\frac{1000\text{ mL H}_2\text{C}_2\text{O}_4\text{ soln}}{1\text{ L H}_2\text{C}_2\text{O}_4\text{ soln}}\right)$$

$$= 0.150\text{ M H}_2\text{C}_2\text{O}_4$$

5.96 a)

$$\# \text{ g NaCl} = (125 \text{ mL NaCl soln})\left(\frac{1 \text{ L NaCl soln}}{1000 \text{ mL NaCl soln}}\right) \times \left(\frac{0.200 \text{ mol NaCl}}{1 \text{ L NaCl soln}}\right)\left(\frac{58.44 \text{ g NaCl}}{1 \text{ mol NaCl}}\right)$$

$$= 1.46 \text{ g NaCl}$$

b)

$$\# \text{ g } C_6H_{12}O_6 = (250 \text{ mL } C_6H_{12}O_6 \text{ soln})\left(\frac{1 \text{ L } C_6H_{12}O_6 \text{ soln}}{1000 \text{ mL } C_6H_{12}O_6 \text{ soln}}\right) \times \left(\frac{0.360 \text{ mol } C_6H_{12}O_6}{1 \text{ L } C_6H_{12}O_6 \text{ soln}}\right)\left(\frac{180.2 \text{ g } C_6H_{12}O_6}{1 \text{ mol } C_6H_{12}O_6}\right)$$

$$= 16.2 \text{ g } C_6H_{12}O_6$$

c)

$$\# \text{ g } H_2SO_4 = (250 \text{ mL } H_2SO_4 \text{ soln})\left(\frac{1 \text{ L } H_2SO_4 \text{ soln}}{1000 \text{ mL } H_2SO_4 \text{ soln}}\right) \times \left(\frac{0.250 \text{ mol } H_2SO_4}{1 \text{ L } H_2SO_4 \text{ soln}}\right)\left(\frac{98.08 \text{ g } H_2SO_4}{1 \text{ mol } H_2SO_4}\right)$$

$$= 6.13 \text{ g } H_2SO_4$$

5.98

$$M_{dil} = \frac{V_{concd} \bullet M_{concd}}{V_{dil}} = \frac{(25.0 \text{ mL})(0.56 \text{ M } H_2SO_4)}{125 \text{ mL}} = 0.11 \text{ M } H_2SO_4$$

5.100

$$V_{dil} = \frac{V_{concd} \bullet M_{concd}}{M_{dil}} = \frac{(25.0 \text{ mL})(18.0 \text{ M } H_2SO_4)}{1.50 \text{ M } H_2SO_4} = 3.00 \times 10^2 \text{ mL}$$

5.102

$$V_{dil} = \frac{V_{concd} \bullet M_{concd}}{M_{dil}} = \frac{(150 \text{ mL})(2.5 \text{ M KOH})}{1.0 \text{ M KOH}} = 380 \text{ mL}$$

This is the total volume we need. In order to have this final volume we must add 230 mL (380 mL – 150 mL) of water to our original solution.

5.104 a) $KOH \rightarrow K^+ + OH^-$

1.25 mol/L x 0.0350 L = 0.0438 mol KOH

$0.0438 \text{ mol KOH} \times 1 \text{ mol OH}^-/\text{mol KOH} = 0.0438 \text{ mol OH}^-$
$0.0438 \text{ mol KOH} \times 1 \text{ mol K}^+/\text{mol KOH} = 0.0438 \text{ mol K}^+$

b) $CaCl_2 \rightarrow Ca^{2+} + 2Cl^-$

$0.45 \text{ mol/L} \times 0.0323 \text{ L} = 0.015 \text{ mol CaCl}_2$
$0.015 \text{ mol CaCl}_2 \times 1 \text{ mol Ca}^{2+}/\text{mol CaCl}_2 = 0.015 \text{ mol Ca}^{2+}$
$0.015 \text{ mol CaCl}_2 \times 2 \text{ mol Cl}^-/\text{mol CaCl}_2 = 0.030 \text{ mol Cl}^-$

c) $AlCl_3 \rightarrow Al^{3+} + 3Cl^-$

$$\#\text{ moles AlCl}_3 = (50 \text{ mL AlCl}_3)\left(\frac{0.40 \text{ mol AlCl}_3}{1000 \text{ mL}}\right) = 2.0\times10^{-2} \text{ moles AlCl}_3$$

$$\#\text{ moles Al}^{3+} = (2.0\times10^{-2} \text{ mol AlCl}_3)\left(\frac{1 \text{ mol Al}^{3+}}{1 \text{ mol AlCl}_3}\right) = 2.0\times10^{-2} \text{ moles Al}^{3+}$$

$$\#\text{ moles Cl}^- = (2.0\times10^{-2} \text{ mol AlCl}_3)\left(\frac{3 \text{ mol Cl}^-}{1 \text{ mol AlCl}_3}\right) = 6.0\times10^{-2} \text{ moles Cl}^-$$

5.106 a) $Cr(NO_3)_2 \rightarrow Cr^{2+} + 2NO_3^-$

$$\text{M Cr}^{2+} = \left(\frac{0.25 \text{ mol Cr(NO}_3)_2}{1 \text{ L Cr(NO}_3)_2 \text{ soln}}\right)\left(\frac{1 \text{ mol Cr}^{2+}}{1 \text{ mol Cr(NO}_3)_2}\right) = 0.25 \text{ M Cr}^{2+}$$

$$\text{M NO}_3^- = \left(\frac{0.25 \text{ mol Cr(NO}_3)_2}{1 \text{ L Cr(NO}_3)_2 \text{ soln}}\right)\left(\frac{2 \text{ mol NO}_3^-}{1 \text{ mol Cr(NO}_3)_2}\right) = 0.50 \text{ M NO}_3^-$$

b) $CuSO_4 \rightarrow Cu^{2+} + SO_4^{2-}$

$$\text{M Cu}^{2+} = \left(\frac{0.10 \text{ mol CuSO}_4}{1 \text{ L CuSO}_4 \text{ soln}}\right)\left(\frac{1 \text{ mol Cu}^{2+}}{1 \text{ mol CuSO}_4}\right) = 0.10 \text{ M Cu}^{2+}$$

$$\text{M SO}_4^{2-} = \left(\frac{0.10 \text{ mol CuSO}_4}{1 \text{ L CuSO}_4 \text{ soln}}\right)\left(\frac{1 \text{ mol SO}_4^{2-}}{1 \text{ mol CuSO}_4}\right) = 0.10 \text{ M SO}_4^{2-}$$

c) $Na_3PO_4 \rightarrow 3Na^+ + PO_4^{3-}$

$$\text{M Na}^+ = \left(\frac{0.16 \text{ mol Na}_3\text{PO}_4}{1 \text{ L Na}_3\text{PO}_4 \text{ soln}}\right)\left(\frac{3 \text{ mol Na}^+}{1 \text{ mol Na}_3\text{PO}_4}\right) = 0.48 \text{ M Na}^+$$

$$\text{M PO}_4^{3-} = \left(\frac{0.16 \text{ mol Na}_3\text{PO}_4}{1 \text{ L Na}_3\text{PO}_4 \text{ soln}}\right)\left(\frac{1 \text{ mol PO}_4^{3-}}{1 \text{ mol Na}_3\text{PO}_4}\right) = 0.16 \text{ M PO}_4^{3-}$$

d) $Al_2(SO_4)_3 \rightarrow 2Al^{3+} + 3SO_4^{2-}$

$$\text{M Al}^{3+} = \left(\frac{0.075 \text{ mol Al}_2(\text{SO}_4)_3}{1 \text{ L Al}_2(\text{SO}_4)_3 \text{ soln}}\right)\left(\frac{2 \text{ mol Al}^{3+}}{1 \text{ mol Al}_2(\text{SO}_4)_3}\right) = 0.15 \text{ M Al}^{3+}$$

$$M\ SO_4^{2-} = \left(\frac{0.075\ mol\ Al_2(SO_4)_3}{1\ L\ Al_2(SO_4)_3\ soln}\right)\left(\frac{3\ mol\ SO_4^{2-}}{1\ mol\ Al_2(SO_4)_3}\right) = 0.23\ M\ SO_4^{2-}$$

5.108 $$M\ Na_3PO_4 = \left(\frac{0.21\ mol\ Na^+}{1\ L\ Na_3PO_4\ soln}\right)\left(\frac{1\ mol\ Na_3PO_4}{3\ mol\ Na^+}\right) = 0.070\ M\ Na_3PO_4$$

5.110

$$\#\ g\ Al_2(SO_4)_3 = (50.0\ mL\ solution)\left(\frac{0.12\ mol\ Al^{3+}}{1000\ mL\ solution}\right)\left(\frac{1\ mol\ Al_2(SO_4)_3}{2\ mol\ Al^{3+}}\right)$$
$$\times\left(\frac{342.14\ g\ Al_2(SO_4)_3}{1\ mol\ Al_2(SO_4)_3}\right) = 1.0\ g\ Al_2(SO_4)_3$$

5.112

$$M\ KOH = \frac{(20.78\ mL\ HCl\ soln)\left(\frac{1\ L\ HCl\ soln}{1000\ mL\ HCl\ soln}\right)\left(\frac{0.116\ mol\ HCl}{1\ L\ HCl\ soln}\right)\left(\frac{1\ mol\ KOH}{1\ mol\ HCl}\right)}{(21.34\ mL\ KOH\ soln)\left(\frac{1\ L\ KOH\ soln}{1000\ mL\ KOH\ soln}\right)}$$
$$= 0.1130\ M\ KOH$$

5.114 $$\#\ mL\ NiCl_2 = (20.0\ mL\ Na_2CO_3)\left(\frac{1\ L\ Na_2CO_3}{1000\ mL\ Na_2CO_3}\right)\left(\frac{0.15\ mol\ Na_2CO_3}{1\ L\ Na_2CO_3}\right)$$
$$\times\left(\frac{1\ mol\ NiCl_2}{1\ mol\ Na_2CO_3}\right)\left(\frac{1\ L\ NiCl_2}{0.25\ mol\ NiCl_2}\right)\left(\frac{1000\ mL\ NiCl_2}{1\ L\ NiCl_2}\right) = 12\ mL\ NiCl_2$$

$$\#\ g\ NiCO_3 = (20.0\ mL\ Na_2CO_3)\left(\frac{1\ L\ Na_2CO_3}{1000\ mL\ Na_2CO_3}\right)\left(\frac{0.15\ mol\ Na_2CO_3}{1\ L\ Na_2CO_3}\right)$$
$$\times\left(\frac{1\ mol\ NiCO_3}{1\ mol\ Na_2CO_3}\right)\left(\frac{118.70\ g\ NiCO_3}{1\ mol\ NiCO_3}\right) = 0.36\ g\ NiCO_3$$

5.116 First, determine the number of moles of H_3PO_4 that are to react:

25.0 mL = 0.0250 L

0.0250 L x 0.250 mol/L = 6.25×10^{-3} mol H_3PO_4

Next, determine the number of moles of NaOH that are required by the balanced chemical equation: $H_3PO_4 + 3NaOH \rightarrow Na_3PO_4 + 3H_2O$

6.25×10^{-3} mol H_3PO_4 x (3 mol NaOH/1 mol H_3PO_4) = 0.0188 mol NaOH

Last, calculate the volume of NaOH solution that will deliver this required number of moles of NaOH:

$$0.0188 \text{ mol NaOH} \times 1000 \text{ mL}/0.100 \text{ mol} = 188 \text{ mL NaOH}$$

5.118

$$\text{M Ba(OH)}_2 = \frac{(20.78 \text{ mL HCl soln})\left(\frac{1 \text{ L HCl soln}}{1000 \text{ mL HCl soln}}\right)\left(\frac{0.116 \text{ mol HCl}}{1 \text{ L HCl soln}}\right)\left(\frac{1 \text{ mol Ba(OH)}_2}{2 \text{ mol HCl}}\right)}{(21.34 \text{ mL Ba(OH)}_2 \text{ soln})\left(\frac{1 \text{ L Ba(OH)}_2 \text{ soln}}{1000 \text{ mL Ba(OH)}_2 \text{ soln}}\right)}$$

$$= 0.0565 \text{ M Ba(OH)}_2$$

5.120 The balanced equation for the reaction is as follows:

$$3AgNO_3(aq) + FeCl_3(aq) \rightarrow 3AgCl(s) + Fe(NO_3)_3$$

$$\# \text{ mL FeCl}_3 = (20.0 \text{ mL AgNO}_3)\left(\frac{1 \text{ L AgNO}_3}{1000 \text{ mL AgNO}_3}\right)\left(\frac{0.0450 \text{ mol AgNO}_3}{1 \text{ L AgNO}_3}\right)$$

$$\times\left(\frac{1 \text{ mol FeCl}_3}{3 \text{ mol AgNO}_3}\right)\left(\frac{1 \text{ L FeCl}_3}{0.150 \text{ mol FeCl}_3}\right)\left(\frac{1000 \text{ mL FeCl}_3}{1 \text{ L FeCl}_3}\right) = 2.00 \text{ mL FeCl}_3$$

$$\# \text{ g AgCl} = (20.0 \text{ mL AgNO}_3)\left(\frac{1 \text{ L AgNO}_3}{1000 \text{ mL AgNO}_3}\right)\left(\frac{0.0450 \text{ mol AgNO}_3}{1 \text{ L AgNO}_3}\right)$$

$$\times\left(\frac{3 \text{ mol AgCl}}{3 \text{ mol AgNO}_3}\right)\left(\frac{143.32 \text{ AgCl}}{1 \text{ mol AgCl}}\right) = 0.129 \text{ g AgCl}$$

5.122 The balanced equation for the reaction is as follows:

$$3Ba(OH)_2(aq) + 2H_3PO_4(aq) \rightarrow Ba_3(PO_4)_2(s) + 6H_2O$$

$$\# \text{ mL Ba(OH)}_2 = (35.0 \text{ mL H}_3\text{PO}_4)\left(\frac{1 \text{ L H}_3\text{PO}_4}{1000 \text{ mL H}_3\text{PO}_4}\right)\left(\frac{0.125 \text{ mol H}_3\text{PO}_4}{1 \text{ L H}_3\text{PO}_4}\right)$$

$$\times\left(\frac{3 \text{ mol Ba(OH)}_2}{2 \text{ mol H}_3\text{PO}_4}\right)\left(\frac{1 \text{ L Ba(OH)}_2}{0.0500 \text{ mol Ba(OH)}_2}\right)\left(\frac{1000 \text{ mL Ba(OH)}_2}{1 \text{ L Ba(OH)}_2}\right) = 131 \text{ mL Ba(OH)}_2$$

5.124 $Ag^+ + Cl^- \rightarrow AgCl(s)$

$$\#\,mL\ AlCl_3 = (20.0 AgC_2H_3O_2)\left(\frac{0.500\ mol\ AgC_2H_3O_2}{1000\ mL\ AgC_2H_3O_2}\right)\left(\frac{1\ mol\ Ag^+}{1\ mol\ AgC_2H_3O_2}\right)$$

$$\left(\frac{1\ mol\ Cl^-}{1\ mol\ Ag^+}\right)\left(\frac{1\ mol\ AlCl_3}{3\ mol\ Cl^-}\right)\left(\frac{1000\ mL\ AlCl_3}{0.250\ moles\ AlCl_3}\right) = 13.3\ mL\ AlCl_3$$

*5.126 $Fe_2O_3 + 6HCl \rightarrow 2FeCl_3 + 3H_2O$

0.0250 L HCl x 0.500 mol/L = 1.25×10^{-2} mol HCl

$$\#\,mol\ Fe^{3+} = (1.25\ X\ 10^{-2}\ mol\ HCl)\left(\frac{1\ mol\ Fe_2O_3}{6\ mol\ HCl}\right)\left(\frac{2\ mol\ Fe^{3+}}{1\ mol\ Fe_2O_3}\right)$$

$$= 4.17\ X\ 10^{-3}\ mol\ Fe^{3+}$$

$$M\ Fe^{3+} = \frac{4.17\ X\ 10^{-3}\ mol\ Fe^{3+}}{0.0250\ L\ soln} = 0.167\ M\ Fe^{3+}$$

$$\#\,g\ Fe_2O_3 = \left(4.17\ X\ 10^{-3}\ mol\ Fe^{3+}\right)\left(\frac{1\ mol\ Fe_2O_3}{2\ mol\ Fe^{3+}}\right)\left(\frac{159.69\ g\ Fe_2O_3}{1\ mol\ Fe_2O_3}\right)$$

$$= 0.333\ g\ Fe_2O_3$$

Therefore, the mass of Fe_2O_3 that remains unreacted is:
(4.00 g – 0.333 g) = 3.67 g

*5.128 The equation for the reaction indicates that the two materials react in equimolar amounts, i.e. the stoichiometry is 1 to 1:

$$AgNO_3(aq) + NaCl(aq) \rightarrow AgCl(s) + NaNO_3(aq)$$

(a) Because this reaction is 1:1, we can see by inspection that the $AgNO_3$ is the limiting reagent. We know this because the concentration of the $AgNO_3$ is lower than the NaCl. Since we start with equal volumes, there are fewer moles of the $AgNO_3$.

$$\#\,mol\ AgCl = (25.0\ mL\ AgNO_3\ soln)\left(\frac{0.320\ mol\ AgNO_3}{1000\ mL\ AgNO_3\ soln}\right)$$

$$\times \left(\frac{1\ mol\ AgCl}{1\ mol\ AgNO_3}\right)$$

$$= 8.00\ X\ 10^{-3}\ mol\ AgCl$$

(b) Assuming that AgCl is essentially insoluble, the concentration of silver ion can be said to be zero since all of the $AgNO_3$ reacted. The number of moles of chloride ion would be reduced by the precipitation of 8.00×10^{-3} mol AgCl, such that the final number of moles of chloride ion would be:

$$0.0250\ \text{L} \times 0.440\ \text{mol/L} - 8.00 \times 10^{-3}\ \text{mol} = 3.0 \times 10^{-3}\ \text{mol Cl}^-$$

The final concentration of Cl^- is, therefore:

$$3.0 \times 10^{-3}\ \text{mol} \div 0.0500\ \text{L} = 0.060\ \text{M Cl}^-$$

All of the original number of moles of NO_3^- and of Na^+ would still be present in solution, and their concentrations would be:

For NO_3^-:

$$\#\,\text{M NO}_3^- = \frac{(25.0\ \text{mL AgNO}_3\ \text{soln})\left(\dfrac{0.320\ \text{mol AgNO}_3}{1000\ \text{mL AgNO}_3\ \text{soln}}\right)\left(\dfrac{1\ \text{mol NO}_3^-}{1\ \text{mol AgNO}_3}\right)}{(50.0\ \text{mL soln})\left(\dfrac{1\ \text{L soln}}{1000\ \text{mL soln}}\right)}$$

$$= 0.160\ \text{M NO}_3^-$$

For Na^+:

$$\#\,\text{M Na}^+ = \frac{(25.0\ \text{mL NaCl soln})\left(\dfrac{0.440\ \text{mol NaCl}}{1000\ \text{mL NaCl soln}}\right)\left(\dfrac{1\ \text{mol Na}^+}{1\ \text{mol NaCl}}\right)}{(50.0\ \text{mL soln})\left(\dfrac{1\ \text{L soln}}{1000\ \text{mL soln}}\right)}$$

$$= 0.220\ \text{M Na}^+$$

5.130

$$\#\,\text{g Pb} = (1.081\ \text{g PbSO}_4)\left(\frac{1\ \text{mol BaSO}_4}{303.27\ \text{g BaSO}_4}\right)\left(\frac{1\ \text{mol Pb}}{1\ \text{mol PbSO}_4}\right)\left(\frac{207.2\ \text{g Pb}}{1\ \text{mol Pb}}\right) = 0.7386\ \text{g Pb}$$

The percentage of Pb in the sample can be calculated as

$$\%\ \text{Pb} = \left(\frac{\text{mass of Pb}}{\text{mass of sample}}\right) \times 100\% = \frac{0.7386\ \text{g Pb}}{1.526\ \text{g PbSO}_4} \times 100\% = 48.40\%\ \text{Pb}$$

5.132 First, calculate the number of moles HCl based on the titration according to the following equation:

$$\text{NaOH(aq)} + \text{HCl(aq)} \rightarrow \text{NaCl(aq)} + \text{H}_2\text{O}$$

$$\#\,\text{mol HCl} = (23.25\ \text{mL NaOH})\left(\frac{0.105\ \text{mol NaOH}}{1000\ \text{mL NaOH}}\right)\left(\frac{1\ \text{mol HCl}}{1\ \text{mol NaOH}}\right)$$

$$= 2.44 \times 10^{-3}\ \text{mol HCl}$$

Next, determine the concentration of the HCl solution:

$$M = mol/L = 2.44 \times 10^{-3} mol \div 0.02145\ L = 0.114\ M\ HCl$$

5.134 Since lactic acid is monoprotic, it reacts with sodium hydroxide on a one to one mole basis:

$$\#\ mol\ HC_3H_5O_3 = (17.25\ mL\ NaOH)\left(\frac{0.155\ mol\ NaOH}{1000\ mL\ NaOH}\right)\left(\frac{1\ mol\ HC_3H_5O_3}{1\ mol\ NaOH}\right)$$

$$= 2.67\ X\ 10^{-3}\ mol\ HC_3H_5O_3$$

5.136 $$\#\ mol\ BaSO_4 = (1.174\ g\ BaSO_4)\left(\frac{1\ mol\ BaSO_4}{233.39\ g\ BaSO_4}\right) = 5.030\ X\ 10^{-3}\ mol\ BaSO_4$$

There were also 5.030×10^{-3}mol $MgSO_4$. We know this because we can see that there is one mole of SO_4^{2-} in both the Ba and Mg compounds. Since both of these elements are in the same family, the reaction to produce the barium salt from the magnesium salt must be 1:1. The balanced equation would be:

$$MgSO_4 + BaCl_2 \rightarrow BaSO_4 + MgCl_2$$

The mass of $MgSO_4$ that was present is:

$$5.030 \times 10^{-3}\ mol\ MgSO_4 \times 120.37\ g/mol = 0.6055\ g\ MgSO_4$$

Subtracting this from the total mass of the sample, we can find the mass of water in the original sample:

$$1.24\ g - 0.6055\ g = 0.63\ g\ H_2O$$

The number of moles of water are:

$$0.63\ g \div 18.0\ g/mol = 3.5 \times 10^{-2}\ mol\ H_2O$$

Finally, the relative mole amounts of water and $MgSO_4$ are found by diving both by the smallest of the two molar amounts above:

For $MgSO_4$, 5.030×10^{-3} moles/5.030×10^{-3} moles = 1.000

For water, $3.5 \times 10^{-2}/5.030 \times 10^{-3} = 7.0$

Hence the formula is $MgSO_4 \cdot 7H_2O$

5.138 First, we calculate how many *moles of NaOH were required for the titration*:

$$\#\ mol\ NaOH = (22.90\ mL\ NaOH)\left(\frac{1\ L\ NaOH}{1000\ mL\ NaOH}\right)\left(\frac{0.100\ mol\ NaOH}{1\ L\ NaOH}\right)$$

$$= 2.29 \times 10^{-3}\ mol\ NaOH$$

This enables us to find how many *moles of unreacted HCl* were titrated. We use the mole ratio in the balanced equation:

$$NaOH + HCl \rightarrow NaCl + H_2O$$

$$\# \text{ mol HCl} = (2.29 \times 10^{-3} \text{ mol NaOH})\left(\frac{1 \text{ mol HCl}}{1 \text{ mol NaOH}}\right)\left(\frac{1 \text{ mol HCl}}{1 \text{ mol NaOH}}\right)$$

$$= 2.29 \times 10^{-3} \text{ mol HCl}$$

The *total number of moles HCl added* were:

$$\# \text{ mol HCl} = (50.00 \text{ mL HCl})\left(\frac{1 \text{ L HCl}}{1000 \text{ mL HCl}}\right)\left(\frac{0.240 \text{ mol HCl}}{1 \text{ L HCl}}\right)$$

$$= 0.0120 \text{ mol HCl}$$

Therefore the number of *moles HCl which reacted* may be found by subtracting moles of unreacted HCl from total moles HCl added:

$$0.0120 \text{ mol HCl} - 2.29 \text{ x } 10^{-3} \text{ mol HCl} = 0.0097 \text{ mol HCl}$$

From this we can calculate how many *grams of Na_2CO_3* were in the original mixture, using the balanced equation for the reaction with HCl:

$$2HCl + Na_2CO_3 \rightarrow 2NaCl + H_2O + CO_2$$

$$\# \text{ g } Na_2CO_3 = (0.0097 \text{ mol HCl})\left(\frac{1 \text{ mol } Na_2CO_3}{2 \text{ mol HCl}}\right)\left(\frac{106.01 \text{ g } Na_2CO_3}{1 \text{ mol } Na_2CO_3}\right)$$

$$= 0.51 \text{ g } Na_2CO_3$$

The mass of the original mixture was 1.243g. If 0.51g was Na_2CO_3, the remainder was NaCl:

$$1.243\text{g} - 0.51\text{g} = 0.73\text{g NaCl}$$

The percentage NaCl by mass in the original mixture was then:

(mass NaCl/total mass)x100 = (0.73g/1.243g) x 100% = 59% NaCl

Additional Exercises

5.140 a) strong electrolyte
b) nonelectrolyte
c) strong electrolyte
d) non electrolyte
e) weak electrolyte
f) nonelectrolyte
g) strong electrolyte
h) weak electrolyte

5.142 a) $CaCO_3(s) + 2H^+(aq) \rightarrow Ca^{2+}(aq) + CO_2(g) + H_2O$
b) $CaCO_3(s) + 2H^+(aq) + SO_4^{2-}(aq) \rightarrow CaSO_4(s) + CO_2(g) + H_2O$
c) $FeS(s) + 2H^+(aq) \rightarrow Fe^{2+}(aq) + H_2S(g)$
d) $Sn^{2+}(aq) + 2OH^-(aq) \rightarrow Sn(OH)_2(s)$

*5.144 The balanced equation is as follows:

$$AgNO_3 + NaCl \rightarrow AgCl + NaNO_3$$

$$\# \text{ g NaCl} = (0.277 \text{ g AgCl})\left(\frac{1 \text{ mol AgCl}}{143.32 \text{ g AgCl}}\right)\left(\frac{1 \text{ mol NaCl}}{1 \text{ mol AgCl}}\right)\left(\frac{58.44 \text{ g NaCl}}{1 \text{ mol NaCl}}\right)$$

$$= 0.113 \text{ g NaCl}$$

Therefore, the entire sample was NaCl.

*5.146 a) $3Ba^{2+}(aq) + 2Al^{3+}(aq) + 6OH^{-}(aq) + 3SO_4^{2-}(aq) \rightarrow 3BaSO_4(s) + 2Al(OH)_3(s)$

b) Because we know the amounts of both starting materials this is a limiting reactant problem. So start by assuming that the barium hydroxide is the limiting reactant.

$$\# \text{ g BaSO}_4 = (40.0 \text{ mL Ba(OH)}_2)\left(\frac{0.270 \text{ mol Ba(OH)}_2}{1000 \text{ mL Ba(OH)}_2}\right)\left(\frac{1 \text{ mol Ba}^{2+}}{1 \text{ mol Ba(OH)}_2}\right)$$

$$\left(\frac{1 \text{ mol BaSO}_4}{1 \text{ mol Ba}^{2+}}\right)\left(\frac{233.39 \text{ g BaSO}_4}{1 \text{ mol BaSO}_4}\right) = 2.52 \text{ g BaSO}_4$$

Now assume $Al_2(SO_4)_3$ is the limiting reactant.

$$\# \text{ g BaSO}_4 = (25.0 \text{ mL Al}_2(\text{SO}_4)_3)\left(\frac{0.330 \text{ mol Al}_2(\text{SO}_4)_3}{1000 \text{ mL Al}_2(\text{SO}_4)_3}\right)\left(\frac{3 \text{ mol SO}_4^{2-}}{1 \text{ mol Al}_2(\text{SO}_4)_3}\right)$$

$$\left(\frac{1 \text{ mol BaSO}_4}{1 \text{ mol SO}_4^{2-}}\right)\left(\frac{233.39 \text{ g BaSO}_4}{1 \text{ mol BaSO}_4}\right) = 5.78 \text{ g BaSO}_4$$

Therefore the barium hydroxide is the limiting reactant. Now we can calculate the mass of aluminum hydroxide that is produced.

$$\# \text{ g Al(OH)}_3 = (40.0 \text{ mL Ba(OH)}_2)\left(\frac{0.270 \text{ mol Ba(OH)}_2}{1000 \text{ mL Ba(OH)}_2}\right)\left(\frac{2 \text{ mol OH}^-}{1 \text{ mol Ba(OH)}_2}\right)$$

$$\left(\frac{1 \text{ mol Al(OH)}_3}{3 \text{ mol OH}^-}\right)\left(\frac{78.00 \text{ g Al(OH)}_3}{1 \text{ mol Al(OH)}_3}\right) = 0.562 \text{ g Al(OH)}_3$$

The total mass of the precipitate is 2.52 g + 0.562 g = 3.08 g

c) All of the barium ion and hydroxide ion are reacted so the concentration of each is 0. We started with the following:

$$\# \text{ mol Al}^{3+} = (25.0\text{mL Al}_2(\text{SO}_4)_3)\left(\frac{0.330 \text{ mol Al}_2(\text{SO}_4)_3}{1000 \text{ mL Al}_2(\text{SO}_4)_3}\right)\left(\frac{2 \text{ mol Al}^{3+}}{1 \text{ mol Al}_2(\text{SO}_4)_3}\right)$$

$$= 1.65 \times 10^{-2} \text{ moles Al}^{3+}$$

$$\# \text{ mol } SO_4^{2-} = (25.0 \text{ mL } Al_2(SO_4)_3)\left(\frac{0.330 \text{ mol } Al_2(SO_4)_3}{1000 \text{ mL } Al_2(SO_4)_3}\right)\left(\frac{3 \text{ mol } SO_4^{2-}}{1 \text{ mol } Al_2(SO_4)_3}\right)$$

$$= 2.48 \times 10^{-2} \text{ moles } SO_4^{2-}$$

In precipitating the $Al(OH)_3$ above, we used 7.2×10^{-3} mol Al leaving $(1.65 \times 10^{-2} - 7.2 \times 10^{-3}) = 9.3 \times 10^{-3}$ mol Al^{3+} in solution.

The resulting concentration of Al^{3+} is:

9.3×10^{-3} mol / (0.0400 + 0.0250) L = 0.143 M Al^{3+}.

Similarly for SO_4^{2-}, the concentration of SO_4^{2-} remaining is

$$\frac{2.48 \times 10^{-2} \text{ mol} - 1.08 \times 10^{-2} \text{ mol}}{(0.0400 + 0.250) \text{ L}} = 0.215 \text{ M } SO_4^{2-},$$

where 1.08×10^{-2} mol SO_4^{2-} represents the amount precipitated as $BaSO_4$.

*5.148 The final desired molarity is 0.25 M. This means:

$$M = \text{moles solute/L solution}$$
$$0.25M = \text{moles solute/L solution}$$

The total moles of solute will be the sum of the moles from each solution:

$$\text{moles solute} = M_1V_1 + M_2V_2 = (0.10 \text{ M})V_1 + (0.40M)(0.050L)$$
$$= 0.10V_1 + 0.020$$

The total volume of solution in liters will be the sum of the two volumes:

$$\text{L solution} = V_1 + V_2$$
$$= V_1 + 0.050$$

Substituting into the top expression we have:

$$0.25 \text{ M} = \text{moles solute/L solution}$$
$$0.25 \text{ M} = (0.10V_1 + 0.020)/(V_1 + 0.050)$$

Multiplying both sides of the equation by $(V_1 + 0.050)$ gives:

$$(V_1 + 0.050)0.25 \text{ M} = (0.10V_1 + 0.020)$$
$$0.25V_1 + 0.0125 = 0.10V_1 + 0.020$$
$$0.15V_1 = 0.0075$$
$$V_1 = 0.050 \text{ L, or } 50 \text{ mL}$$

Chapter 6

Practice Exercises

6.1 $2Al(s) + 3Cl_2(g) \rightarrow 2AlCl_3(aq)$
Aluminum is oxidized and is, therefore, the reducing agent.
Chlorine is reduced and is, therefore, the oxidizing agent.

6.2 If H_2O_2 acts as an oxidizing agent, it gets reduced itself in the process.
Examining the oxidation numbers:

H_2O_2: H = +1, O = -1
H_2O: H = +1, O = -2
O_2: O = 0

If H_2O_2 is reduced it must form water, since the oxidation number of oxygen drops from -1 to -2 in the formation of water (a reduction).

The product is therefore water.

6.3 a) Ni +2; Cl –1
b) Mg +2; Ti +4; O –2
c) K +1; Cr +6; O –2
d) H +1; P +5, O –2
e) V +3; C 0; H +1; O –2

6.4 There is a total charge of +8, divided over three atoms, so the average charge is +8/3.

6.5 First the oxidation numbers of all atoms must be found.

$$Cl_2 + 2NaClO_2 \rightarrow 2ClO_2 + 2NaCl$$

Reactants:	Products:
Cl = 0	Cl = +4
	O = –2
Na = +1	
Cl = +3	Na = +1
O = –2	Cl = –1

The oxidation numbers for O and Na do not change. However, the oxidation numbers for all chlorine atoms change. There is no simple way to tell which chlorines are reduced and which are oxidized in this reaction.

One analysis would have the Cl in Cl_2 end up as the Cl in NaCl, while the Cl in $NaClO_2$ ends up as the Cl in ClO_2. In this case Cl_2 is reduced and is the oxidizing agent, while $NaClO_2$ is oxidized and is the reducing agent.

6.6 First the oxidation numbers of all atoms must be found.

$$KClO_3 + 6HNO_2 \rightarrow KCl + HNO_3$$

Reactants:	Products:
K = +1	K = +1
Cl = +5	Cl = -1
O = -2	
	H = +1
H = +1	N = +5
N = +3	O = -2
O = -2	

The oxidation numbers for K and H do not change. However, the oxidation numbers for all chlorines atoms drop. The oxidation numbers for nitrogen increase.

Therefore, $KClO_3$ is reduced and HNO_2 is oxidized.
This means $KClO_3$ is the oxidizing agent and HNO_2 is the reducing agent.

6.7 $Al(s) + Cu^{2+}(aq) \rightarrow Al^{3+}(aq) + Cu(s)$

First, we break the reaction above into half-reactions:

$$Al(s) \rightarrow Al^{3+}(aq)$$
$$Cu^{2+}(aq) \rightarrow Cu(s)$$

Each half-reaction is already balanced with respect to atoms, so next we add electrons to balance the charges on both sides of the equations:

$$Al(s) \rightarrow Al^{3+}(aq) + 3e^-$$
$$2e^- + Cu^{2+}(aq) \rightarrow Cu(s)$$

Next, we multiply both equations so that the electrons gained equals the electrons lost,

$$2(Al(s) \rightarrow Al^{3+}(aq) + 3e^-)$$
$$3(2e^- + Cu^{2+}(aq) \rightarrow Cu(s))$$

which gives us:

$$2Al(s) \rightarrow 2Al^{3+}(aq) + 6e^-$$
$$6e^- + 3Cu^{2+}(aq) \rightarrow 3Cu(s)$$

Now, by adding the half-reactions back together, we have our balanced equation:

$$2Al(s) + 3Cu^{2+}(aq) \rightarrow 2Al^{3+}(aq) + 3Cu(s)$$

6.8 $TcO_4^- + Sn^{2+} \rightarrow Tc^{4+} + Sn^{4+}$

First, we break the reaction above into half-reactions:

$$TcO_4^- \rightarrow Tc^{4+}$$
$$Sn^{2+} \rightarrow Sn^{4+}$$

Each half-reaction is already balanced with respect to atoms other than O and H, so next we balance the O atoms by using water:

$$TcO_4^- \rightarrow Tc^{4+} + 4H_2O$$
$$Sn^{2+} \rightarrow Sn^{4+}$$

Now we balance H by using H^+:

$$8H^+ + TcO_4^- \rightarrow Tc^{4+} + 4H_2O$$
$$Sn^{2+} \rightarrow Sn^{4+}$$

Next, we add electrons to balance the charges on both sides of the equations:

$$3e^- + 8H^+ + TcO_4^- \rightarrow Tc^{4+} + 4H_2O$$
$$Sn^{2+} \rightarrow Sn^{4+} + 2e^-$$

We multiply the equations so that the electrons gained equals the electrons lost,

$$2(3e^- + 8H^+ + TcO_4^- \rightarrow Tc^{4+} + 4H_2O)$$
$$3(Sn^{2+} \rightarrow Sn^{4+} + 2e^-)$$

which gives us:

$$6e^- + 16H^+ + 2TcO_4^- \rightarrow 2Tc^{4+} + 8H_2O$$
$$3Sn^{2+} \rightarrow 3Sn^{4+} + 6e^-$$

Now, by adding the half-reactions back together, we have our balanced equation:

$$3Sn^{2+} + 16H^+ + 2TcO_4^- \rightarrow 2Tc^{4+} + 8H_2O + 3Sn^{4+}$$

6.9 $(Cu \rightarrow Cu^{2+} + 2e^-) \times 4$

$2NO_3^- + 10H^+ + 8e^- \rightarrow N_2O + 5H_2O$

$4Cu + 2NO_3^- + 10H^+ \rightarrow 4Cu^{2+} + N_2O + 5H_2O$

6.10 $(MnO_4^- + 4H^+ + 3e^- \rightarrow MnO_2 + 2H_2O) \times 2$

$(C_2O_4^{2-} + 2H_2O \rightarrow 2CO_3^{2-} + 4H^+ + 2e^-) \times 3$

$2MnO_4^- + 3C_2O_4^{2-} + 2H_2O \rightarrow 2MnO_2 + 6CO_3^{2-} + 4H^+$

Adding $4OH^-$ to both sides of the above equation we get:

$$2MnO_4^- + 3C_2O_4^{2-} + 2H_2O + 4OH^- \rightarrow 2MnO_2 + 6CO_3^{2-} + 4H_2O$$

which simplifies to give:

$$2MnO_4^- + 3C_2O_4^{2-} + 4OH^- \rightarrow 2MnO_2 + 6CO_3^{2-} + 2H_2O$$

6.11 a) molecular: $Mg(s) + 2HCl(aq) \rightarrow MgCl_2(aq) + H_2(g)$

ionic: $Mg(s) + 2H^+(aq) + 2Cl^-(aq) \rightarrow Mg^{2+}(aq) + 2Cl^-(aq) + H_2(g)$

net ionic: $Mg(s) + 2H^+(aq) \rightarrow Mg^{2+}(aq) + H_2(g)$

b) molecular: $2Al(s) + 6HCl(aq) \rightarrow 2AlCl_3(aq) + 3H_2(g)$

ionic: $2Al(s) + 6H^+(aq) + 6Cl^-(aq) \rightarrow 2Al^{3+}(aq) + 6Cl^-(aq) + 3H_2(g)$

net ionic: $2Al(s) + 6H^+(aq) \rightarrow 2Al^{3+}(aq) + 3H_2(g)$

6.12 a) $2Al(s) + 3Cu^{2+}(aq) \rightarrow 2Al^{3+}(aq) + 3Cu(s)$

b) N.R.

6.13 $2C_4H_{10}(\ell) + 13O_2(g) \rightarrow 8CO_2(g) + 10H_2O(g)$

6.14 $C_2H_5OH(\ell) + 3O_2(g) \rightarrow 2CO_2(g) + 3H_2O(g)$

6.15 $4Fe(s) + 3O_2(g) \rightarrow 2Fe_2O_3(s)$

6.16 First we need a balanced equation:

$Cl_2 + 2e^- \rightarrow 2Cl^-$

$S_2O_3^{2-} + 5H_2O \rightarrow 2SO_4^{2-} + 10H^+ + 8e^-$

$4Cl_2 + S_2O_3^{2-} + 5H_2O \rightarrow 8Cl^- + 2SO_4^{2-} + 10H^+$

$$\#\,g\,Na_2S_2O_3 = (4.25\,g\,Cl_2)\left(\frac{1\,mol\,Cl_2}{70.906\,g\,Cl_2}\right)\left(\frac{1\,mol\,Na_2S_2O_3}{4\,mol\,Cl_2}\right)\left(\frac{158.132\,g\,Na_2S_2O_3}{1\,mol\,Na_2S_2O_3}\right)$$

$$= 2.37\,g\,Na_2S_2O_3$$

6.17 a) $(Sn^{2+} \rightarrow Sn^{4+} + 2e^-) \times 5$

$(MnO_4^- + 8H^+ + 5e^- \rightarrow Mn^{2+} + 4H_2O) \times 2$

$5Sn^{2+} + 2MnO_4^- + 16H^+ \rightarrow 5Sn^{4+} + 2Mn^{2+} + 8H_2O$

b)

$$\#\,g\,Sn = (8.08\,mL\,KMnO_4\,soln)\left(\frac{0.0500\,mol\,KMnO_4}{1000\,mL\,KMnO_4}\right)\left(\frac{1\,mol\,MnO_4^-}{1\,mol\,KMnO_4}\right)$$

$$\times\left(\frac{5\,mol\,Sn^{2+}}{2\,mol\,MnO_4^-}\right)\left(\frac{1\,mol\,Sn}{1\,mol\,Sn^{2+}}\right)\left(\frac{118.71\,g\,Sn}{1\,mol\,Sn}\right)$$

$$= 0.120\,g\,Sn$$

c) $$\%\,Sn = \frac{0.120\,g\,Sn}{0.300\,g\,sample} \times 100\,\% = 40.0\,\%\,Sn$$

d)

$$\#\,g\,Sn = (8.08\,mL\,KMnO_4\,soln)\left(\frac{0.0500\,mol\,KMnO_4}{1000\,mL\,KMnO_4}\right)\left(\frac{1\,mol\,MnO_4^-}{1\,mol\,KMnO_4}\right)$$

$$\times\left(\frac{5\,mol\,Sn^{2+}}{2\,mol\,MnO_4^-}\right)\left(\frac{1\,mol\,SnO_2}{1\,mol\,Sn^{2+}}\right)\left(\frac{150.71\,g\,SnO_2}{1\,mol\,SnO_2}\right)$$

$$= 0.152\,g\,SnO_2$$

$$\%\,SnO_2 = \frac{0.152\,g\,SnO_2}{0.300\,g\,sample} \times 100\,\% = 50.7\,\%\,SnO_2$$

Review Problems

6.25 a) substance reduced (and oxidizing agent): HNO_3

substance oxidized (and reducing agent): H_3AsO_3

b) substance reduced (and oxidizing agent): HOCl
substance oxidized (and reducing agent): NaI
c) substance reduced (and oxidizing agent): $KMnO_4$
substance oxidized (and reducing agent): $H_2C_2O_4$
d) substance reduced (and oxidizing agent): H_2SO_4
substance oxidized (and reducing agent): Al

6.27 Recall that the sum of the oxidation numbers must equal the charge on the molecule or ion.
a) –2
b) +4; we know that O is usually in a –2 oxidation state.
c) The oxidation state of an element is always zero, by definition.
d) –3; hydrogen is usually in a +1 oxidation state.

6.29 The sum of the oxidation numbers should be zero:

a) Na: +1; H: +1; P: +5; O: –2

b) Ba: +2; Mn: +6; O: –2

c) Na: +1; S: +2.5; O: –2

d) Cl: +3; F: –1

6.31 The sum of the oxidation numbers should be zero:
a) +2
b) +5
c) –1
d) +4
e) –2

6.33 The sum of the oxidation numbers should be zero:

a) O: –2; Na: +1; Cl: +1

b) O: –2; Na: +1; Cl: +3

c) O: –2; Na: +1; Cl: +5

d) O: –2; Na: +1; Cl: +7

6.35 The sum of the oxidation numbers should be zero:

a) S: –2; Pb: +2

b) Cl: –1; Ti: +4

c) O: –2; Sr: +2; I: +5

d) S: –2; Cr: +3

6.37 $Cl_2(aq) + H_2O \rightarrow H^+(aq) + Cl^-(aq) + HOCl(aq)$

In the forward direction: The oxidation number of the chlorine atoms decreases from 0 to -1. Therefore Cl_2 is reduced. However, in HOCl, chlorine has an oxidation number of +1, so Cl_2 also oxidized! (One atom is reduced, the other is oxidized.)

In the reverse direction: The Cl^- ion begins with an oxidation number of -1 and ends with an oxidation number of 0. Therefore the Cl^- ion is oxidized: This means Cl^- is the reducing agent. Since the oxidation number of H^+ does not change, HOCl must be the oxidizing agent.

6.39 a) $BiO_3^- + 6H^+ + 2e^- \rightarrow Bi^{3+} + 3H_2O$; this is a reduction of BiO_3^{2-}.
b) $Pb^{2+} + 2H_2O \rightarrow PbO_2 + 4H^+ + 2e^-$; this is an oxidation of Pb^{2+}.

6.41 a) $Fe + 2OH^- \rightarrow Fe(OH)_2 + 2e^-$; this is an oxidation of Fe.
b) $2e^- + 2OH^- + SO_2Cl_2 \rightarrow SO_3^{2-} + 2Cl^- + H_2O$; this is a reduction of SO_2Cl_2.

6.43 a)
$2S_2O_3^{2-} \rightarrow S_4O_6^{2-} + 2e^-$
$OCl^- + 2H^+ + 2e^- \rightarrow Cl^- + H_2O$
$OCl^- + 2S_2O_3^{2-} + 2H^+ \rightarrow S_4O_6^{2-} + Cl^- + H_2O$

b)
$(NO_3^- + 2H^+ + e^- \rightarrow NO_2 + H_2O)^2$
$Cu \rightarrow Cu^{2+} + 2e^-$
$2NO_3^- + Cu + 4H^+ \rightarrow 2NO_2 + Cu^{2+} + 2H_2O$

c)
$IO_3^- + 6H^+ + 6e^- \rightarrow I^- + 3H_2O$
$(H_2O + AsO_3^{3-} \rightarrow AsO_4^{3-} + 2H^+ + 2e^-) \times 3$
$IO_3^- + 3AsO_3^{3-} + 6H^+ + 3H_2O \rightarrow I^- + 3AsO_4^{3-} + 3H_2O + 6H^+$
which simplifies to give:
$3AsO_3^{3-} + IO_3^- \rightarrow I^- + 3AsO_4^{3-}$

d)
$SO_4^{2-} + 4H^+ + 2e^- \rightarrow SO_2 + 2H_2O$
$Zn \rightarrow Zn^{2+} + 2e^-$
$Zn + SO_4^{2-} + 4H^+ \rightarrow Zn^{2+} + SO_2 + 2H_2O$

e)
$NO_3^- + 10H^+ + 8e^- \rightarrow NH_4^+ + 3H_2O$
$(Zn \rightarrow Zn^{2+} + 2e^-) \times 4$
$NO_3^- + 4Zn + 10H^+ \rightarrow 4Zn^{2+} + NH_4^+ + 3H_2O$

f)
$2Cr^{3+} + 7H_2O \rightarrow Cr_2O_7^{2-} + 14H^+ + 6e^-$
$(BiO_3^- + 6H^+ + 2e^- \rightarrow Bi^{3+} + 3H_2O) \times 3$
$2Cr^{3+} + 3BiO_3^- + 18H^+ + 7H_2O \rightarrow Cr_2O_7^{2-} + 14H^+ + 3Bi^{3+} + 9H_2O$
which simplifies to give:
$2Cr^{3+} + 3BiO_3^- + 4H^+ \rightarrow Cr_2O_7^{2-} + 3Bi^{3+} + 2H_2O$

g) $I_2 + 6H_2O \rightarrow 2IO_3^- + 12H^+ + 10e^-$
$(OCl^- + 2H^+ + 2e^- \rightarrow Cl^- + H_2O) \times 5$
$I_2 + 5OCl^- + H_2O \rightarrow 2IO_3^- + 5Cl^- + 2H^+$

h) $(Mn^{2+} + 4H_2O \rightarrow MnO_4^- + 8H^+ + 5e^-) \times 2$
$(BiO_3^- + 6H^+ + 2e^- \rightarrow Bi^{3+} + 3H_2O) \text{ x } 5$
$2Mn^{2+} + 5BiO_3^- + 30H^+ + 8H_2O \rightarrow 2MnO_4^- + 5Bi^{3+} + 16H^+ + 15H_2O$
which simplifies to:
$2Mn^{2+} + 5BiO_3^- + 14H^+ \rightarrow 2MnO_4^- + 5Bi^{3+} + 7H_2O$

i) $(H_3AsO_3 + H_2O \rightarrow H_3AsO_4 + 2H^+ + 2e^-) \times 3$
$Cr_2O_7^{2-} + 14H^+ + 6e^- \rightarrow 2Cr^{3+} + 7H_2O$
$3H_3AsO_3 + Cr_2O_7^{2-} + 3H_2O + 14H^+ \rightarrow 3H_3AsO_4 + 2Cr^{3+} + 6H^+ + 7H_2O$
which simplifies to give:
$3H_3AsO_3 + Cr_2O_7^{2-} + 8H^+ \rightarrow 3H_3AsO_4 + 2Cr^{3+} + 4H_2O$

j) $2I^- \rightarrow I_2 + 2e^-$
$HSO_4^- + 3H^+ + 2e^- \rightarrow SO_2 + 2H_2O$
$2I^- + HSO_4^- + 3H^+ \rightarrow I_2 + SO_2 + 2H_2O$

6.45 For redox reactions in basic solution, we proceed to balance the half reactions as if they were in acid solution, and then add enough OH^- to each side of the resulting equation in order to neutralize (titrate) all of the H^+. This gives a corresponding amount of water ($H^+ + OH^- \rightarrow H_2O$) on one side of the equation, and an excess of OH^- on the other side of the equation, as befits a reaction in basic solution.

a) $(CrO_4^{2-} + 4H^+ + 3e^- \rightarrow CrO_2^- + 2H_2O) \times 2$
$(S^{2-} \rightarrow S + 2e^-) \times 3$
$2CrO_4^{2-} + 3S^{2-} + 8H^+ \rightarrow 2CrO_2^- + 2S + 4H_2O$
Adding $8OH^-$ to both sides of the above equation we obtain:
$2CrO_4^{2-} + 3S^{2-} + 8H_2O \rightarrow 2CrO_2^- + 8OH^- + 3S + 4H_2O$
which simplifies to:
$2CrO_4^{2-} + 3S^{2-} + 4H_2O \rightarrow 2CrO_2^- + 3S + 8OH^-$

b) $(C_2O_4^{2-} \rightarrow 2CO_2 + 2e^-) \times 3$
$(MnO_4^- + 4H^+ + 3e^- \rightarrow MnO_2 + 2H_2O) \times 2$
$3C_2O_4^{2-} + 2MnO_4^- + 8H^+ \rightarrow 6CO_2 + 2MnO_2 + 4H_2O$
Adding $8OH^-$ to both sides of the above equation we get:
$3C_2O_4^{2-} + 2MnO_4^- + 8H_2O \rightarrow 6CO_2 + 2MnO_2 + 4H_2O + 8OH^-$
which simplifies to give:
$3C_2O_4^{2-} + 2MnO_4^- + 4H_2O \rightarrow 6CO_2 + 2MnO_2 + 8OH^-$

c) $(ClO_3^- + 6H^+ + 6e^- \rightarrow Cl^- + 3H_2O) \times 4$
$(N_2H_4 + 2H_2O \rightarrow 2NO + 8H^+ + 8e^-) \times 3$

$4ClO_3^- + 3N_2H_4 + 24H^+ + 6H_2O \rightarrow 4Cl^- + 6NO + 12H_2O + 24H^+$
which needs no OH^-, because it simplifies directly to:
$4ClO_3^- + 3N_2H_4 \rightarrow 4Cl^- + 6NO + 6H_2O$

d) $NiO_2 + 2H^+ + 2e^- \rightarrow Ni(OH)_2$
$2Mn(OH)_2 \rightarrow Mn_2O_3 + H_2O + 2H^+ + 2e^-$
$NiO_2 + 2Mn(OH)_2 \rightarrow Ni(OH)_2 + Mn_2O_3 + H_2O$

e) $(SO_3^{2-} + H_2O \rightarrow SO_4^{2-} + 2H^+ + 2e^-) \times 3$
$(MnO_4^- + 4H^+ + 3e^- \rightarrow MnO_2 + 2H_2O) \times 2$
$3SO_3^{2-} + 3H_2O + 8H^+ + 2MnO_4^- \rightarrow 3SO_4^{2-} + 6H^+ + 2MnO_2 + 4H_2O$
Adding $8OH^-$ to both sides of the equation we obtain:
$3SO_3^{2-} + 11H_2O + 2MnO_4^- \rightarrow 3SO_4^{2-} + 10H_2O + 2MnO_2 + 2OH^-$
which simplifies to:
$3SO_3^{2-} + 2MnO_4^- + H_2O \rightarrow 3SO_4^{2-} + 2MnO_2 + 2OH^-$

6.47 First, we write the half-reactions for the equation:

$$NaOCl \rightarrow NaCl$$
$$Na_2S_2O_3 \rightarrow Na_2SO_4$$

Now, we write them as ionic equations:

$$Na^+ + OCl^- \rightarrow Na^+ + Cl^-$$
$$2Na^+ + S_2O_3^{2-} \rightarrow 2Na^+ + SO_4^{2-}$$

Eliminating spectator ions, gives us the net ionic half-reaction equations:

$$OCl^- \rightarrow Cl^-$$
$$S_2O_3^{2-} \rightarrow SO_4^{2-}$$

At this point, we may begin balancing each half-reaction (we will assume a basic solution). First we balance all atoms other than O and H:

$$OCl^- \rightarrow Cl^-$$
$$S_2O_3^{2-} \rightarrow 2SO_4^{2-}$$

Next, we balance oxygen atoms using OH^- ions:

$$OCl^- \rightarrow Cl^- + OH^-$$
$$OH^- + S_2O_3^{2-} \rightarrow 2SO_4^{2-}$$

Now H atoms are balanced by putting OH^- ions on one side and water molecules on the other:

$$H_2O + OCl^- \rightarrow Cl^- + 2OH^-$$
$$2OH^- + S_2O_3^{2-} \rightarrow 2SO_4^{2-} + H_2O$$

Next, we add electrons to balance the charges on both sides of the equations:

$$2e^- + H_2O + OCl^- \rightarrow Cl^- + 2OH^-$$
$$2OH^- + S_2O_3^{2-} \rightarrow 2SO_4^{2-} + H_2O + 2e^-$$

Here, electrons gained in the first equation equal electrons lost in the second, so we may simply add them:

$$2e^- + H_2O + OCl^- + 2OH^- + S_2O_3^{2-} \rightarrow 2SO_4^{2-} + H_2O + 2e^- + Cl^- + 2OH^-$$

Canceling like species on each side gives us the balanced equation:

$$OCl^- + S_2O_3^{2-} \rightarrow 2SO_4^{2-} + Cl^-$$

6.49 First, we write the half-reactions for the equation:

$$O_3 \rightarrow H_2O$$
$$Br^- \rightarrow BrO_3^-$$

Now we may begin balancing each half-reaction, assume an acidic solution. All atoms other than O and H are already balanced, so we begin by balancing oxygen atoms using water:

$$O_3 \rightarrow 3H_2O$$
$$3H_2O + Br^- \rightarrow BrO_3^-$$

Next, we balance hydrogen atoms using H^+ ions:

$$6H^+ + O_3 \rightarrow 3H_2O$$
$$3H_2O + Br^- \rightarrow BrO_3^- + 6H^+$$

At this point, we add electrons to balance the charges on both sides of the equations:

$$6e^- + 6H^+ + O_3 \rightarrow 3H_2O$$
$$3H_2O + Br^- \rightarrow BrO_3^- + 6H^+ + 6e^-$$

Here, electrons gained in the first equation equal electrons lost in the second, so we may simply add them:

$$6e^- + 6H^+ + O_3 + 3H_2O + Br^- \rightarrow BrO_3^- + 6H^+ + 6e^- + 3H_2O$$

Canceling like species on each side gives us the balanced equation:

$$O_3 + Br^- \rightarrow BrO_3^-$$

6.51 a) Molecular: $Mn(s) + 2HCl(aq) \rightarrow MnCl_2(aq) + H_2(g)$
Ionic: $Mn(s) + 2H^+(aq) + 2Cl^-(aq) \rightarrow Mn^{2+}(aq) + 2Cl^-(aq) + H_2(g)$
Net Ionic: $Mn(s) + 2H^+(aq) \rightarrow Mn^{2+}(aq) + H_2(g)$

b) Molecular: $Cd(s) + 2HCl(aq) \rightarrow CdCl_2(aq) + H_2(g)$
Ionic: $Cd(s) + 2H^+(aq) + 2Cl^-(aq) \rightarrow Cd^{2+}(aq) + Cl^-(aq) + H_2(g)$
Net Ionic: $Cd(s) + 2H^+(aq) \rightarrow Cd^{2+}(aq) + H_2(g)$

c) Molecular: $Sn(s) + 2HCl(aq) \rightarrow SnCl_2(aq) + H_2(g)$
Ionic: $Sn(s) + 2H^+(aq) + 2Cl^-(aq) \rightarrow Sn^{2+}(aq) + 2Cl^-(aq) + H_2(g)$
Net Ionic: $Sn(s) + 2H^+(aq) \rightarrow Sn^{2+}(aq) + H_2(g)$

d) Molecular: $Ni(s) + 2HCl(aq) \rightarrow NiCl_2(aq) + H_2(g)$
Ionic: $Ni(s) + 2H^+(aq) + 2Cl^-(aq) \rightarrow Ni^{2+}(aq) + 2Cl^-(aq) + H_2(g)$
Net Ionic: $Ni(s) + 2H^+(aq) \rightarrow Ni^{2+}(aq) + H_2(g)$

e) Molecular: $2Cr(s) + 6HCl(aq) \rightarrow 2CrCl_3(aq) + 3H_2(g)$
Ionic: $2Cr(s) + 6H^+(aq) + 6Cl^-(aq) \rightarrow 2Cr^{3+}(aq) + 6Cl^-(aq) + 3H_2(g)$
Net Ionic: $2Cr(s) + 6H^+(aq) \rightarrow 2Cr^{3+}(aq) + 3H_2(g)$

6.53 a) $3Ag(s) + 4HNO_3(aq) \rightarrow 3AgNO_3(aq) + 2H_2O + NO(g)$
b) $Ag(s) + 2HNO_3(aq) \rightarrow AgNO_3(aq) + H_2O + NO_2(g)$

6.55 First, we write the half-reactions for the equation:

$$Sn^{2+} \rightarrow Sn^{4+}$$
$$BrO_3^- \rightarrow Br^-$$

Now we may begin balancing each half-reaction, assume an acidic solution. All atoms other than O and H are already balanced, so we begin by balancing oxygen atoms using water:

$$Sn^{2+} \rightarrow Sn^{4+}$$
$$BrO_3^- \rightarrow Br^- + 3H_2O$$

Next, we balance hydrogen atoms using H^+ ions:

$$Sn^{2+} \rightarrow Sn^{4+}$$
$$6H^+ + BrO_3^- \rightarrow Br^- + 3H_2O$$

At this point, we add electrons to balance the charges on both sides of the equations:

$$Sn^{2+} \rightarrow Sn^{4+} + 2e^-$$
$$6e^- + 6H^+ + BrO_3^- \rightarrow Br^- + 3H_2O$$

We multiply the first equation by 3 so that the electrons gained equals the electrons lost,

$$3(Sn^{2+} \rightarrow Sn^{4+} + 2e^-)$$
$$6e^- + 6H^+ + BrO_3^- \rightarrow Br^- + 3H_2O$$

which gives us:

$$3Sn^{2+} \rightarrow 3Sn^{4+} + 6e^-$$
$$6e^- + 6H^+ + BrO_3^- \rightarrow Br^- + 3H_2O$$

Adding the two equations together produces the following:

$$3Sn^{2+} + 6e^- + 6H^+ + BrO_3^- \rightarrow Br^- + 3H_2O + 3Sn^{4+} + 6e^-$$

Canceling like species on each side gives us the final, balanced equation:

$$3Sn^{2+} + 6H^+ + BrO_3^- \rightarrow Br^- + 3H_2O + 3Sn^{4+}$$

6.57 a) $Fe + Mg^{2+} \rightarrow NR$
b) $2Cr + 3Pb^{2+} \rightarrow 3Pb + 2Cr^{3+}$
c) $Fe + 2Ag^+ \rightarrow 2Ag + Fe^{2+}$
d) $3Ag + Au^{3+} \rightarrow Au + 3Ag^+$

6.59 Since Ru reduces Pt^{2+}, it is more active than Pt.

$$Pt < Ru$$

Since Tl reduces Ru^{2+}, it is more active than Ru.

$$Ru < Tl$$

Since Pu reduces Tl^{2+}, it is more active than Tl.

$$Tl < Pu$$

Placing these together, we arrive at the following:

$$Pt < Ru < Tl < Pu$$

6.61 The equation given shows that Cd is more active than Ru. Coupled with the information in Review Problem 6.59, we also see that Cd is more active than Pt. This means that in a mixture of Cd and Pt, Cd will be oxidized and Pt will be reduced:

$$Cd(s) + PtCl_2(aq) \rightarrow CdCl_2(aq) + Pt(s)$$

(The Pt(s) and the $Cd(NO_3)_2(aq)$ will not react.)

6.63 a) $2C_6H_6(\ell) + 15O_2(g) \rightarrow 12CO_2(g) + 6H_2O(g)$
b) $C_3H_8(g) + 5O_2(g) \rightarrow 3CO_2(g) + 4H_2O(g)$
c) $C_{21}H_{44}(s) + 32O_2(g) \rightarrow 21CO_2(g) + 22H_2O(g)$

6.65 a) $2C_6H_6(\ell) + 9O_2(g) \rightarrow 12CO(g) + 6H_2O(g)$
$2C_3H_8(g) + 7O_2(g) \rightarrow 6CO(g) + 8H_2O(g)$
$2C_{21}H_{44}(s) + 43O_2(g) \rightarrow 42CO(g) + 44H_2O(g)$

b) $2C_6H_6(\ell) + 3O_2(g) \rightarrow 12C(s) + 6H_2O(g)$
$C_3H_8(g) + 2O_2(g) \rightarrow 3C(s) + 4H_2O(g)$
$C_{21}H_{44}(s) + 11O_2(g) \rightarrow 21C(s) + 22H_2O(g)$

6.67 $2CH_3OH(\ell) + 3O_2(g) \rightarrow 2CO_2(g) + 4H_2O(g)$

6.69 a) $IO_3^- + 6H^+ + 6e^- \rightarrow I^- + 3H_2O$
$[SO_3^{2-} + H_2O \rightarrow SO_4^{2-} + 2H^+ + 2e^-] \times 3$
$IO_3^- + 3SO_3^{2-} + 6H^+ + 3H_2O \rightarrow I^- + 3SO_4^{2-} + 3H_2O + 6H^+$
Which simplifies to:
$IO_3^- + 3SO_3^{2-} \rightarrow I^- + 3SO_4^{2-}$

b)

$$\#\,g\,Na_2SO_3 = (5.00\,g\,NaIO_3)\left(\frac{1\,mol\,NaIO_3}{197.9\,g\,NaIO_3}\right) \times \left(\frac{3\,mol\,Na_2SO_3}{1\,mol\,NaIO_3}\right)\left(\frac{126.0\,g\,Na_2SO_3}{1\,mol\,Na_2SO_3}\right)$$

$$= 9.55\,g\,Na_2SO_3$$

6.71 $Cu + 2Ag^+ \rightarrow Cu^{2+} + Ag$

$$\#\,g\,Cu = (12.0\,g\,Ag)\left(\frac{1\,mol\,Ag}{107.868\,g\,Ag}\right)\left(\frac{1\,mol\,Cu}{2\,mol\,Ag}\right)\left(\frac{63.54\,g\,Cu}{1\,mol\,Cu}\right) = 3.53\,g\,Cu$$

6.73 a) $[MnO_4^- + 8H^+ + 5e^- \rightarrow Mn^{2+} + 4H_2O] \times 2$
$[Sn^{2+} \rightarrow Sn^{4+} + 2e^-] \times 5$
$2MnO_4^- + 5Sn^{2+} + 16H^+ \rightarrow 2Mn^{2+} + 5Sn^{4+} + 8H_2O$

b)

$$\#\,\text{mL KMnO}_4 = (40.0\text{ mL SnCl2})\left(\frac{0.250\text{ mol SnCl}_2}{1000\text{ mL SnCl}_2}\right)\left(\frac{1\text{ mol Sn}^{2+}}{1\text{ mol SnCl}_2}\right)$$

$$\left(\frac{2\text{ mol MnO}_4^-}{5\text{ mol Sn}^{2+}}\right)\left(\frac{1\text{ mol KMnO}_4}{1\text{ mol MnO}_4^-}\right)\left(\frac{1000\text{ mL KMnO}_4}{0.230\text{ mol kMnO}_4}\right) = 17.4\text{ mL}$$

6.75 a) M = mol solute/liter of solution
L of solution is given as 0.250 L (250 mL)

$$\#\,\text{mol I}_3^- = (0.462\text{ g KIO}_3)\left(\frac{1\text{ mol KIO}_3}{214.0\text{ g KIO}_3}\right)\left(\frac{1\text{ mol IO}_3^-}{1\text{ mol KIO}_3}\right)\left(\frac{3\text{ mol I}_3^-}{1\text{ mol IO}_3^-}\right)$$

$$= 0.006{,}48\text{ mol I}_3^-$$

M = moles/L = 0.006,48 mol/0.250 L = 0.0259 M I_3^-

b)

$$\#\,\text{g (NH}_4)_2\text{S}_2\text{O}_3 = (27.99\text{ mL I}_3^-\text{ solution})\left(\frac{1\text{ L I}_3^-\text{ solution}}{1000\text{ mL I}_3^-\text{ solution}}\right)\left(\frac{0.0259\text{ mol I}_3^-}{1\text{ L I}_3^-\text{ solution}}\right)$$

$$\times\left(\frac{2\text{ mol S}_2\text{O}_3^{2-}}{1\text{ mol I}_3^-}\right)\left(\frac{1\text{ mol (NH}_4)_2\text{S}_2\text{O}_3}{1\text{ mol S}_2\text{O}_3^{2-}}\right)\left(\frac{148.3\text{ g (NH}_4)_2\text{S}_2\text{O}_3}{1\text{ mol (NH}_4)_2\text{S}_2\text{O}_3}\right)$$

$$= 0.2150\text{ g (NH}_4)_2\text{S}_2\text{O}_3$$

c) $$\%\,(\text{NH}_4)_2\text{S}_2\text{O}_3 = \frac{0.2150\text{ g (NH}_4)_2\text{S}_2\text{O}_3}{0.218\text{ g fertilizer}} \times 100\% = 98.6\%$$

6.77 a)

$$\#\,\text{mol Cu}^{2+} = (29.96\text{ mL S}_2\text{O}_3^{2-})\left(\frac{0.02100\text{ mol S}_2\text{O}_3^{2-}}{1000\text{ mL S}_2\text{O}_3^{2-}}\right)$$

$$\times\left(\frac{1\text{ mol I}_3^-}{2\text{ mol S}_2\text{O}_3^{2-}}\right)\left(\frac{2\text{ mol Cu}^{2+}}{1\text{ mol I}_3^-}\right)$$

$$= 6.292\text{ X }10^{-4}\text{ mol Cu}^{2+}$$

g Cu = (6.292 X 10^{-4} mol Cu) × (63.546 g Cu/mol Cu)
= 3.998 X 10^{-2} g Cu

% Cu = (3.998 X 10^{-2} g Cu/0.4225 g sample) × 100 % = 9.463 %

b)

$$\# \text{g CuCO}_3 = (6.292 \text{ X } 10-4 \text{ mol CuCO}_3)\left(\frac{123.56 \text{ g CuCO}_3}{1 \text{ mol CuCO}_3}\right)$$

$$= 0.07774 \text{ g CuCO}_3$$

$$\% \text{ CuCO}_3 = \left(\frac{0.07774 \text{ g CuCO}_3}{0.4225 \text{ g sample}}\right) \times 100\,\% = 18.40\,\%$$

6.79 a)

$$\# \text{g H}_2\text{O}_2 = (17.60 \text{ mL KMnO}_4)\left(\frac{0.02000 \text{ mol KMnO}_4}{1000 \text{ mL KMnO}_4}\right)\left(\frac{1 \text{ mol MnO}_4^-}{1 \text{ mol KMnO}_4}\right)$$

$$\times \left(\frac{5 \text{ mol H}_2\text{O}_2}{2 \text{ mol MnO}_4^-}\right)\left(\frac{34.02 \text{ g H}_2\text{O}_2}{1 \text{ mol H}_2\text{O}_2}\right)$$

$$= 0.02994 \text{ g H}_2\text{O}_2$$

b) 0.02994 g/1.000 g × 100 % = 2.994 % H_2O_2

6.81 a) $2CrO_4^{2-} + 3SO_3^{2-} + H_2O \rightarrow 2CrO_2^- + 3SO_4^{2-} + 2OH^-$

b)

$$\# \text{mol CrO}_4^{2-} = (3.18 \text{ g Na}_2\text{SO}_3)\left(\frac{1 \text{ mol Na}_2\text{SO}_3}{126.04 \text{ g Na}_2\text{SO}_3}\right)$$

$$\times \left(\frac{1 \text{ mol SO}_3^{2-}}{1 \text{ mol Na}_2\text{SO}_3}\right)\left(\frac{2 \text{ mol CrO}_4^{2-}}{3 \text{ mol SO}_3^{2-}}\right)$$

$$= 1.68 \text{ X } 10^{-2} \text{ mol CrO}_4^{2-}$$

Since there is one mole of Cr in each mole of CrO_4^{2-}, then the above number of moles of CrO_4^{2-} is also equal to the number of moles of Cr that were present:

0.0168 mol Cr × 52.00 g/mol = 0.875 g Cr in the original alloy.

c) (0.875 g/3.450 g) × 100 = 25.4 % Cr

6.83 a)

$$\# \text{mol C}_2\text{O}_4^{2-} = (21.62 \text{ mL KMnO}_4)\left(\frac{0.1000 \text{ mol KMnO}_4}{1000 \text{ mL KMnO}_4}\right)\left(\frac{5 \text{ mol C}_2\text{O}_4^{2-}}{2 \text{ mol KMnO}_4}\right)$$

$$= 5.405 \times 10^{-3} \text{ mol C}_2\text{O}_4^{2-}$$

b) The stoichiometry for calcium is as follows:

mol $C_2O_4^{2-}$ = mol Ca^{2+} = mol $CaCl_2$

Thus the number of grams of $CaCl_2$ is given simply by:

5.405×10^{-3} mol $CaCl_2 \times 110.98$ g/mol = 0.5999 g $CaCl_2$

c) (0.5999 g/2.463 g) × 100 % = 24.35 % $CaCl_2$

Additional Exercises

6.85 Total charge = −2 = (charge of sulfur atoms) + (charge of oxygen atoms)
Charge of oxygens = 6(−2) = −12
Charge of sulfur atoms = −2 −(−12) = +10
+10 spread out over 4 sulfur atoms gives a charge of: +10/4 per S atom, or:
Oxidation number of S = +2.5

6.87 For each problem, first find the total of the oxidation number of all non-carbon atoms. Since each species is neutral, the carbon atoms must then have a total oxidation number opposite to that total:
a) −4/2 = −2
b) 0 (the oxidation number of O and H cancel eachother)
c) +4
d) +4

6.89 No, the first, fourth and fifth reactions were not necessary.

6.91 We choose the metal that is lower (more reactive) in the activity series shown in Table 6.2: a) aluminum b) zinc c) magnesium

6.93 In each case, the reaction should proceed to give the less reactive of the two metals, together with the ion of the more reactive of the two metals. The reactivity is taken from the reactivity series.
a) $Zn(s) + Sn^{2+}(aq) \rightarrow Zn^{2+}(aq) + Sn(s)$
b) $2Cr(s) + 6H^{+}(aq) \rightarrow 2Cr^{3+}(aq) + 3H_2(g)$
c) N.R.
d) $Mn(s) + Pb^{2+}(aq) \rightarrow Mn^{2+}(aq) + Pb(s)$
e) $Zn(s) + Co^{2+}(aq) \rightarrow Zn^{2+}(aq) + Co(s)$

6.95 a) $2Zn(s) + O_2(g) \rightarrow 2ZnO(s)$
b) $4Al(s) + 3O_2(g) \rightarrow 2Al_2O_3(s)$
c) $2Mg(s) + O_2(g) \rightarrow 2MgO(s)$
d) $2Fe(s) + O_2(g) \rightarrow 2FeO(s)$

Alternatively we have: $4Fe(s) + 3O_2(g) \rightarrow 2Fe_2O_3(s)$

e) $2Ca(s) + O_2(g) \rightarrow 2CaO(s)$

6.97

$$\# \text{ g PbO}_2 = (15.0 \text{ g Cl}_2)\left(\frac{1 \text{ mol Cl}_2}{70.91 \text{ g Cl}_2}\right)\left(\frac{1 \text{ mol PbO}_2}{1 \text{ mol Cl}_2}\right)\left(\frac{239.2 \text{ g PbO}_2}{1 \text{ mol PbO}_2}\right)$$

$$= 50.6 \text{ g PbO}_2$$

*6.99 $Cu + 2Ag^+ \rightarrow 2Ag + Cu^{2+}$

The number of moles of Ag^+ available for the reaction is
$0.125 \text{ M} \times 0.255 \text{ L} = 0.0319 \text{ mol } Ag^+$

Since the stoichiometry is 2/1, the number of moles of Cu^{2+} ion that are consumed is $0.0319 \div 2 = 0.0159$ mol. The mass of copper consumed is 0.0159 mol x 63.546 g/mol = 1.01 g. The amount of unreacted copper is thus: 12.340 g – 1.01 g = 11.33 g Cu. The mass of Ag that is formed is: 0.0319 mol × 108 g/mol = 3.45 g Ag. The final mass of the bar is: 11.33 g + 3.45 g = 14.78 g.

*6.101 The balanced equation for the oxidation-reduction reaction is:
$3H_2C_2O_4 + Cr_2O_7^{2-} + 8H^+ \rightarrow 6CO_2 + 2Cr^{3+} + 7H_2O$

$$\# \text{ moles } H_2C_2O_4 = (6.25 \text{ mL } K_2Cr_2O_7)\left(\frac{0.200 \text{ moles } K_2Cr_2O_7}{1000 \text{ mL } K_2Cr_2O_7}\right)$$

$$\times\left(\frac{3 \text{ moles } H_2C_2O_4}{1 \text{ mole } K_2Cr_2O_7}\right) = 3.75\times10^{-3} \text{ moles } H_2C_2O_4$$

So, if we titrate the same oxalic acid solution using NaOH we will need:

$$\# \text{ ml NaOH} = (3.75\times10^{-3} \text{ moles } H_2C_2O_4)\left(\frac{2 \text{ moles NaOH}}{1 \text{ mole } H_2C_2O_4}\right)$$

$$\times\left(\frac{1000 \text{ mL NaOH}}{0.450 \text{ moles NaOH}}\right) = 16.7 \text{ ml NaOH}$$

Chapter 7

Practice Exercises

7.1 $\text{\# dietary calories} = (54 \text{ kJ})\left(\dfrac{1 \text{ Cal}}{4.184 \text{ kJ}}\right) = 13 \text{ Cal per pound of body weight}$

7.2 For 50.0 g water, the energy needed is 50 times that for 1.0 g water, or 50 cal:

$$\text{\# cal} = (100\ °\text{C})\left(\frac{50 \text{ cal}}{1°\text{C}}\right) = 5.00 \times 10^3 \text{ cal}$$

In joules, this is:

$$\text{\# J} = (5.00 \times 10^3 \text{ cal})\left(\frac{4.184 \text{ J}}{1 \text{ cal}}\right) = 20{,}900 \text{ J}$$

7.3 $\text{q gained by water} = (5.0\ °\text{C})\left(\dfrac{250 \text{ cal}}{1°\text{C}}\right)\left(\dfrac{4.184 \text{ J}}{1 \text{ cal}}\right) = 5230 \text{ J}$

q lost by ball bearing = − q gained by water = −5,230 J

$C = q/\Delta T$

$C = -5{,}230 \text{ J}/(30.0\ °\text{C} - 220\ °\text{C}) = 27.5 \text{ J}/°\text{C}$

7.4 The amount of heat transferred into the water is:

$$\text{\# J} = (250 \text{ g } H_2O)(4.184\ ^{\text{J}}/_{\text{g °C}})(30.0\ °\text{C} - 25.0\ °\text{C}) = 5200 \text{ J}$$

$$\text{\# kJ} = (5200 \text{ J})\left(\frac{1 \text{ kJ}}{1000 \text{ J}}\right) = 5.2 \text{ kJ}$$

$$\text{\# cal} = (5200 \text{ J})\left(\frac{1 \text{ cal}}{4.184 \text{ J}}\right) = 1200 \text{ cal}$$

$$\text{\# kcal} = (1200 \text{ cal})\left(\frac{1 \text{ kcal}}{1000 \text{ cal}}\right) = 1.2 \text{ kcal}$$

7.5 Supplied. It takes energy to separate particles which are attracted to one another.

7.6 $\text{Heat absorbed by calorimeter} = (25.51°\text{C} - 20.00°\text{C})\left(\dfrac{8.930 \text{ kJ}}{1°\text{C}}\right) = 49.2 \text{ kJ}$

$$\text{\# mol C} = (1.50 \text{ g C})\left(\frac{1 \text{ mol C}}{12.01 \text{ g C}}\right) = 0.125 \text{ mol C}$$

ΔE = energy/mol = 49.2 kJ/0.125 mol C = 394 kJ/mol C

7.7 q = specific heat x mass x temperature change
= 4.184 J/g °C x (175 g + 4.90 g) x (14.9 °C – 10.0 °C)
= 3.7 x 10^3 J = 3.7 kJ of heat released by the process.

This should then be converted to a value representing kJ per mole of reactant, remembering that the sign of ΔH is to be negative, since the process releases heat energy to surroundings. The number of moles of sulfuric acid is:

$$\#\,\text{moles}\ H_2SO_4 = (4.90\ \text{g}\ H_2SO_4)\left(\frac{1\ \text{mole}\ H_2SO_4}{98.08\ \text{g}\ H_2SO_4}\right) = 5.00\times10^{-2}\ \text{moles}\ H_2SO_4$$

and the enthalpy change in kJ/mole is given by:

3.7 kJ ÷ 0.0500 moles = 74 kJ/mole

7.8 We can proceed by multiplying both the equation and the thermochemical value of Example 7.9 by the fraction 2.5/2:

$2.5\ H_2(g) + 2.5/2\ O_2(g) \rightarrow 2.5\ H_2O(g)$, ΔH = (–517.8 kJ) x 2.5/2
$2.5\ H_2(g) + 1.25\ O_2(g) \rightarrow 2.5\ H_2O(g)$, ΔH = –647.3 kJ

7.9

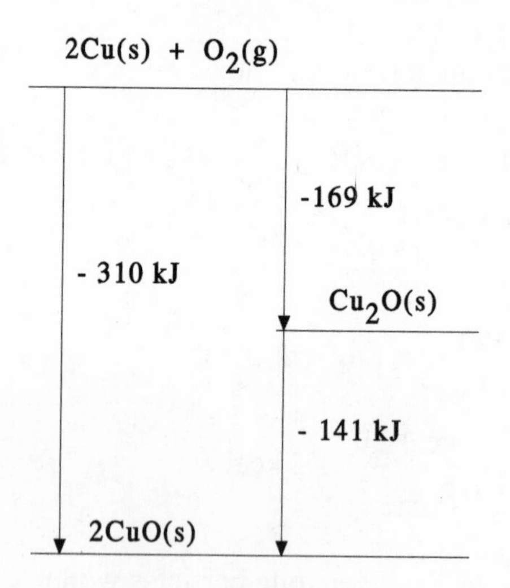

7.10 This problem requires that we add the reverse of the second equation (remembering to change the sign of the associated ΔH value) to the first equation:

$C_2H_4(g) + 3O_2(g) \rightarrow 2CO_2(g) + 2H_2O(\ell)$, $\Delta H°$ = –1411.1 kJ

$2CO_2(g) + 3H_2O(\ell) \rightarrow C_2H_5OH(\ell) + 3O_2(g)$, $\Delta H°$ = +1367.1 kJ which gives the following net equation and value for $\Delta H°$:

$C_2H_4(g) + H_2O(\ell) \rightarrow C_2H_5OH(\ell)$ $\Delta H° = -44.0$ kJ

7.11 $\# kJ = (480 \text{ moles } C_8H_{18})\left(\frac{5450.5 \text{ kJ/mol}}{1 \text{ moles } C_8H_{18}}\right) = 2.62 \times 10^6 \text{ kJ}$

7.12 $Na(s) + 1/2H_2(g) + C(s) + 3/2O_2(g) \rightarrow NaHCO_3(s)$, $\Delta H_f° = -947.7$ kJ/mol

7.13 a) $\Delta H°$ = sum $\Delta H_f°$[products] – sum $\Delta H_f°$[reactants]
= $2\Delta H_f°[NO_2(g)] - \{2\Delta H_f°[NO(g)] + \Delta H_f°[O_2(g)]\}$
= 2 mol x 33.8 kJ/mol – [2 mol x 90.37 kJ/mol + 1 mol x 0 kJ/mol]
= –113.1 kJ

b) $\Delta H° = \{\Delta H_f°[H_2O(\ell)] + \Delta H_f°[NaCl(s)]\} - \{\Delta H_f°[NaOH(s)] + \Delta H_f°[HCl(g)]\}$
= [(–285.9 kJ/mol) + (–411.0 kJ/mol)]
– [(–426.8 kJ/mol) + (–92.30 kJ/mol)]
= –177.8 kJ

Review Problems

7.47 $KE = \frac{1}{2} mv^2$

$$= 1/2(2150 \text{ kg})(80 \frac{\text{km}}{\text{hr}})^2 \left(\frac{1 \text{ hr}}{3600\text{s}}\right)^2 \left(\frac{1000\text{m}}{1 \text{ km}}\right)^2$$

$$= (5.31 \times 10^5 \text{ kgm}^2/\text{s}^2)\left(\frac{1 \text{ kJ}}{1000 \text{ kgm}^2/\text{s}^2}\right)$$

$$= 531 \text{ kJ}$$

7.49 a) $\# \text{kcal} = (347 \text{ kJ})\left(\frac{1 \text{ kcal}}{4.184 \text{ kJ}}\right) = 82.9 \text{ kJ}$

b) $\# \text{kJ} = (308 \text{ kcal})\left(\frac{4.184 \text{ kJ}}{1 \text{ kcal}}\right) = 1290 \text{ kJ}$

7.51 $\# \text{cal} = (1 \text{ hr})\left(\frac{60 \text{ min}}{1 \text{ hl}}\right)\left(\frac{60 \text{ s}}{1 \text{ min}}\right)\left(\frac{100 \text{ J}}{1 \text{ s}}\right)\left(\frac{1 \text{ cal}}{4.184 \text{ J}}\right) = 86{,}000 \text{ cal}$

A dietary calorie is 1,000 calories, so this = 86 dietary calories.

7.53 $\Delta E = q + w = 28\text{J} - 45\text{J} = -17 \text{ J}$

7.55 Here, ΔE must = 0 in order for there to be no change in energy for the cycle.

$$\Delta E = q + w$$
$$0 = q + (-100J)$$
$$q = +100J$$

7.57 $\# kJ = (175\ g\ H_2O)(4.184\ \frac{J}{g\ ^\circ C})(25.0\ ^\circ C - 15.0\ ^\circ C)\left(\frac{1\ kJ}{1000\ J}\right) = 7.32\ kJ$

7.59 $\# J = (0.4498\ J\ g^{-1}\ ^\circ C^{-1})(15.0\ g)(20.0\ ^\circ C) = 135\ J$

7.61 a) $\# J = 4.184\ J\ g^{-1}\ ^\circ C^{-1} \times 100\ g \times 4.0\ ^\circ C = 1.67 \times 10^3\ J$
b) $1.67 \times 10^3\ J$
c) $1.67 \times 10^3\ J/(100 - 28.0)\ ^\circ C = 23.2\ J\ ^\circ C^{-1}$
d) $23.2\ J\ ^\circ C^{-1} \div 5.00\ g = 4.64\ J\ g^{-1}\ ^\circ C^{-1}$

7.63 a) Since fat tissue is 85% fat, there are only 0.85 lbs of fat lost for every pound of tissue lost in the weight reduction program. The water, which is a part of the fat tissue, is not lost. Therefore;

$$\# kcal = (0.85\ lb\ fat)\left(\frac{456\ g}{1\ lb}\right)\left(\frac{9.0\ kcal}{1\ g\ fat}\right) = 3.5 \times 10^3\ kcal$$

b) In part (a) we determined that we would need to expend 3500 kcal in order to burn off 1 lb of fat tissue.

$$\# miles = (1.0\ lb\ fat\ tissue)\left(\frac{3.5\ X\ 10^3\ kcal}{1.0\ lb\ fat\ tissue}\right)\left(\frac{1\ hr}{5.0 \times 10^2\ kcal}\right)\left(\frac{8.0\ miles}{hr}\right)$$
$$= 56\ miles$$

7.65 $\frac{\# J}{mol\ ^\circ C} = \left(\frac{0.4498\ J}{g\ ^\circ C}\right)\left(\frac{55.847\ g\ Fe}{1\ mol\ Fe}\right) = 25.12\ J/mol\ ^\circ C$

7.67 $(4.18\ J\ g^{-1}\ ^\circ C^{-1})(4.54 \times 10^3\ g)(58.65 - 60.25)\ ^\circ C = -3.04 \times 10^4\ J = -30.4\ kJ$

7.69 Keep in mind that the total mass must be considered in this calculation, and that both liquids, once mixed, undergo the same temperature increase:

heat $= (4.18\ J/g\ ^\circ C)(55.0\ g + 55.0\ g)(31.8\ ^\circ C - 23.5\ ^\circ C)$
$= 3.8 \times 10^3$ J of heat energy released

Next determine the number of moles of reactant involved in the reaction:
0.0550 L x 1.3 mol/L = 0.072 mol of acid and of base.

Thus the enthalpy change is: $\frac{\# kJ}{mol} = \frac{(3.8 \times 10^3\ J)\left(\frac{1\ kJ}{1000\ J}\right)}{(0.072\ mol)} = 53\ kJ/mol$

7.71 a) $\#J = (97.1\ kJ/^\circ C)(27.282\ ^\circ C - 25.000\ ^\circ C) = 222\ kJ = 2.22 x 10^5\ J$

b) $\Delta H^\circ = -222$ kJ/mol

7.73 a) Multiply the given equation by the fraction 2/3.

$$2CO(g) + O_2(g) \rightarrow 2CO_2(g), \quad \Delta H^\circ = -566\ kJ$$

b) To determine ΔH for 1 mol, simply multiply the original ΔH by 1/3; –283 kJ/mol.

7.75 $4Al(s) + 2Fe_2O_3(s) \rightarrow 2Al_2O_3(s) + 4Fe(s) \quad \Delta H^\circ = -1708\ kJ$

7.77 $$\#kJ = (6.54\ g\ Mg)\left(\frac{-1203\ kJ}{2\ mol\ Mg}\right)\left(\frac{1\ mol\ Mg}{24.305\ g\ Mg}\right) = -162\ kJ$$

(162 kJ of heat are evolved)

7.79

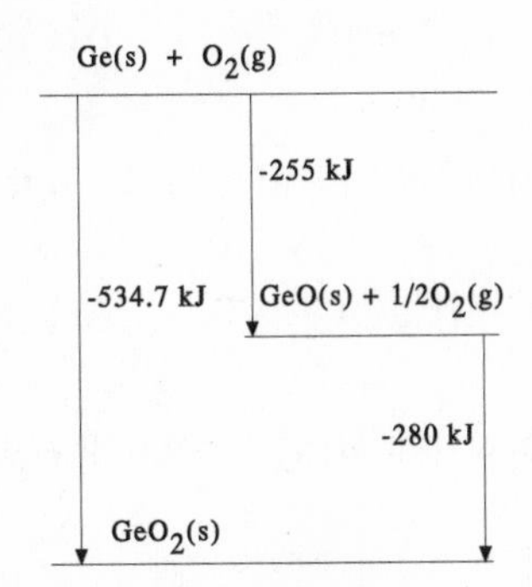

The enthalpy change for the reaction $GeO(s) + 1/2O_2(g) \rightarrow GeO_2(s)$ is –280 kJ as seen in the figure above.

7.81 Since NO_2 does not appear in the desired overall reaction, the two steps are to be manipulated in such a manner so as to remove it by cancellation. Add the second equation to the inverse of the first, remembering to change the sign of the first equation, since it is to be reversed:

$2NO_2(g) \rightarrow N_2O_4(g), \quad \Delta H^\circ = -57.93\ kJ$

$2NO(g) + O_2(g) \rightarrow 2NO_2(g), \quad \Delta H^\circ = -113.14\ kJ$

Adding, we have:

$2NO(g) + O_2(g) \rightarrow N_2O_4(g), \quad \Delta H^\circ = -171.07\ kJ$

7.83 If we label the four known thermochemical equations consecutively, 1, 2, 3, and 4, then the sum is made in the following way: Divide equation #3 by two, and reverse all of the other equations (#1, #2, and #4), while also dividing each by two:

$\frac{1}{2}Na_2O(s) + HCl(g) \rightarrow \frac{1}{2}H_2O(\ell) + NaCl(s), \quad \Delta H° = -253.66 \text{ kJ}$

$NaNO_2(s) \rightarrow \frac{1}{2}Na_2O(s) + \frac{1}{2}NO_2(g) + \frac{1}{2}NO(g), \quad \Delta H° = 213.57 \text{ kJ}$

$\frac{1}{2}NO(g) + \frac{1}{2}NO_2(g) \rightarrow \frac{1}{2}N_2O(g) + \frac{1}{2}O_2(g), \quad \Delta H° = -21.34 \text{ kJ}$

$\frac{1}{2}H_2O(\ell) + \frac{1}{2}O_2(g) + \frac{1}{2}N_2O(g) \rightarrow HNO_2(\ell), \quad \Delta H° = -17.18 \text{ kJ}$

Adding gives:

$HCl(g) + NaNO_2(s) \rightarrow NaCl(s) + HNO_2(\ell), \; \Delta H° = -78.61 \text{ kJ}$

7.85 Reverse the second equation, and then divide each by two before adding:

$CO(g) + \frac{1}{2}O_2(g) \rightarrow CO_2(g), \quad \Delta H° = -283.0 \text{ kJ}$

$CuO(s) \rightarrow Cu(s) + \frac{1}{2}O_2(g), \quad \Delta H° = 155.2 \text{ kJ}$

$CuO(s) + CO(g) \rightarrow Cu(s) + CO_2(g), \quad \Delta H° = -127.8 \text{ kJ}$

7.87 Multiply all of the equations by ½ and and them together.

$1/2CaO(s) + 1/2Cl_2(g) \rightarrow 1/2CaOCl_2(s) \quad \Delta H° = \frac{1}{2}(-110.9\text{kJ})$

$1/2H_2O(\ell) + 1/2CaOCl_2(s) + NaBr(s) \rightarrow NaCl(s) + 1/2Ca(OH)_2(s) + 1/2Br_2(\ell)$

$\Delta H° = \frac{1}{2}(-60.2 \text{ kJ})$

$1/2Ca(OH)_2(s) \rightarrow 1/2CaO(s) + 1/2H_2O(\ell) \quad \Delta H° = \frac{1}{2}(+65.1 \text{ kJ})$

$1/2Cl_2(g) + NaBr(s) \rightarrow NaCl(s) + 1/2Br_2(\ell) \; \Delta H° = \frac{1}{2}(-106 \text{ kJ})$

7.89 We need to eliminate the NO_2 from the two equations. To do this, multiply the first reaction by 3 and the second reaction by two and add them together.

$12NH_3(g) + 21O_2(g) \rightarrow 12NO_2(g) + 18H_2O(g) \quad \Delta H° = 3(-1132 \text{ kJ})$

$12NO_2(g) + 16NH_3(g) \rightarrow 14N_2(g) + 24H_2O(g) \quad \Delta H° = 2(-2740 \text{ kJ})$

$28NH_3(g) + 21O_2(g) \rightarrow 14N_2(g) + 42O_2(g) \quad \Delta H° = -8876 \text{ kJ}$

Now divide this equation by 7 to get

$4NH_3(g) + 3O_2(g) \rightarrow 2N_2(g) + 6O_2(g) \quad \Delta H° = 1/7(-8876 \text{ kJ}) = -1268 \text{ kJ}$

7.91 The equation we want is:

$3Mg(s) + N_2(g) + 3O_2(g) \rightarrow Mg(NO_3)_2(s)$

Reverse all three reactions AND multiply the third equation by three

$Mg_3N_2(s) + 6MgO(s) \rightarrow 8Mg(s) + Mg(NO_3)_2(s)$ $\Delta H° = +3884$ kJ

$3Mg(s) + N_2(g) \rightarrow Mg_3N_2(s)$ $\Delta H° = -463$ kJ

$6Mg(s) + 3O_2(g) \rightarrow Mg(NO_3)_2(s)$ $\Delta H° = 3(-1203$ kJ)

$Mg(s) + N_2(g) + 3O_2(g) \rightarrow Mg(NO_3)_2(s)$ $\Delta H° = -188$ kJ

7.93 Only (b) should be labeled with $\Delta H_f°$.

7.95 a) $2C(graphite) + 2H_2(g) + O_2(g) \rightarrow HC_2H_3O_2(\ell)$, $\Delta H_f° = -487.0$ kJ

b) $Na(s) + 1/2H_2(g) + C(graphite) + 3/2O_2(g) \rightarrow NaHCO_3(s)$$\Delta H_f° = -947.7$ kJ

c) $Ca(s) + 8S(s) + 3O_2(g) + 2H_2(g) \rightarrow CaSO_4.2H_2O(s)$ $\Delta H_f° = -2021.1$ kJ

7.97 a) $\Delta H° = \Delta H_f°[O_2(g)] + 2\Delta H_f°[H_2O(\ell)] - 2\Delta H_f°[H_2O_2(\ell)]$

$\Delta H° = 0$ kJ/mol + 2 mol × (–285.9 kJ/mol) – 2 × (–187.6 kJ/mol)
= –196.6 kJ

b) $\Delta H° = \Delta H_f°[H_2O(\ell)] + \Delta H_f°[NaCl(s)] - \Delta H_f°[HCl(g)] - \Delta H_f°[NaOH(s)]$
= 1 mol × (–285.9 kJ/mol) + 1 mol × (–411.0 kJ/mol)
– 1 mol × (–92.30 kJ/mol) – 1 mol × (–426.8 kJ/mol)
= –177.8 kJ

7.99 a) $½H_2(g) + ½Cl_2(g) \rightarrow HCl(g)$, $\Delta H_f° = -92.30$ kJ/mol

b) $½N_2(g) + 2H_2(g) + ½Cl_2(g) \rightarrow NH_4Cl(s)$, $\Delta H_f° = -315.4$ kJ/mol

7.101 $C_{12}H_{22}O_{11}(s) + 12O_2(g) \rightarrow 12CO_2(g) + 11H_2O$ $\Delta H° = -5.65 \times 10^3$ kJ/mol

$\Delta H° = \Sigma\Delta H_f°(\text{products}) - \Sigma\Delta H_f°(\text{reactants})$
$=[12\ \Delta H_f°(CO_2(g)) + 11\ \Delta H_f°(H_2O)] - [\Delta H_f°(C_{12}H_{22}O_{11}(s)) + 12\Delta H_f°(O_2(g))]$

Rearranging and realizing the $\Delta H_f°O_2(g) = 0$ we get
$\Delta H_f°[C_{12}H_{22}O_{11}(s)] = 12\Delta H_f°(CO_2(g)) + 11\Delta H_f°(H2O(\ell)) - \Delta H°$
$= 12(-393kJ) + 11(-285.9\ kJ) - (-5.65 \times 10^3\ kJ) = -2.22 \times 10^3\ kJ$

Additional Exercises

7.103 a) Either heat was subtracted from the system or work was done by the system.

b) $\# \text{kJ} = (750\text{g})(4.184\text{J/g}^\circ\text{C})(19.50 - 25.50)^\circ\text{C}\left(\frac{1\,\text{kJ}}{1000\,\text{J}}\right) = -18.8\,\text{kJ}$

7.105 $\Delta H^\circ = \Delta H_f^\circ[H_2SO_4(\ell)] - [\Delta H_f^\circ(SO_3(g)) + \Delta H_f^\circ(H_2O(\ell))]$
$= -811.32\text{ kJ} - [(-395.2\text{ kJ}) + (-285.9\text{ kJ})]$
$= -130.2\text{ kJ}$

7.107 Multiply the first reaction by ½:
$\frac{1}{2}Fe_2O_3(s) + 3/2CO(g) \rightarrow Fe(s) + 3/2CO_2(g)$ $\Delta H^\circ = \frac{1}{2}(-28\text{ kJ})$

Reverse the second reaction AND multiply by 1/6:
$1/3Fe_3O_4(s) + 1/6CO_2(g) \rightarrow 1/3\ Fe_3O_4(s) + 1/6CO(g)$ $\Delta H^\circ = 1/6(+59\text{ kJ})$

Reverse the third reaction AND multiply by 1/3:
$FeO(s) + 1/3CO_2(g) \rightarrow 1/3Fe_3O_4(s) + 1/3CO(g)$ $\Delta H^\circ = 1/3(-38\text{ kJ})$

Now add the equations together:
$FeO(s) + CO(g) \rightarrow Fe(s) + CO_2(g)$ $\Delta H^\circ = -16.8\text{ kJ}$

7.109 These two thermochemical equations are added along with six times that for the formation of liquid water:

$P_4O_{10}(s) + 6H_2O(\ell) \rightarrow 4H_3PO_4(\ell),$ $\Delta H^\circ = -257.2\text{ kJ}$
$4P(s) + 5O_2(g) \rightarrow P_4O_{10}(s),$ $\Delta H^\circ = -3062\text{ kJ}$
$6H_2(g) + 3O_2(g) \rightarrow 6H_2O(\ell),$ $\Delta H^\circ = -1715.4\text{ kJ}$

Adding gives:
$4P(s) + 6H_2(g) + 8O_2(g) \rightarrow 4H_3PO_4(\ell),$ $\Delta H^\circ = -5035\text{ kJ}$

The above result should then be divided by four:
$P(s) + 3/2H_2(g) + 2O_2(g) \rightarrow H_3PO_4(\ell),$ $\Delta H^\circ = -1259\text{ kJ/mol}$

7.111 The desired net equation is obtained by adding together the reverse of the two thermochemical equations:

$Zn(NO_3)_2(aq) + Cu(s) \rightarrow Cu(NO_3)_2(aq) + Zn(s),$ $\Delta H^\circ = 258\text{ kJ}$
$Cu(NO_3)_2(aq) + 2Ag(s) \rightarrow 2AgNO_3(aq) + Cu(s),$ $\Delta H^\circ = 106\text{ kJ}$

$2Ag(s) + Zn(NO_3)_2(aq) \rightarrow Zn(s) + 2AgNO_3(aq),$ $\Delta H^\circ = 364\text{ kJ}$

Since this ΔH° has a positive value, it is the reverse reaction that occurs spontaneously.

7.113 The equation may be written as: 1/2 $H_2(g)$ + 1/2 $Br_2(\ell) \rightarrow HBr(g)$; $\Delta H_f° = -36$ kJ

To obtain ΔH, combine the equations in the following manner:

$Br_2(aq) + 2KCl(aq)$	$\rightarrow Cl_2(g) + 2KBr(aq)$	$\Delta H° = 96.2$ kJ
$H_2(g) + Cl_2(g)$	$\rightarrow 2HCl(g)$	$\Delta H° = -184$ kJ
$2HCl(aq) + 2KOH(aq)$	$\rightarrow 2KCl(aq) + 2H_2O(\ell)$	$\Delta H° = -115$ kJ
$2KBr(aq) + 2H_2O(\ell)$	$\rightarrow 2HBr(aq) + 2KOH(aq)$	$\Delta H° = 115$ kJ
$2HCl(g)$	$\rightarrow 2HCl(aq)$	$\Delta H° = -154$ kJ
$2HBr(aq)$	$\rightarrow 2HBr(g)$	$\Delta H° = 160$ kJ
$Br_2(\ell)$	$\rightarrow Br_2(aq)$	$\Delta H° = -4.2$ kJ

Add all of the above to get;

$H_2(g) + Br_2(\ell) \rightarrow 2HBr(g)$; $\Delta H = -86$ kJ

Now divide this equation by two to give the thermochemical equation for the formation of 1 mol of HBr(g):

1/2 $H_2(g)$ + 1/2 $Br_2(\ell) \rightarrow HBr(g)$; $\Delta H = -43$ kJ

Comparing this value to the $\Delta H_f°$ value listed in Appendix E and at the outset of this problem, we see that this experimental data indicates a value that is close to the reported value.

7.115 $C_{12}H_{22}O_{11}(s) + 12O_2(g) \rightarrow 12CO_2(g) + 11H_2O(\ell)$ $\Delta H° = -5.63 \times 10^3$ kJ/mol

$\Delta H° = \Sigma\Delta H_f°(\text{products}) - \Sigma\Delta H_f°(\text{reactants})$

$= [12\ \Delta H_f°(CO_2(g)) + 11\ \Delta H_f°(H_2O(\ell))] - [\Delta H_f°(C_{12}H_{22}O_{11}(s)) + 12\Delta H_f°(O_2(g))]$

$= [12(-393\text{kJ}) + 11(-285.9\text{ kJ})] - [(-2230\text{ kJ}) + 12(0\text{ kJ})] = -5.63 \times 10^3$ J/mol

This is the amount of heat liberated for 1 mol of sucrose. Thus, for 28.4 g we have,

$$\#\text{kJ} = (28.4\text{ g})\left(\frac{1\text{ mol}}{342.3\text{ g}}\right)\left(\frac{-5630\text{ kJ}}{1\text{ mol}}\right) = -467\text{ kJ}$$

7.117 $1/2HCHO_2(\ell) + 1/2H_2O(\ell) \rightarrow 1/2CH_3OH(\ell) + 1/2O_2(g)$ $\Delta H° = +206$kJ

$1/2CO(g) + H_2(g) \rightarrow 1/2CH_3OH(\ell)$ $\Delta H° = -64$ kJ

$1/2HCHO_2(\ell) \rightarrow 1/2CO(g) + 1/2H_2O(\ell)$ $\Delta H° = -17$ kJ

Add these together:

$HCHO_2(\ell) + H_2(g) \rightarrow CH_3OH(\ell) + 1/2O_2(g)$ $\Delta H° = +125$ kJ

7.119 Add together the fourth equation, the first equation, the second equation and the reverse of the third equation:

$4CuS(s) + 2CuO(s) \rightarrow 3Cu_2S(s) + SO_2(g)$ $\Delta H° = -13.1$ kJ

$2Cu(s) + O_2(g) \rightarrow 2CuO(s)$ $\Delta H° = -155$ kJ

$Cu(s) + S(s) \rightarrow CuS(s)$ $\Delta H° = -53.1$ kJ

$SO_2(g) \rightarrow S(s) + O_2(g)$ $\Delta H° = +297$ kJ

The net reaction is:

$3CuS(s) + 3Cu(s) \rightarrow 3Cu_2S(s)$ $\Delta H° = +76$ kJ

We want 1/3 of this:

$CuS(s) + Cu(s) \rightarrow Cu_2S(s)$ $\Delta H° = +25$ kJ

Chapter 8

Practice Exercises

8.1 $\nu = 10.0\,\mu\text{m}\left(\dfrac{1\times10^{-6}\,\text{m}}{1\,\mu\text{m}}\right) = 1.00\times10^{-5}\,\text{m}$

$\nu = c/\lambda = (3.00 \times 10^{8}\text{ m/s})/(1.00 \times 10^{-5}\text{ m}) = 3.00 \times 10^{13}\text{ s}^{-1} = 3.00 \times 10^{13}\text{ Hz}$

8.2 $\lambda = \dfrac{c}{\nu} = \dfrac{2.9979\times10^{8}\text{ m s}^{-1}}{104.3\times10^{6}\text{ s}^{-1}} = 2.874\text{ m}$

8.3

$$\frac{1}{\lambda} = 109{,}678\text{ cm}^{-1}\times\left(\frac{1}{2^2}-\frac{1}{3^2}\right) = 109{,}678\text{ cm}^{-1}\times(0.2500 - 0.1111)$$

$$\frac{1}{\lambda} = 1.523\text{ x }10^{4}\text{ cm}^{-1}$$

$\lambda = 6.566\text{ x }10^{-5}\text{ cm} = 656.6\text{ nm}$, which is red.

8.4 When $n = 3$, $\ell = 0, 1, 2$. Thus we have s, p and d subshells.

When $n = 4$, $\ell = 0, 1, 2, 3$. Thus we have s, p, d and f subshells.

8.5 For the g subshell, $\ell = 4$ and there are 9 possible values of m_ℓ; –4, –3, –2, –1, 0, +1, +2, +3, +4. There are therefore 9 orbitals.

8.6 a) Mg: $1s^2 2s^2 2p^6 3s^2$
b) Ge: $1s^2 2s^2 2p^6 3s^2 3p^6 3d^{10} 4s^2 4p^2$
c) Cd: $1s^2 2s^2 2p^6 3s^2 3p^6 3d^{10} 4s^2 4p^6 4d^{10} 5s^2$
d) Gd: $1s^2 2s^2 2p^6 3s^2 3p^6 3d^{10} 4s^2 4p^6 4d^{10} 4f^7 5s^2 5p^6 5d^1 6s^2$

8.7

(a) Na:

1s	2s	2p	3s	3p	4s	3d
(↑↓)	(↑↓)	(↑↓)(↑↓)(↑↓)	(↑)	( )( )( )	( )	( )( )( )( )( )

(b) S:

1s	2s	2p	3s	3p	4s	3d
(↑↓)	(↑↓)	(↑↓)(↑↓)(↑↓)	(↑↓)	(↑↓)(↑)(↑)	( )	( )( )( )( )( )

(c) Fe:

1s	2s	2p	3s	3p	4s	3d
(↑↓)	(↑↓)	(↑↓)(↑↓)(↑↓)	(↑↓)	(↑↓)(↑↓)(↑↓)	(↑↓)	(↑↓)(↑)(↑)(↑)(↑)

8.8 a) P: $[Ne]3s^23p^3$

[Ne] (↑↓) (↑)(↑)(↑)
3s 3p (3 unpaired electrons)

b) Sn: $[Kr]4d^{10}5s^25p^2$

[Kr] (↑↓)(↑↓)(↑↓)(↑↓)(↑↓) (↑↓) (↑)(↑)( )
4d 5s 5p

(2 unpaired electrons)

8.9 a) Se: $4s^24p^4$ b) Sn: $5s^25p^2$ c) I: $5s^25p^5$

8.10 a) Sn b) Ga c) Cr d) S^{2-}

8.11 a) Be b) C

Review Problems

8.73 $\nu = \frac{c}{\lambda} = \frac{3.00\times10^8 \text{ m/s}}{430\times10^{-9} \text{ m}} = 6.98\times10^{14} \text{ s}^{-1} = 6.98\times10^{14} \text{ Hz}$

8.75 $\nu = \frac{c}{\lambda} = \frac{3.00\times10^8 \text{ m/s}}{6.85\times10^{-6} \text{ m}} = 4.38\times10^{13} \text{ s}^{-1} = 4.38\times10^{13} \text{ Hz}$

8.77 $295 \text{ nm} = 295 \times 10^{-9} \text{ m}$

$\nu = \frac{c}{\lambda} = \frac{3.00\times10^8 \text{ m/s}}{295\times10^{-9} \text{ m}} = 1.02\times10^{15} \text{ s}^{-1} = 1.02\times10^{15} \text{ Hz}$

8.79 $101.1 \text{ MHz} = 101.1 \times 10^6 \text{ Hz} = 101.1 \times 10^6 \text{ s}^{-1}$

$\lambda = \frac{c}{\nu} = \frac{3.00\times10^8 \text{ m/s}}{101.1\times10^6 \text{ s}^{-1}} = 2.98 \text{ m}$

8.81 $\lambda = \frac{c}{\nu} = \frac{3.00\times10^8 \text{ m/s}}{60 \text{ s}^{-1}} = 5.0\times10^6 \text{ m} = 5.0\times10^3 \text{ km}$

8.83 $E = h\nu = 6.63 \text{ x } 10^{-34} \text{ J s} \times 4.0 \times 10^{14} \text{ s}^{-1} = 2.7 \times 10^{-19} \text{ J}$

$\frac{\#\text{J}}{\text{mol}} = \left(\frac{2.7\times10^{-19} \text{ J}}{1 \text{ photon}}\right)\left(\frac{6.022\times10^{23} \text{ photons}}{1 \text{ mol}}\right) = 1.6\times10^5 \text{ J mol}^{-1}$

8.85 a) violet (see Figure 8.8)
b) $\nu = c/\lambda = (3.00 \times 10^8 \text{ m s}^{-1})/(410.3 \times 10^{-9} \text{ m}) = 7.31 \times 10^{14} \text{ s}^{-1}$
c) $E = h\nu = (6.63 \times 10^{-34} \text{ J s})(7.31 \times 10^{14} \text{ s}^{-1}) = 4.85 \times 10^{-19} \text{ J}$

8.87 $\frac{1}{\lambda} = 109{,}678\ \text{cm}^{-1} \times \left(\frac{1}{3^2} - \frac{1}{6^2}\right) = 109{,}678\ \text{cm}^{-1} \times (0.1111 - 0.02778)$

$\frac{1}{\lambda} = 9.140 \times 10^3\ \text{cm}^{-1}$

$\lambda = 1.094 \times 10^{-4}$ cm = 1090 nm, which is not in the visible region. (We would not see the line.)

8.89 $\frac{1}{\lambda} = 109{,}678\ \text{cm}^{-1} \times \left(\frac{1}{4^2} - \frac{1}{10^2}\right)$

$\frac{1}{\lambda} = 5.758 \times 10^3\ \text{cm}^{-1}$

$\lambda = 1.737 \times 10^{-6}$ m

This is in the infrared region.

8.91 a) *p* b) *f*

8.93 a) n = 3, $\ell = 0$ b) n = 5, $\ell = 2$

8.95 $\ell = 0, 1, 2, 3, 4, 5$

8.97 a) $m_\ell = 1, 0, \text{or} -1$ b) $m_\ell = 3, 2, 1, 0, -1, -2, \text{or} -3$

8.99 When $m_\ell = -4$, the minimum value of ℓ is 4 and the minimum value of n is 5.

8.101

n	ℓ	m_ℓ	m_s
2	1	−1	+1/2
2	1	−1	−1/2
2	1	0	+1/2
2	1	0	−1/2
2	1	+1	+1/2
2	1	+1	−1/2

8.103 21 electrons have $\ell = 1$, 20 electrons have $\ell = 2$

8.105
a) S $1s^22s^22p^63s^23p^4$
b) K $1s^22s^22p^63s^23p^64s^1$
c) Ti $1s^22s^22p^63s^23p^63d^24s^2$
d) Sn $1s^22s^22p^63s^23p^64s^23d^{10}4p^64d^{10}5s^25p^2$

8.107 a) Mn is $[Ar]4s^2 3d^5$, ∴ five unpaired electrons. PARAMAGNETIC
b) As is $[Ar]\ 3d^{10}4s^2 4p^3$, ∴ three unpaired electrons. PARAMAGNETIC
c) S is $[Ne]3s^2 3p^4$, ∴ two unpaired electrons. PARAMAGNETIC
d) Sr is $[Kr]5s^2$, ∴ zero unpaired electrons
e) Ar is $1s^2 2s^2 2p^6 3s^2 3p^6$, ∴ zero unpaired electrons

8.109 a) Mg is $1s^2 2s^2 2p^6 3s^2$, ∴ zero unpaired electrons
b) P is $1s^2 2s^2 2p^6 3s^2 3p^3$, ∴ three unpaired electrons
c) V is $1s^2 2s^2 2p^6 3s^2 3p^6 3d^3 4s^2$, ∴ three unpaired electrons

8.111 a) $[Ar]\ 3d^8 4s^2$
b) $[Xe]\ 6s^1$
c) $[Ar]\ 3d^{10}4s^2 4p^2$
d) $[Ar]\ 3d^{10}4s^2 4p^5$
e) $[Xe]\ 4f^{14}5d^{10}6s^2 6p^3$

8.113

(a) Mg: (↑↓) (↑↓) (↑↓)(↑↓)(↑↓) (↑↓) ( )( )( ) ( ) ( )( )( )( )( )
1s 2s 2p 3s 3p 4s 3d

(b) Ti: (↑↓) (↑↓) (↑↓)(↑↓)(↑↓) (↑↓) (↑↓)(↑↓)(↑↓) (↑↓) (↑)(↑)( )( )( )
1s 2s 2p 3s 3p 4s 3d

8.115

(a) Ni: [Ar] (↑↓) (↑↓)(↑↓)(↑↓)(↑)(↑)
4s 3d

(b) Cs: [Xe] (↑)
6s

(c) Ge: [Ar] (↑↓) (↑↓)(↑↓)(↑↓)(↑↓)(↑↓) (↑)(↑)( )
4s 3d 4p

(d) Br: [Ar] (↑↓) (↑↓)(↑↓)(↑↓)(↑↓)(↑↓) (↑↓)(↑↓)(↑)
4s 3d 4p

8.117 The value corresponds to the row in which the element resides:
a) 5 b) 4 c) 4 d) 6

8.119 a) $3s^1$ b) $3s^2 3p^1$ c) $4s^2 4p^2$ d) $3s^2 3p^3$

8.121

(a) Na: ↑ (3s)

(b) Al: ↑↓ (3s) ↑ _ _ (3p)

(c) Ge: ↑↓ (4s) ↑ ↑ _ (4p)

(d) P: ↑↓ (3s) ↑ ↑ ↑ (3p)

8.123 a) 1 b) 6 c) 7

8.125 a) Na b) Sb

8.127 Sb. (Based upon trends, Sn would be predicted to be larger, but this is one area in which an exception to the trend exists. See Figure 8.30.)

8.129 Cations are generally smaller than the corresponding atom, and anions are generally larger than the corresponding atom:
a) Na b) Co^{2+} c) Cl^{-}

8.131 a) C b) O c) Cl

8.133 a) Cl b) Br

8.135 Mg

Additional Exercises

8.137 a) In air:

$$\lambda = \frac{c}{\nu} = \frac{330\ \text{m s}^{-1}}{20\ \text{s}^{-1}} = 17\ \text{m (longest wavelength)}$$

$$\lambda = \frac{c}{\nu} = \frac{330\ \text{m s}^{-1}}{20{,}000\ \text{s}^{-1}} = 0.017\ \text{m (1.7 cm) (shortest wavelength)}$$

b) In water:

$$\lambda = \frac{c}{\nu} = \frac{1500\ \text{m s}^{-1}}{20\ \text{s}^{-1}} = 75\ \text{m (longest wavelength)}$$

$$\lambda = \frac{c}{\nu} = \frac{1500\ \text{m s}^{-1}}{20{,}000\ \text{s}^{-1}} = 0.075\ \text{m (7.5 cm) (shortest wavelength)}$$

8.139

$$\frac{1}{\lambda} = 109{,}678\ \text{cm}^{-1} \times \left(\frac{1}{n_2^{\ 2}} - \frac{1}{n_1^{\ 2}}\right)$$

$$= \left(\frac{1}{410.3\ \text{nm}}\right)\left(\frac{1\ \text{nm}}{1 \times 10^{-9}\ \text{m}}\right)\left(\frac{1\ \text{m}}{100\ \text{cm}}\right)$$

$$= 2.437 \times 10^{4}\ \text{cm}^{-1} = 109{,}678\ \text{cm}^{-1} \times \left(\frac{1}{2^2} - \frac{1}{x^2}\right)$$

$$\frac{2.437 \times 10^{4}\ \text{cm}^{-1}}{109678\ \text{cm}^{-1}} = \frac{1}{4} - \frac{1}{x^2} = 0.222$$

$$\frac{1}{x^2} = \frac{1}{4} - 0.222$$

$$x = 6$$

8.141 A transition from high energy to low energy may result in light emission. The transition from 4p→3d, 5f→4d, and 4d→2p are the only possibilities.

*8.143 a) We must first calculate the energy in joules of a mole of photons.

$$E = \frac{hc}{\lambda} = \frac{(6.63 \times 10^{-34}\ \text{J s})(3.00 \times 10^{8}\ \text{m/s})}{600 \times 10^{-9}\ \text{m}} = 3.32 \times 10^{-19}\ \text{J/photon}$$

$$3.32 \times 10^{-19}\ \text{J/photon} \times 6.02 \times 10^{23}\ \text{photons/mol} = 2.00 \times 10^{5}\ \text{J/mol}$$

Next, we calculate the heat transfer problem as in Chapter 7:

Heat = (specific heat)(mass)(change in temperature)

2.00×10^{5} J = (4.18 J g^{-1} °C^{-1})(m)(5.0 °C)

mass = 9.57×10^{3} g

b) E = 6.63×10^{-19} J/photon

E = 3.99×10^{5} J/mol

mass = 1.91×10^{4} g

8.145 a) Start by calculating λ:

$$\frac{1}{\lambda} = 109{,}678\ \text{cm}^{-1} \times \left(\frac{1}{1^2} - \frac{1}{5^2}\right) = 109{,}678\ \text{cm}^{-1} \times (0.1000 - 0.04000)$$

$$\frac{1}{\lambda} = 1.053 \text{ x } 10^5\ \text{cm}^{-1}$$

$\lambda = 9.498 \times 10^{-6}$ cm = 94.98 nm, which is in the ultraviolet region of the visible spectrum.

b)

$$\frac{1}{\lambda} = 109{,}678\ \text{cm}^{-1} \times \left(\frac{1}{2^2} - \frac{1}{4^2}\right) = 109{,}678\ \text{cm}^{-1} \times (0.2500 - 0.0625)$$

$$\frac{1}{\lambda} = 2.056 \text{ x } 10^4\ \text{cm}^{-1}$$

$\lambda = 4.863 \times 10^{-5}$ cm = 486.3 nm, which is in the visible region of the spectrum.

c)

$$\frac{1}{\lambda} = 109{,}678\ \text{cm}^{-1} \times \left(\frac{1}{4^2} - \frac{1}{6^2}\right) = 109{,}678\ \text{cm}^{-1} \times (0.0625 - 0.0278)$$

$$\frac{1}{\lambda} = 3.808 \text{ x } 10^3\ \text{cm}^{-1}$$

$\lambda = 2.626 \times 10^{-4}$ cm = 2626 nm, which is in the infrared region of the spectrum.

8.147 a) This diagram violates the aufbau principle. Specifically, the s-orbital should be filled before filling the higher energy p-orbitals.

b) This diagram violates the aufbau principle. Specifically, the lower energy s orbital should be filled completely before filling the p-orbitals.

c) This diagram violates the aufbau principle. Specifically, the lower energy s orbital should be filled completely before filling the p-orbitals.

d) This diagram violates the Pauli Exclusion Principle since 2 electrons have the same set of quantum numbers

8.149 The 4*s* electrons are lost; $n = 4$, $\ell = 0$, $m_\ell = 0$, $m_s = \pm 1/2$

8.151

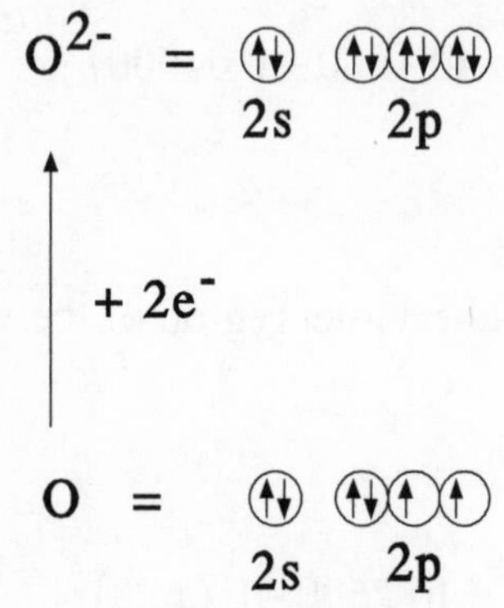

Thus the oxide ion, O^{2-}, has the electron configuration of [Ne].

Since the oxide ion has a closed shell electron configuration, the addition of a third electron is extremely difficult.

8.153 In problem 8.152 parts b and c we can determine that:

$$O(g) + 2e^- \rightarrow O^{2-}(g) \qquad \Delta H = +703kJ$$

From Table 8.2 we can see that $O(g) \rightarrow O^+(g) + e^-$ $\Delta H = +1314kJ/mol$.

It takes more energy to ionize oxygen than to create $O^{2-}(g)$.

Chapter 9

Practice Exercises

9.1 Cr: [Ar] $3d^4 4s^2$
a) Cr^{2+}: [Ar]$3d^4$
b) Cr^{3+}: [Ar]$3d^3$
c) Cr^{6+}: [Ar]

9.2 S^{2-}: [Ne] $3s^2 3p^6$
Cl^-: [Ne] $3s^2 3p^6$
The electron configurations are identical.

9.3

(a) :Se: (b) :I· (c) ·Ca·

9.4

·Mg· :O: ⟶ Mg^{2+} [:O:]$^{2-}$

9.5

R–C(=O)–H aldehyde   R–C(=O)–OH acid   R–O–H alcohol

R–N(H)–H amine   R–C(=O)–R ketone

9.6 a) Br b) Cl c) Cl

9.7

SO_2 O S O

NO_3^- O
O N O

$HClO_3$ O
H O Cl O

O
H O P O H
H_3PO_4 O
H

9.8 SO_2 has 18 valence electrons
PO_4^{3-} has 32 valence electrons
NO^+ has 10 valence electrons

9.9

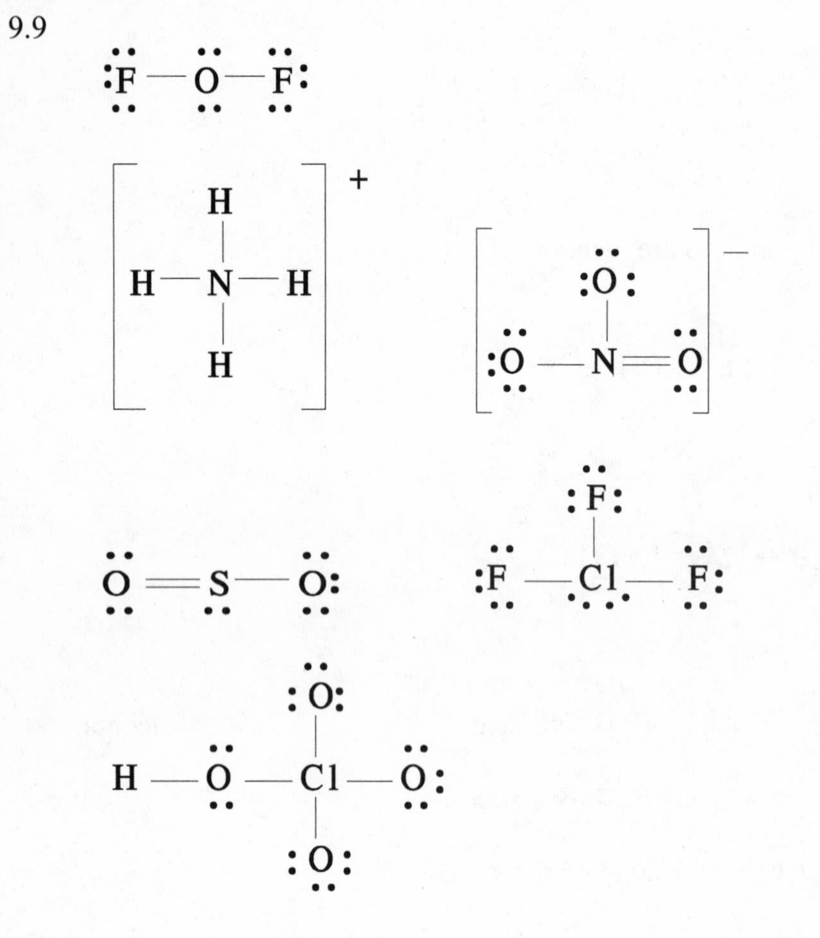

9.10 a) b)

-2 N—N≡O +1 +1

[S=C=N]$^-$ 0 0 -1

9.11 In each of these problems, we try to minimize the formal charges in order to determine the preferred Lewis structure. This frequently means violating the octet rule by expanding the octet. Of course, this can only be done for atoms beyond the second period as the atoms in the first and second periods will never expand the octet.

(a) 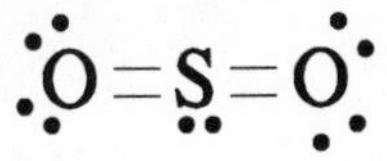

(b) 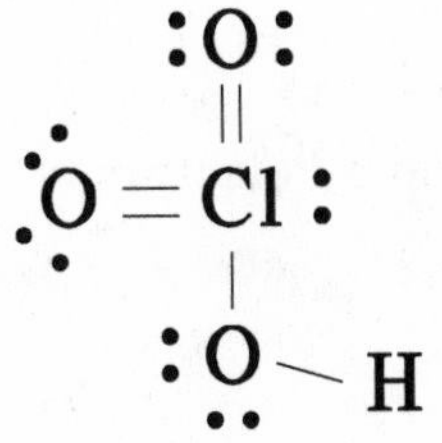

(c) 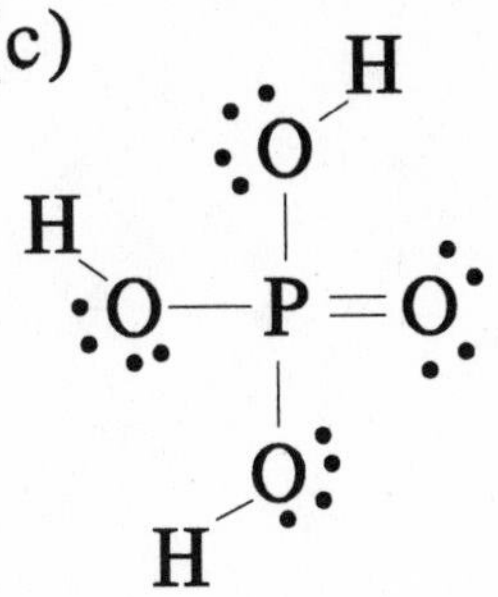

9.12

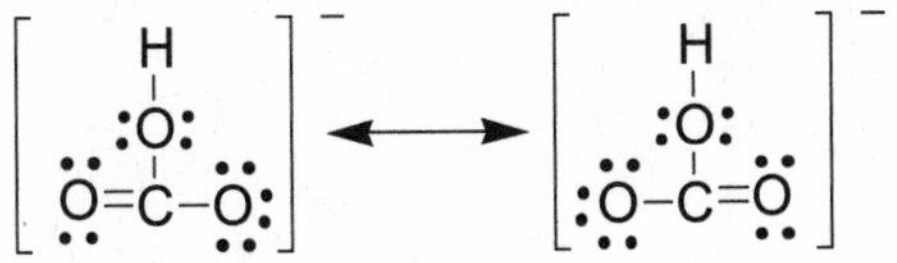

9.13

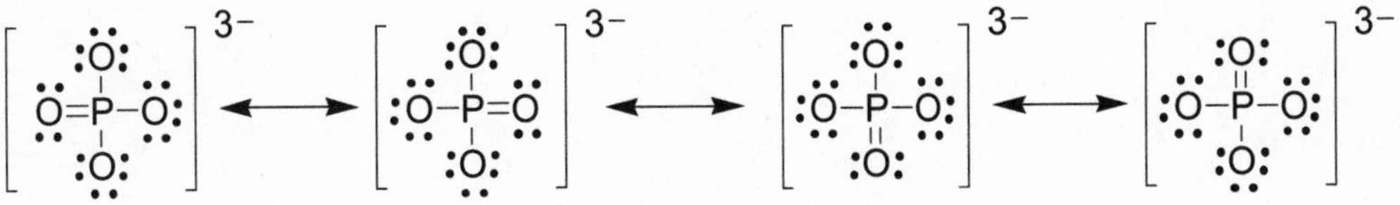

9.14

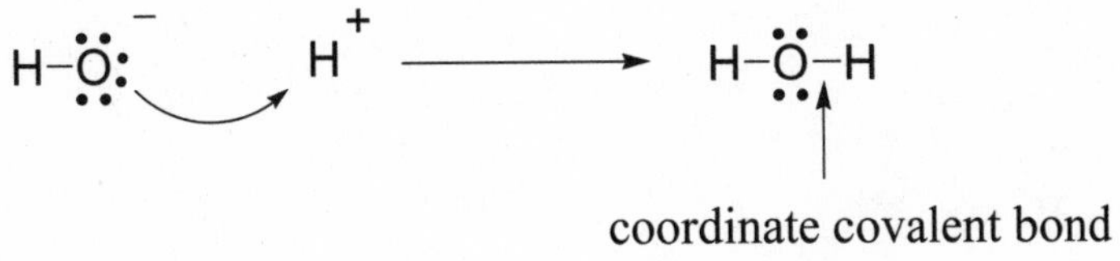

Review Problems

9.68 Magnesium loses two electrons:
$Mg \rightarrow Mg^{2+} + 2e^{-}$
$[Ne]3s^{2} \rightarrow [Ne]$

Bromine gains an electron:
$Br + e^{-} \rightarrow Br^{-}$

$[Ar]3d^{10}4s^24p^5 \rightarrow [Kr]$

9.70 Pb^{2+}: $[Xe]\,4f^{14}5d^{10}6s^2$
Pb^{4+}: $[Xe]4f^{14}5d^{10}$

9.72 Mn^{3+}: $[Ar]3d^4$ 4 unpaired electrons

9.74

(a) ·Si· (b) ·Sb·

(c) ·Ba· (d) ·Al·

(e) :S·

9.76

(a) $[K]^+$ (b) $[Al]^{3+}$

(c) $[:S:]^{2-}$ (d) $[:Si:]^{4-}$

(e) $[Mg]^{2+}$

9.78

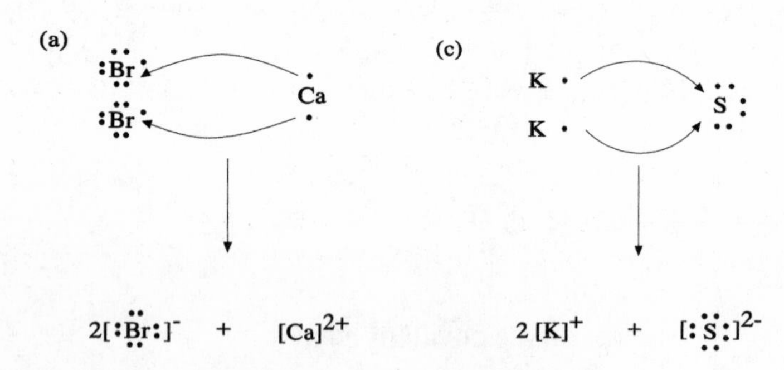

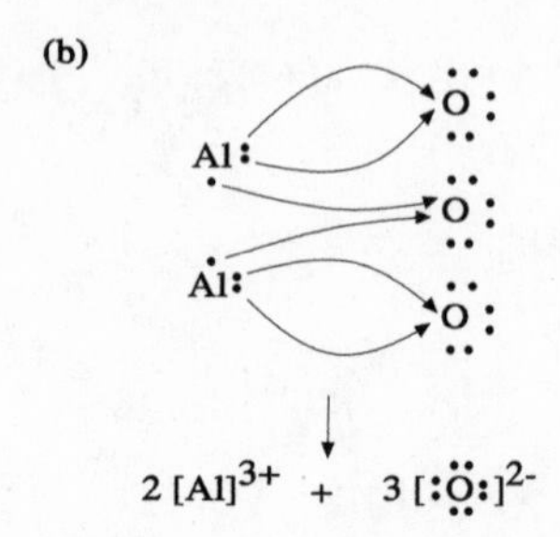

9.80 The dipole moment of a molecule may be found by:

$$\mu = q \cdot r,$$

where q is the charge at either end of the molecule and r is the distance between the charges. First, we must convert the units for the data in Table 9.3:

$$\mu = 0.16 \text{ D } (3.34 \times 10^{-30} \text{ C·m/D}) = 5.3 \times 10^{-31} \text{ C·m}$$

$$r = 1.15 \text{ Å}(1 \times 10^{-10} \text{ m/1 Å}) = 1.15 \times 10^{-10} \text{ m}$$

Now, we may use the equation above:

$$5.3 \text{ x } 10^{-31} \text{ C·m} = q \cdot (1.15 \times 10^{-10} \text{ m})$$

$$q = 4.6 \times 10^{-21} \text{ C}$$

Finally, we may convert to electronic charge units:

$$4.6 \text{ x } 10^{-21} \text{ C } (1\ e/1.6 \times 10^{-19} \text{ C}) = 0.029\ e$$

This suggests a charge of +0.029 on the nitrogen atom (less electronegative) and −0.029 for the oxygen atom.

9.82 Let q be the amount of energy released in the formation of 1 mol of H_2 molecules from H atoms: 435 kJ/mol, the single bond energy for hydrogen.

q = (specific heat)(mass)(ΔT)

∴ mass = q ÷ [(specific heat)(ΔT)]

$$\# \text{g H}_2\text{O} = \frac{(435 \times 10^3 \text{ J})}{\left(4.184 \ {}^{\text{J}}\!/_{\text{g °C}}\right)\left(100 \text{ °C} - 25 \text{ °C}\right)} = 1.4 \times 10^3 \text{ g}$$

9.84 First, we determine the energy of a single C-C bond:

$$348 \text{ kJ/mol C-C bonds}(1 \text{ mol bond}/6.022 \times 10^{23} \text{ C-C bonds})$$

$$= 5.78 \times 10^{-22} \text{ kJ}$$

$$= 5.78 \times 10^{-19} \text{ J}$$

Next, we determine the frequency of photon which carries this minimum energy:

$$E = h\nu$$

$$5.78 \times 10^{-19} \text{ J} = (6.626 \times 10^{-34} \text{ J·s})\nu$$

$$\nu = 8.72 \times 10^{14} \text{ Hz}$$

Finally, we calculate the wavelength to which this frequency of light corresponds:

$$c = \lambda\nu$$

$$3.00 \times 10^{8} \text{ m/s} = \lambda\ (8.72 \times 10^{14} \text{ Hz})$$

$$\lambda = 3.44 \times 10^{-7} \text{ m} = 344 \text{ nm}$$

This is in the ultraviolet (UV) region of the electromagnetic spectrum.

9.86

(a) $:\ddot{Br}\cdot + \cdot\ddot{Br}: \longrightarrow :\ddot{Br}-\ddot{Br}:$

(b) $2\ H\cdot + \cdot\ddot{O}\cdot \longrightarrow H-\ddot{O}:$ (with a second H bonded below O)

(c) $3\ H\cdot + \cdot\ddot{N}\cdot \longrightarrow H-\ddot{N}-H$ (with a third H bonded below N)

9.88 a) We predict the formula H_2Se because selenium, being in Group VIA, needs only two additional electrons (one each from two hydrogen atoms) in order to complete its octet.

b) Arsenic, being in Group VA, needs three electrons from hydrogen atoms in order to complete its octet, and we predict the formula H_3As.

c) Silicon is in Group IVB, and it needs four electrons (and hence four hydrogen atoms) to complete its octet: SiH_4.

9.90 Here we choose the atom with the smaller electronegativity:

a) S b) Si c) Br d) C

9.92 Here we choose the linkage that has the greatest difference in electronegativities between the atoms of the bond: N—S.

9.94

(a)

```
    Cl
Cl  Si  Cl
    Cl
```

(b)

```
F  P  F
   F
```

(c)

```
H  P  H
   H
```

(d)

```
Cl  S  Cl
```

9.96 a) 32 b) 26 c) 8 d) 20

9.98 (a)

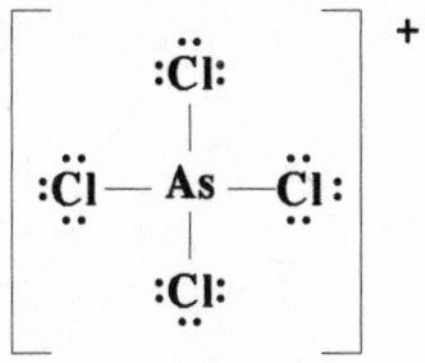

$$\left[\begin{array}{c} :\ddot{Cl}: \\ | \\ :\ddot{Cl}-As-\ddot{Cl}: \\ | \\ :\ddot{Cl}: \end{array}\right]^+$$

(b)

$$\left[:\ddot{O}-\ddot{Cl}-\ddot{O}:\right]^-$$

(c)

$$H-\ddot{O}-\ddot{N}=\ddot{O}$$

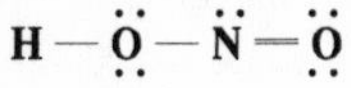

(d)

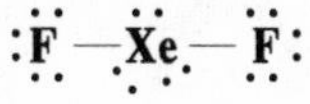

$$:\ddot{F}-\ddot{Xe}-\ddot{F}:$$

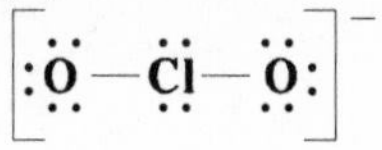

9.100

(a) $:\ddot{Cl}-Si-\ddot{Cl}:$ (with $:\ddot{Cl}:$ above and below Si)

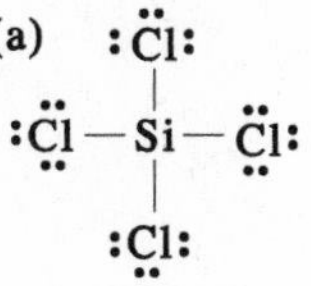

(b) $:\ddot{F}-\ddot{P}-\ddot{F}:$ (with $:\ddot{F}:$ below P)

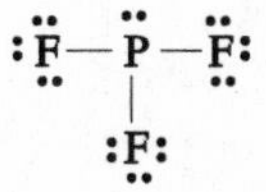

(c) $H-\ddot{P}-H$ (with H below P)

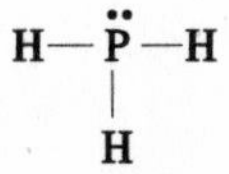

(d) $:\ddot{Cl}-\ddot{S}-\ddot{Cl}:$

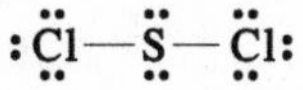

9.102 (a)

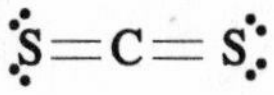

$$\ddot{S}=C=\ddot{S}$$

(b)

$$[:C\equiv N:]^-$$

9.104 (a)

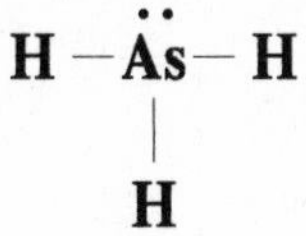

$H-\ddot{As}-H$ (with H below As)

(b)

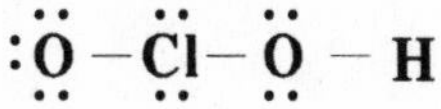

$$:\ddot{O}-\ddot{Cl}-\ddot{O}-H$$

(c)

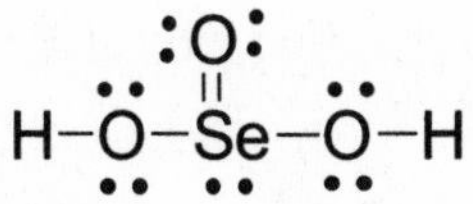

$H-\ddot{O}-\ddot{Se}-\ddot{O}-H$ (with $:\ddot{O}:$ double-bonded above Se)

(d)

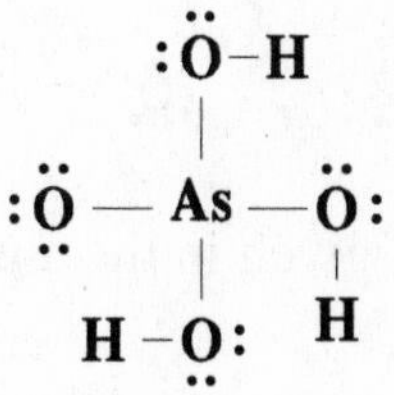

$:\ddot{O}-As-\ddot{O}:$ (with $:\ddot{O}-H$ above As, $H-\ddot{O}:$ below As, and H on the right O)

9.106 a)

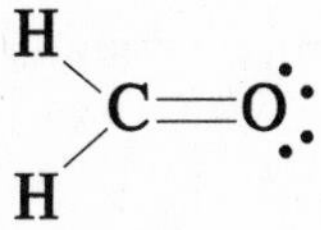

$H_2C=\ddot{O}:$

b)

$:\ddot{Cl}-\ddot{S}-\ddot{Cl}:$ (with $:\ddot{O}:$ double-bonded above S)

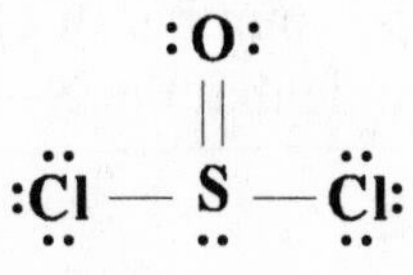

9.108

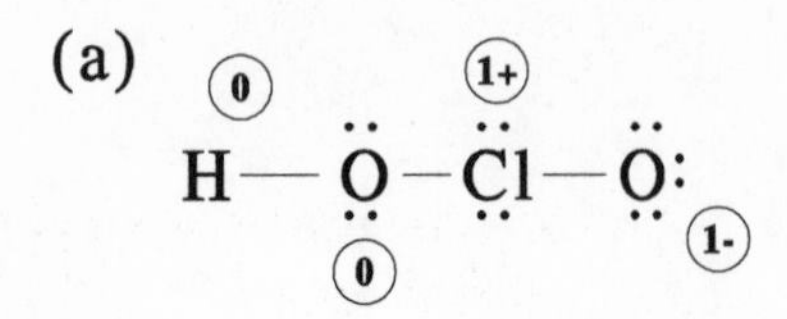

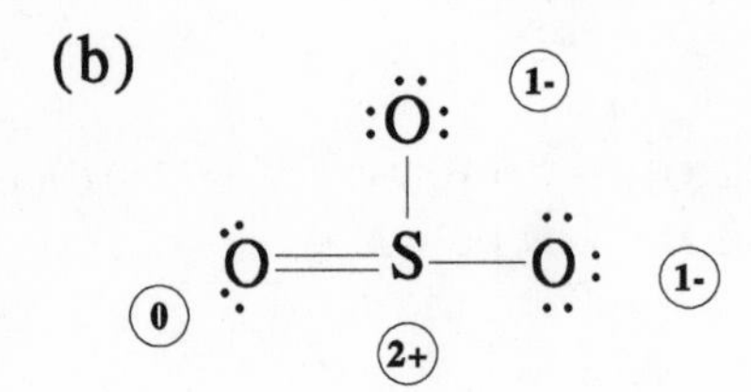

(c) O=S—O: (formal charges: O 0, S 1+, O 1-)

9.110

H—O—Cl(—O:)(—O:)—O: (formal charges: H 0, O 0, Cl 3+, O 1-, O 1-, O 1-)

H—O—Cl(=O)(=O)—O: (formal charges: Cl 1+, O 1-)

9.112 The formal charges on all of the atoms of the left structure are zero. Also, the non-preferred structure places a positive formal charge on chlorine.

9.114 The average bond order is 4/3.

$$[CO_3]^{2-} \longleftrightarrow [CO_3]^{2-} \longleftrightarrow [CO_3]^{2-}$$

9.116 The Lewis structure for NO_3^- is given in the answer to practice exercise 9, and that for NO_2^- is below.

$$[O=N-O:]^- \longleftrightarrow [:O-N=O]^-$$

Resonance causes the average number of bonds in each N—O linkage of NO_3^- to be 1.33. Resonance causes the average number of electron pair bonds in each linkage of NO_2^- to be 1.5. We conclude that the N—O bond in NO_2^- should be shorter than that in NO_3^-.

9.118

$$:O\equiv C-\ddot{\underset{..}{O}}: \;\leftrightarrow\; :\ddot{\underset{..}{O}}-C\equiv O:$$

These are not preferred structures because in each Lewis diagram one oxygen bears a formal charge of +1 whereas the other bears a formal charge of –1. The structure with the formal charges of zero has a lower potential energy and is more stable.

9.120 The Lewis structure that is obtained using the rules of Figure 9.8 is shown at top. This structure contains more formal charges than the one below.

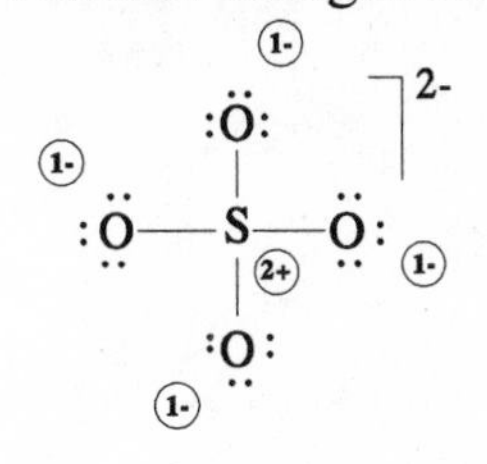

$$\left[\begin{array}{c} :\ddot{O}:^{(1-)} \\ | \\ {}^{(0)}\ddot{O}=S^{(0)}=\ddot{O}^{(0)} \\ | \\ {}^{(1-)}:\ddot{O}: \end{array} \right]^{2-} \longleftrightarrow \left[\begin{array}{c} \ddot{O}^{(0)} \\ \| \\ {}^{(1-)}:\ddot{O}-S^{(1-)}-\ddot{O}: \\ \|^{(0)} \\ {}^{(0)}\ddot{O} \end{array} \right]^{2-}$$

(Many other such resonance structures may be drawn as well, each containing two S=O double bonds.) Formal charges are indicated. The average bond order is 1.5.

9.122

$$H-\ddot{O}: \longrightarrow H^+$$
$$\;\;|$$
$$\;\;H$$

$$\downarrow$$

$$\left[\begin{array}{c} H-\ddot{O}-H \\ | \\ H \end{array} \right]^+$$

Additional Exercises

9.124 $Na(g) \rightarrow Na^+(g) + e^-$ IE = 496 kJ/mol
$Cl(g) + e^- \rightarrow Cl^-(g)$ EA = –348 kJ/mol
$Na(g) + Cl(g) \rightarrow Na^+(g) + Cl^-(g)$ ΔH = 148 kJ/mol

$Na(g) \rightarrow Na^+(g) + e^-$ IE = 496 kJ/mol
$Na^+(g) \rightarrow Na^{2+}(g) + e^-$ IE = 4563 kJ/mol
$2Cl(g) + 2e^- \rightarrow 2Cl^-$ EA = 2(–348kJ/mol)

$Na(g) + 2Cl(g) \rightarrow Na^{2+}(g) + 2Cl^{-}(g) \qquad \Delta H = 4.36 \times 10^{3}$ kJ/mol

In order for $NaCl_2$ to be more stable than NaCl, the lattice energy should be almost 30 times larger 4360 kJ/148 kJ = 29.5.

*9.126

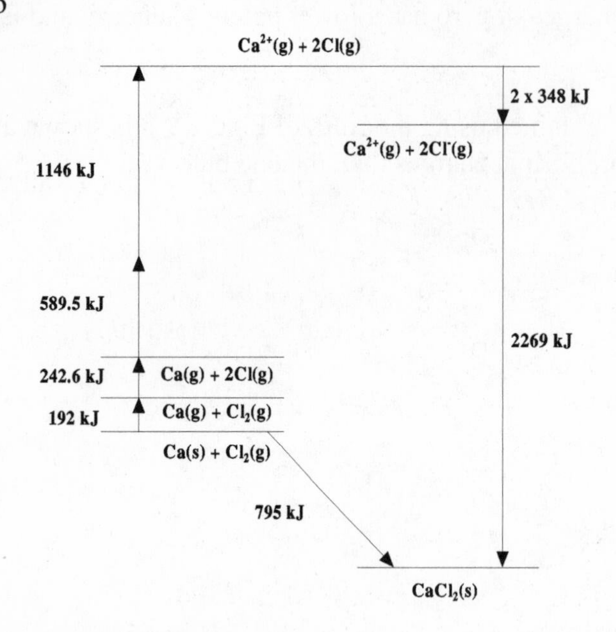

9.128

:C̈l:
|
:C̈l— Sn— C̈l:
|
:C̈l:

9.130 Eighteen electrons ($3s^2$, $3p^6$, $3d^{10}$)

9.132 Given: Better (lower energy):

Given: H—N≡N—N̈: with formal charges H (0), first N (1+), second N (1+), terminal N (2-)

Better: H—N̈=N=N̈ with formal charges H (0), first N (0), central N (1+), terminal N (1-)

9.134 Of the four resonance structures, the one in the upper left is the best possible structure. Each of the terminal nitrogen atoms has a –1 formal charge and the central atom has a formal charge of +1. The structure in the upper right is also a good structure except that the formal charge on the rightmost nitrogen atom is –2 and the central nitrogen is +1 while the leftmost nitrogen has a formal charge of zero. The bottom two resonance structures are both poor because each violates the octet rule. The left bottom structure has too many electrons around the central nitrogen atom while the right bottom structure has too few electrons on the central atom.

9.136

$$:\ddot{\underset{..}{Cl}}-\ddot{\underset{..}{S}}-\ddot{\underset{..}{S}}-\ddot{\underset{..}{Cl}}:$$

*9.138 We need to calculate the energy necessary to carry out this process, then find the photon of minimum wavelength which carries this energy. We can think of this process as occurring in three steps:

Step 1: The cleavage of the H-H bond to produce two H atoms

Step 2: The removal of an electron from one H atom, and

Step 3: The capture of that electron by the second H atom.

The process may or may not occur in these distinct steps, but the sum of all the steps gives us the net energy for the change.

Step 1

First, we determine the energy of a single H-H bond. Section 9.3 tells that the bond energy for H-H is 435 kJ/mol.

$$435 \text{ kJ/mol H-H bonds}(1 \text{ mol bond}/6.022 \times 10^{23} \text{ H-H bonds})$$
$$= 7.22 \times 10^{-22} \text{ kJ}$$
$$= 7.22 \times 10^{-19} \text{ J}$$

Step 2

This is given as the ionization energy of Hydrogen from chapter 8:

$$1312 \text{ kJ/mol H-H bonds}(1 \text{ mol bond}/6.022 \times 10^{23} \text{ H-H bonds})$$
$$= 2.18 \times 10^{-21} \text{ kJ}$$
$$= 2.18 \times 10^{-18} \text{ J}$$

Step 3

This is given as the electron affinity of Hydrogen from chapter 8. It is a negative number, which means energy is given off:

$$-73 \text{ kJ/mol H-H bonds}(1 \text{ mol bond}/6.022 \times 10^{23} \text{ H-H bonds})$$
$$= -1.21 \times 10^{-22} \text{ kJ}$$
$$= -1.21 \times 10^{-19} \text{ J}$$

Adding all of the energies from the steps above, we get:

Total energy needed per H atom = 2.78×10^{-18} J.

Next, we determine the frequency of photon which carries this minimum energy:

$$E = h\nu$$
$$2.78 \text{ x } 10^{-18} \text{ J} = (6.626 \times 10^{-34} \text{ J}\cdot\text{s})\nu$$
$$\nu = 4.20 \times 10^{15} \text{ Hz}$$

Finally, we calculate the wavelength to which this frequency of light corresponds:

$$c = \lambda\nu$$
$$3.00 \times 10^{8} \text{ m/s} = \lambda\ (4.20 \times 10^{15} \text{ Hz})$$
$$\lambda = 7.15 \times 10^{-8} \text{ m} = 71.5 \text{ nm}$$

(These are very high energy photons, in the X-ray region of the electromagnetic spectrum.)

9.140

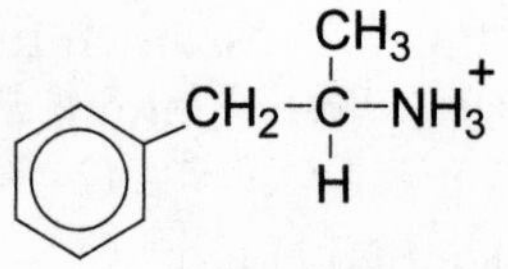

$$C_6H_5-CH_2-\underset{\displaystyle H}{\overset{\displaystyle CH_3}{\underset{|}{\overset{|}{C}}}}-NH_3^+$$

Chapter 10

Practice Exercises

10.1 Each bond below corresponds to a bonding domain, including the double bond, which is considered a single bonding domain. (Carbon has 3 bonding domains.)

:O:
||
H–Ö–C–Ö–H

The lone pairs on the oxygens correspond to non-bonding domains.

10.2 $SbCl_5$ should have a trigonal bipyramidal shape (Figure 10.2) because, like PCl_5, it has five electron pairs around the central atom.

10.3 I_3^- should have a linear shape (Figure 10.4) because although it has five electron pairs around the central I atom, only two are being used for bonding.

10.4 HArF should have a linear shape (Figure 10.4) because although it has five electron pairs around the central Ar atom, only two are being used for bonding.

10.5 XeF_4 should have a square planar shape (Figure 10.5) because although it has six electron pairs around the central Xe atom, only four are being used for bonding.

10.6 In SO_3^{2-}, there are three bond pairs and one lone pair of electrons at the sulfur atom, and as shown in Figure 10.3, this ion has a trigonal pyramidal shape.

In CO_3^{2-}, there are three bonding domains (2 single bonds and 1 double bond) at the carbon atom, and as shown in Figure 10.2, this ion has a planar triangular shape.

In XeO_4, there are four bond pairs of electrons around the Xe atom, and as shown in Figure 10.3, this molecule is tetrahedral.

In OF_2, there are two bond pairs and two lone pairs of electrons around the oxygen atom, and as shown in Figure 10.3, this molecule is bent.

10.7 a) SF_6 is octahedral, and it is not polar.
b) SO_2 is bent, and it is polar.
c) BrCl is polar because there is a difference in electronegativity between Br and Cl.
d) AsH_3, like NH_3, is pyramidal, and it is polar.
e) CF_2Cl_2 is polar, because there is a difference in electronegativity between F and Cl.

10.8 The H–Cl bond is formed by the overlap of the half–filled 1s atomic orbital of a H atom with the half–filled 3p valence orbital of a Cl atom:

Cl atom in HCl (x = H electron):

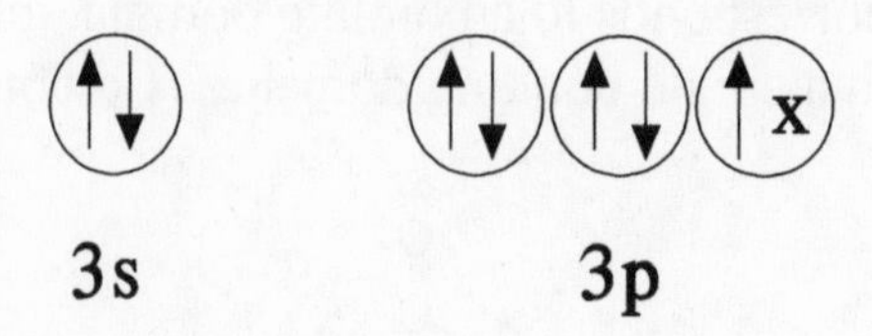

The overlap that gives rise to the H–Cl bond is that of a 1s orbital of H with a 3p orbital of Cl:

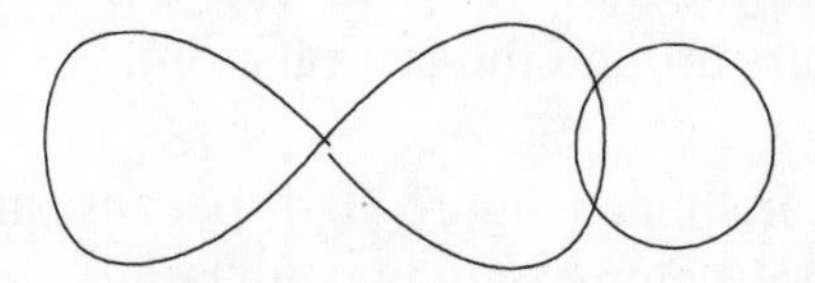

10.9 The half–filled 1s atomic orbital of each H atom overlaps with a half–filled 3p atomic orbital of the P atom, to give three P–H bonds. This should give a bond angle of 90°.

P atom in PH_3 (x = H electron):

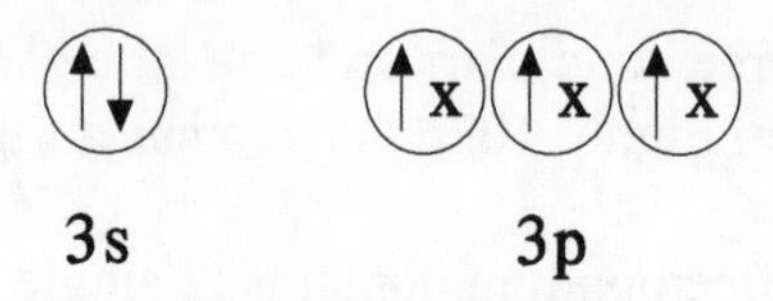

The orbital overlap that forms the P–H bond combines a 1s orbital of hydrogen with a 3p orbital of phosphorus (note: only half of each p orbital is shown):

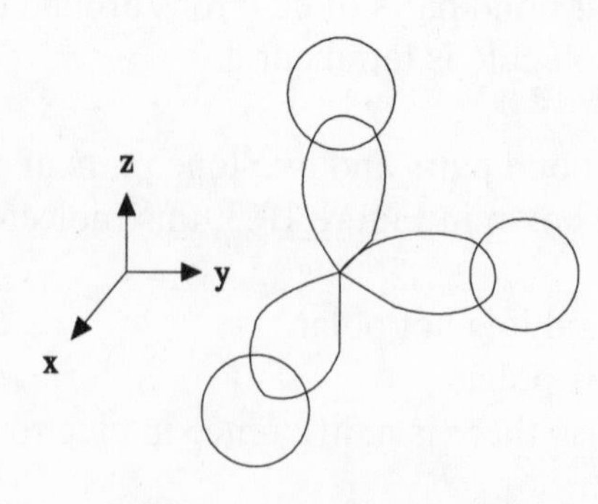

10.10 Since there are five bonding pairs of electrons on the central arsenic atom, we choose sp^3d hybridization for the As atom. Each of arsenic's five sp^3d hybrid orbitals overlaps with a 3p atomic orbital of a chlorine atom to form a total of five As–Cl single bonds. Four of the 3d atomic orbitals of As remain unhybridized.

10.11 a) sp^3 b) sp^3d

10.12 a) sp^3 b) sp^3d

10.13 sp^3d^2, since six atoms are bonded to the central atom.

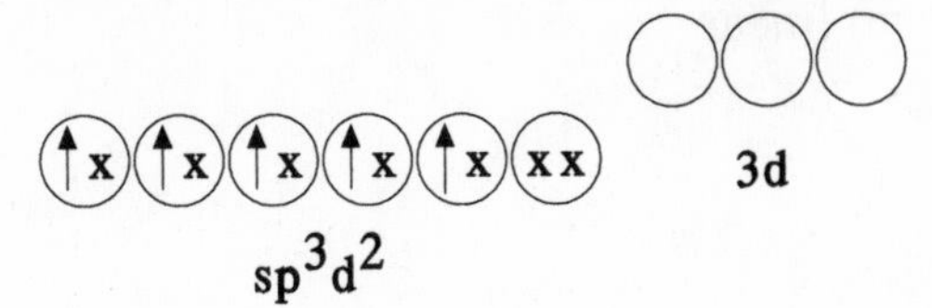

The ion is octahedral because six atoms and no lone pairs surround the central atom.

10.14 NO has 11 valence electrons, and the MO diagram is similar to that shown in Table 10.1 for O_2, except that one fewer electron is employed at the highest energy level

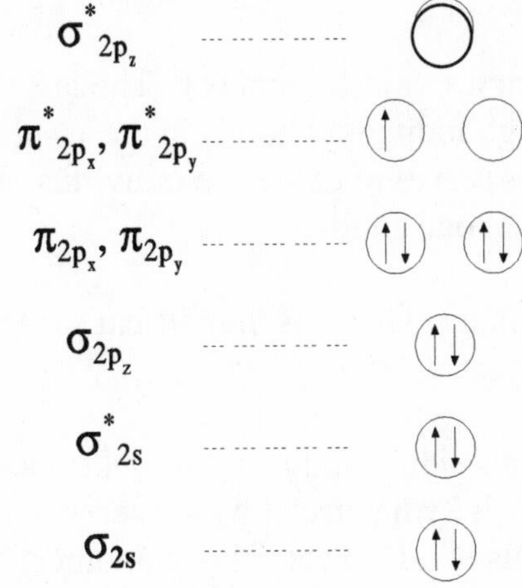

The bond order is calculated to be 5/2:

$$\text{Bond Order} = \frac{(8 \text{ bonding e}^-) - (3 \text{ antibonding e}^-)}{2} = \frac{5}{2}$$

Review Problems

10.54 a) bent (central N atom has two single bonds and two lone pairs)
b) planar triangular (central C atom has three bonding domains—a double bond and a single bond)
c) T-shaped (central I atom has three single bonds and two lone pairs)

d) Linear (central Br atom has two single bonds and three lone pairs)
e) planar triangular (central Ga atom has three single bonds and no lone pairs)

10.56 a) nonlinear
b) trigonal bipyramidal
c) trigonal pyramidal
d) trigonal pyramidal
e) nonlinear

10.58 a) tetrahedral
b) square planar
c) octahedral
d) tetrahedral
e) linear

10.60 180°

10.62 (a), (b), (c), and (e). Actually, all of these compounds contain nonbonding domains. Compound (d) contains no nonbonding domains on the central atom, but contains them on the outer atoms.

10.64 a) 109.5°
b) 109.5°
c) 120°
d) 180°
e) 109.5°

10.66 The ones that are polar are (a), (b), and (c). The last two have symmetrical structures, and although individual bonds in these substances are polar bonds, the geometry of the bonds serves to cause the individual dipole moments of the various bonds to cancel one another.

10.68 All are polar. (a)-(c) and (e) have asymmetrical structures, and (d) only has one bond, which is polar.

10.70 In SF_6, although the individual bonds in this substance are polar bonds, the geometry of the bonds is symmetrical which serves to cause the individual dipole moments of the various bonds to cancel one another.

In SF_5Br, one of the six bonds has a different polarity so the individual dipole moments of the various bonds do not cancel one another.

10.72 The 1s atomic orbitals of the hydrogen atoms overlap with the mutually perpendicular p atomic orbitals of the selenium atom.

Se atom in H_2Se (x = H electron):

↑↓ ↑x ↑x

4s **4p**

10.74

Atomic Be:

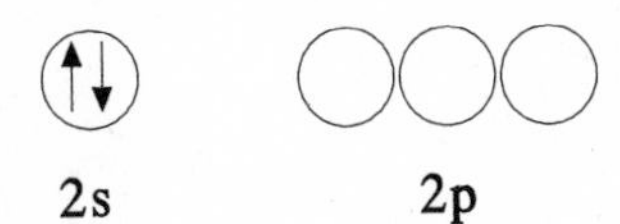

Hybridized Be:

(x = a Cl electron)

10.76 a) There are three bonds to the central Cl atom, plus one lone pair of electrons. The geometry of the electron pairs is tetrahedral so the Cl atom is to be sp^3 hybridized:

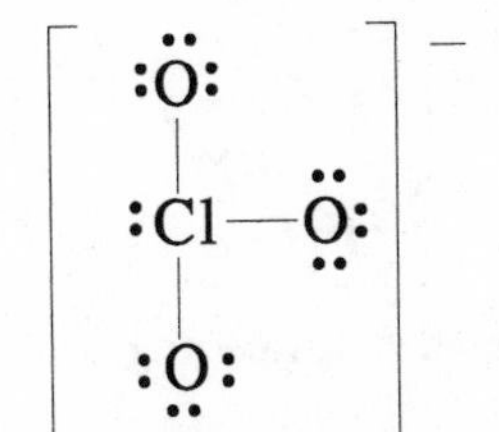

b) There are three atoms bonded to the central sulfur atom, and no lone pairs on the central sulfur. The geometry of the electron pairs is that of a planar triangle, and the hybridization of the S atom is sp^2:

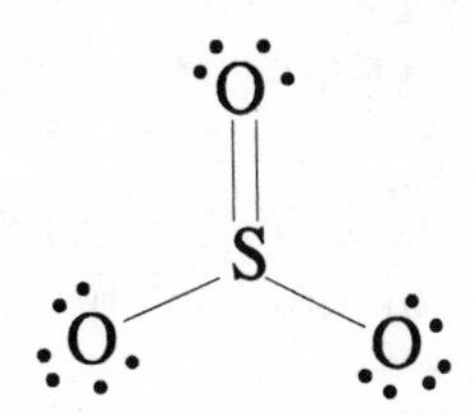

Two other resonance structures should also be drawn for SO_3.

c) There are two bonds to the central O atom, as well as two lone pairs. The O atom is to be sp^3 hybridized, and the geometry of the electron pairs is tetrahedral.

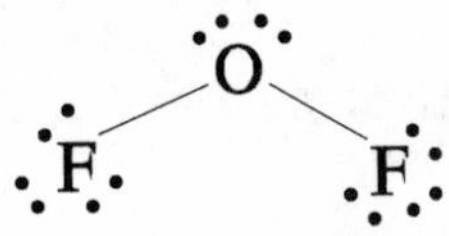

10.78 a) There are three bonds to As and one lone pair at As, requiring As to be sp^3 hybridized.

The Lewis diagram:

:C̈l—Äs—C̈l:
|
:C̈l:

The hybrid orbital diagram for As:

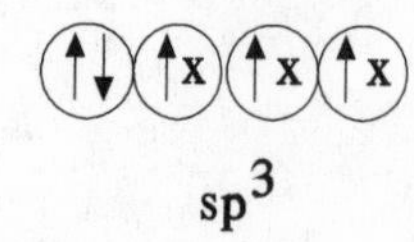

(x = a Cl electron)

b) There are three atoms bonded to the central Cl atom, and it also has two lone pairs of electrons. The hybridization of Cl is thus sp^3d.

The Lewis diagram:

:F̈—C̈l—F̈:
|
:F̈:

The hybrid orbital diagram for Cl:

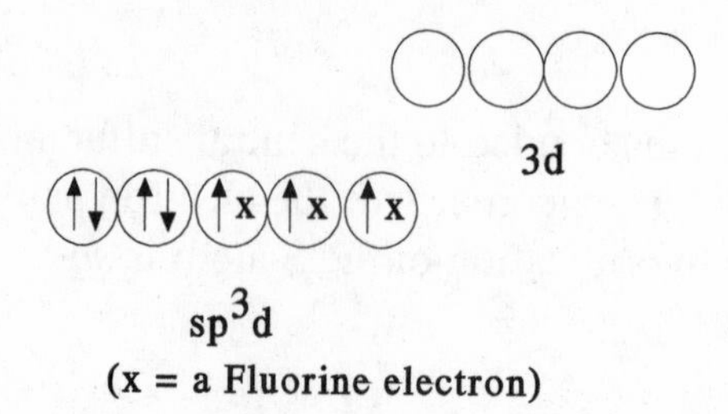

(x = a Fluorine electron)

10.80 We can consider that this ion is formed by reaction of SbF_5 with F^-. The antimony atom accepts a pair of electrons from fluoride:

Sb in SbF_6^-:

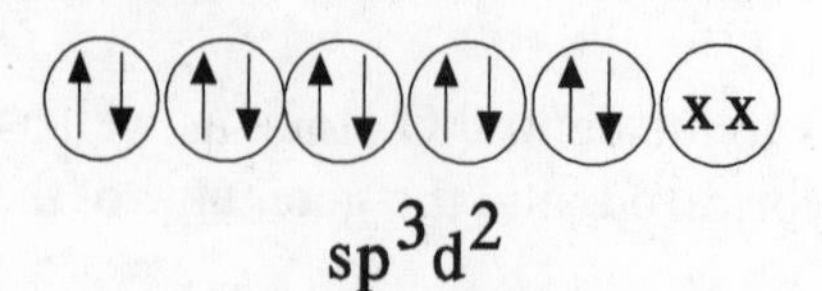

(xx = an electron pair from the donor F^-)

10.82 a)

N in the C=N system:

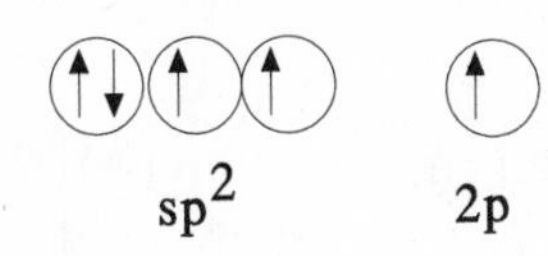

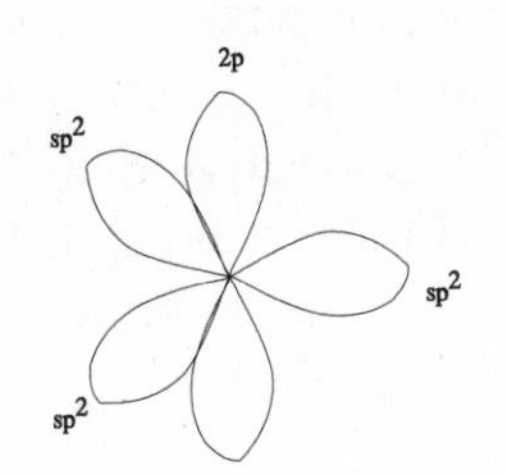

(b)

sigma bond

C N

pi bond

C N

(c)

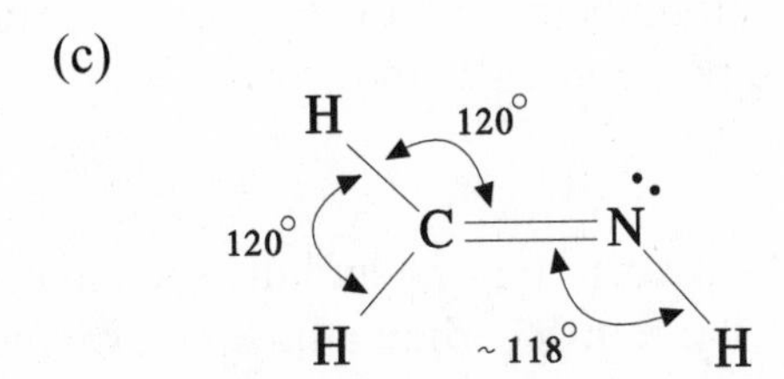

10.84 Each carbon atom is sp^2 hybridized, and each C–Cl bond is formed by the overlap of an sp^2 hybrid of carbon with a p atomic orbital of a chlorine atom. The C=C double bond consists first of a C–C σ bond formed by "head on" overlap of sp^2 hybrids from each C atom. Secondly, the C=C double bond consists of a side–to–side overlap of unhybridized p orbitals of each C atom, to give one π bond. The molecule is planar, and the expected bond angles are all 120°.

10.86 1. sp^3 2. sp 3. sp^2 4. sp^2

10.88 Here we pick the one with the higher bond order.
a) O_2^+ b) O_2 c) N_2

10.90 Using Fig. 10.34b, and filling in the appropriate number of valence electrons, we see that (a)-(d) all have unpaired electrons, and hence are paramagnetic. (e) does not have any unpaired electrons, and hence is not paramagnetic.

Additional Exercises

10.92 Planar triangular

10.94 The normal C–C–C angle for an sp^3 hybridized carbon atom is 109.5°. The 60° bond angle in cyclopropane is much less than this optimum bond angle. This means that the bonding within the ring cannot be accomplished through the desirable "head on" overlap of hybrid orbitals from each C atom. As a result, the overlap of the hybrid orbitals in cyclopropane is less effective than that in the more normal, noncyclic propane molecule, and this makes the C–C bonds in cyclopropane comparatively weaker than those in the noncyclic molecule. We can also say that there is a severe "ring strain" in the molecule.

10.96 a) PF_3 is a pyramidal molecule and uses sp^3 hybrid orbitals. The expected bond angle is 109.5°.

b) Using the unhybidized p orbitals, we would anticipate a bond angle of 90°.

c) The observed bond angle is almost exactly the average of the bond angles listed in parts (a) and (b) above. So, neither hybid orbitals nor unhybridized atomic orbitals explain the observed bond angle.

*10.98 a) The C–C single bonds are formed from head–to–head overlap of C atom sp^2 hybrids. This leaves one unhybridized atomic p orbital on each carbon atom, and each such atomic orbital is oriented perpendicular to the plane of the molecule.

b) Sideways or π type overlap is expected between the first and the second carbon atoms, as well as between the third and the fourth carbon atoms. However, since all of these atomic p orbitals are properly aligned, there can be continuous π type overlap between all four carbon atoms.

c) The situation described in part (b) is delocalized. We expect completely delocalized π type bonding among the carbon atoms.

d) Double bonds are shorter, and this bond has some double bond character.

*10.100

The arrangement of the atoms is trigonal bipyramidal.

Recall that the bond angle between equatorial atoms is 120°. The bond angle from the equatorial position to the axial position is 90°. Due to the smaller bond angles, the atoms in the axial positions create more repulsions. The structure with the least amount of total repulsion is

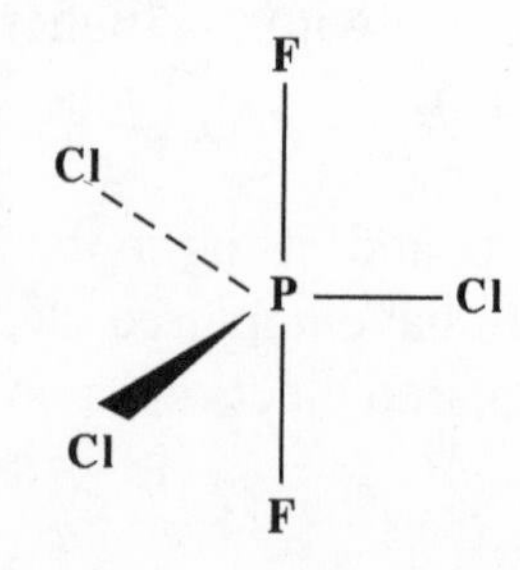

preferred; the statement implies that the more electronegative atoms create less repulsion, therefore the more electronegative atom should be placed in the axial position. Since fluorine is more electronegatove then chlorine, the F atoms will be in the axial positions and the Cl atoms will be in the equatorial positions. The molecule is non–polar.

*10.102

Double bonds are predicted to be between S and O atoms. Hence, the Cl–S–Cl angle diminishes under the influence of the S=O double bonds.

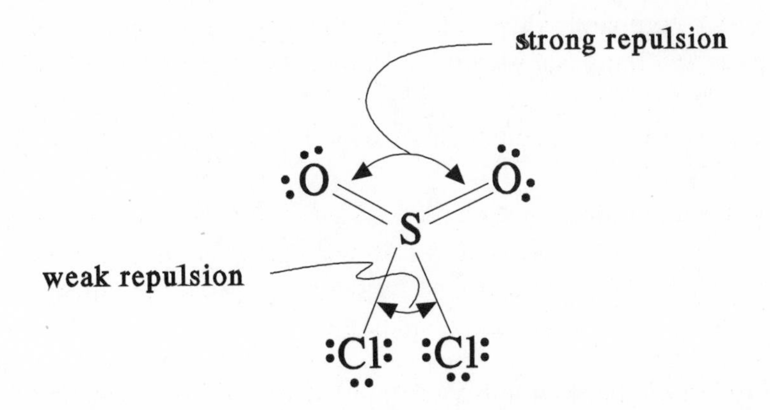

10.104 This is a π bond, since overlap is *side to side* rather than the *end to end.* Also, consider that no bond rotation is possible here without breaking the bond since overlap occurs both above and below the bond axis.

10.106 a) The O-O-N bond angle is about 109.5° and the O-N-O bond angle is around 120°.

b) sp^2

c) The nitrate ion benefits from three stable resonance structures (or electron delocalization) which stabilize it. The peroxynitrite ion does not.

Chapter 11

Practice Exercises

11.1 $\# \text{torr} = 887 \text{ mbar}\left(\frac{1 \text{ bar}}{1000 \text{ mbar}}\right)\left(\frac{0.9868 \text{ atm}}{1 \text{ bar}}\right)\left(\frac{760 \text{ torr}}{1 \text{ atm}}\right) = 665 \text{ torr}$

11.2 Assuming the pressure of the atmosphere in the room is 756 torr, the pressure in the container would be 756 − 87 = 669 torr.

11.3 $h_B = h_A \times \frac{d_A}{d_B}$

$$h_B = 760 \text{ mm} \times \frac{13.6 g/mL}{1.00 g/mL} = 10{,}300 mm$$

$$\# \text{ft} = 10{,}300 \text{ mm}\left(\frac{1 \text{ cm}}{10\text{mm}}\right)\left(\frac{1 \text{ inch}}{2.54\text{cm}}\right)\left(\frac{1 \text{ ft}}{12 \text{ inches}}\right) = 33.8 \text{ ft}$$

11.4 Since volume is to decrease, pressure must increase, and we multiply the starting pressure by a volume ratio that is larger than one. Also, since $P_1V_1 = P_2V_2$, we can solve for P_2:

$$P_2 = \frac{P_1V_1}{V_2} = \frac{(740 \text{ torr})(880 \text{ mL})}{(870 \text{ mL})} = 750 \text{ torr}$$

11.5 In general the combined gas law equation is: $\frac{P_1V_1}{T_1} = \frac{P_2V_2}{T_2}$, and in particular, for this problem, we have:

$$P_2 = \frac{P_1V_1T_2}{T_1V_2} = \frac{(745 \text{ torr})(950 \text{ m}^3)(333.2 \text{ K})}{(1150 \text{ m}^3)(298.2 \text{ K})} = 688 \text{ torr}$$

11.6 $n = \frac{PV}{RT} = \frac{(57.8 \text{ atm})(12.0 \text{ L})}{(0.0821 \frac{\text{L atm}}{\text{mol K}})(298 \text{ K})} = 28.3 \text{ moles gas}$

28.3 mol Ar (39.95 g Ar/mol) = 1,130 g Ar = 1.13 kg Ar

11.7 Since PV = nRT, then n = PV/RT

$$n = \frac{PV}{RT} = \frac{(685 \text{ torr})\left(\frac{1 \text{ atm}}{760 \text{ torr}}\right)(0.300 \text{ L})}{(0.0821 \frac{\text{L atm}}{\text{mol K}})(300.2 \text{ K})} = 0.0110 \text{ moles gas}$$

$$\text{molar mass} = \frac{1.45 \text{ g}}{0.0110 \text{ mol}} = 132 \text{ g mol}^{-1}$$

The gas must be Xenon.

11.8 $d = m/V$

Taking 1.00 mol SO2:

$m = 64.1$ g

$$V = \frac{nRT}{P} = \frac{(1.00\ \text{mol})(0.0821\ \frac{\text{L atm}}{\text{mol K}})(253.15\ \text{K})}{\left(96.5\ \text{kPa}\frac{1\ \text{atm}}{101.325\ \text{kPa}}\right)} = 21.8\ \text{L} = 21{,}800\ \text{mL}$$

$$\text{density} = \frac{64.1\ \text{g}}{21{,}800\ \text{mL}} = 2.94\times10^{-3}\ \text{g/mL}$$

11.9 In general PV = nRT, where n = mass ÷ formula mass. Thus

$$PV = \frac{\text{mass}}{\text{formula mass}}RT$$

We can rearrange this equation to get;

$$\text{formula mass} = \frac{(\text{mass/V})RT}{P} = \frac{dRT}{P}$$

$$\text{formula mass} = \frac{(5.60\ \text{g L}^{-1})(0.0821\ \frac{\text{L atm}}{\text{mol K}})(295.2\ \text{K})}{(750\ \text{torr})\left(\frac{1\ \text{atm}}{760\ \text{torr}}\right)} = 138\ \text{g mol}^{-1}$$

The empirical mass is 69 g mol^{-1}. The ratio of the molecular mass to the empirical mass is 138 g mol^{-1}/69 g mol^{-1} = 2. Therefore, the molecular formula is 2 times the empirical formula, i.e., P_2F_4.

11.10 We can determine the pressure due to the oxygen since $P_{total} = P_{N_2} + P_{O_2}$. $P_{O_2} = P_{total} - P_{N_2} = 30.0\ \text{atm} - 15.0\ \text{atm} = 15.0\ \text{atm}$. We can now use the ideal gas law to determine the number of moles of O_2:

$$n = \frac{PV}{RT} = \frac{(15.0\ \text{atm})(5.00\ \text{L})}{\left(0.0821\frac{\text{L atm}}{\text{mol K}}\right)(298\ \text{K})} = 3.06\ \text{moles}\ O_2$$

$$\#\,\text{g}\ O_2 = (3.06\ \text{moles}\ O_2)\left(\frac{32.0\ \text{g}\ O_2}{1\ \text{mol}\ O_2}\right) = 98.1\ \text{g}\ O_2$$

11.11 First we find the partial pressure of nitrogen, using the vapor pressure of water at 15 °C:

$$P_{N_2} = P_{total} - P_{water} = 745 \text{ torr} - 12.79 \text{ torr} = 732 \text{ torr.}$$

To calculate the volume of the nitrogen we can use the combined gas law

$$\frac{P_1V_1}{T_1} = \frac{P_2V_2}{T_2}$$

For this problem,

$$V_2 = \frac{P_1V_1T_2}{P_2T_1} = \frac{(732 \text{ torr})(310 \text{ mL})(273 \text{ K})}{(760 \text{ torr})(288 \text{ K})} = 283 \text{ mL}$$

11.12 The mole fraction is defined in Equation 11.8:

$$X_{O_2} = \frac{P_{O_2}}{P_{total}} = \frac{116 \text{ torr}}{760 \text{ torr}} = 0.153 \text{ or } 15.3\%$$

11.13 When gases are held at the same temperature and pressure, and dispensed in this fashion during chemical reactions, then they react in a ratio of volumes that is equal to the ratio of the coefficients (moles) in the balanced chemical equation for the given reaction. We can, therefore, directly use the stoichiometry of the balanced chemical equation to determine the combining ratio of the gas volumes:

$$\# \text{L } O_2 = (4.50 \text{ L } CH_4)\left(\frac{2 \text{ volume } O_2}{1 \text{ volume } CH_4}\right) = 9.00 \text{ L } O_2$$

11.14 First, we determine the number of moles of NO using the ideal gas law:

$$n = \frac{PV}{RT} = \frac{[720 \text{ torr}(1 \text{ atm}/760 \text{ torr})][0.180 \text{ L } CH_4]}{(0.0821 \frac{\text{L atm}}{\text{mol K}})(318 \text{ K})} = 0.00653 \text{ mol NO}$$

Now we calculate the number of moles of O_2 which will be consumed, using the mole ratio from the balanced equation:

$$\# \text{mol } O_2 = 0.00653 \text{ mol NO} \frac{1 \text{ mol } O_2}{2 \text{ mol NO}} = 0.003265 \text{ mol } O_2$$

Finally, we use the ideal gas law to determine the volume of O_2:

$$V = \frac{nRT}{P} = \frac{(0.003265 \text{ mol } O_2)(0.0821 \frac{\text{L atm}}{\text{mol K}})(293 \text{ K})}{[755 \text{ torr}(1 \text{ atm}/760 \text{ torr})]} = 0.0791 \text{ L } O_2 = 79.1 \text{ mL } O_2$$

11.15 First lets determine the number of moles of CO_2 that are produced:

$$n = \frac{PV}{RT} = \frac{(738\ \text{torr})\left(\frac{1\ \text{atm}}{760\ \text{torr}}\right)(0.250\ \text{L})}{\left(0.0821\frac{\text{L atm}}{\text{mol K}}\right)(296\ \text{K})} = 9.99 \times 10^{-3}\ \text{moles CO}_2$$

The stoichiometry of the reaction indicates that one mole of $Na_2CO_3(s)$ will produce one mole $CO_2(g)$, so we will need to use 9.99 X 10^{-3} mol of $Na_2CO_3(s)$.

$$\#\,\text{g Na}_2\text{CO}_3 = (9.99 \times 10^{-3}\ \text{mol Na}_2\text{CO}_3)\left(\frac{106\ \text{g Na}_2\text{CO}_3}{1\ \text{mol Na}_2\text{CO}_3}\right) = 1.06\ \text{g Na}_2\text{CO}_3$$

11.16 Use Equation 11.10;

$$\frac{\text{effusion rate (HX)}}{\text{effusion rate (HCl)}} = \sqrt{\frac{M_{HCl}}{M_{HX}}}$$

$$M_{HX} = M_{HCl} \times \left(\frac{\text{effusion rate (HX)}}{\text{effusion rate (HCl)}}\right)^2 = 36.46\ \text{g mol}^{-1} \times (1.88)^2 = 128.9\ \text{g mol}^{-1}$$

The unknown gas must be HI.

Review Problems

11.36 a) # torr = (1.26 atm)(760 torr / 1 atm) = 958 torr
b) # atm = (740 torr)(1 atm / 760 torr) = 0.974 atm
c) 738 torr = 738 mm Hg
d) # torr = (1.45 X 10^3 Pa)(760 torr / 1.01325×10^5 Pa) = 10.9 torr

11.38 a) # torr = (0.329 atm)(760 torr / 1 atm) = 250 torr
b) # torr = (0.460 atm)(760 torr / 1 atm) = 350 torr

11.40 765 torr – 720 torr = 45 torr 45 torr = 45 mm Hg

$$\#\,\text{cm Hg} = 45\ \text{mm Hg}\left(\frac{1\ \text{cm}}{10\ \text{mm}}\right) = 4.5\ \text{cm Hg}$$

11.42 65 mm Hg = 65 torr, 748 torr + 65 torr = 813 torr

11.44 The reaction produced gas. (The pressure inside the container was higher than that outside, as evidenced by the fact that the Hg level was lower.)

When the mercury level in the manometer arm nearest the bulb goes down by 4.00 cm (12.50 to 8.50) the mercury in the other arm goes up by 4.00 cm. Hence, the difference in the heights of the two arms is 8.00 cm = 80.0 mm.

P_{gas} = P_{atm} + (height difference)
= 746 mm + 80.0 mm = 826 mm = 826 torr

11.46 In closed-end manometer the difference in height of the mercury levels in the two arms corresponds to the pressure of the gas:

$$\#\text{ torr} = (12.5\text{ cm Hg})\left(\frac{10\text{ mm}}{1\text{ cm}}\right)\left(\frac{1\text{ torr}}{1\text{ mm Hg}}\right) = 125\text{ torr}$$

11.48 First convert the pressure in torr to mm Hg:

$$\#\text{ mm Hg} = (755\text{ torr})\left(\frac{1\text{ mm Hg}}{1\text{ torr}}\right) = 755\text{ mm Hg}$$

Then use Equation 10.3 to convert height of the mercury column (liquid A) to height of the fluid column with a density of 1.22 g/mL (liquid B).

$$h_B = h_A\left(\frac{d_A}{d_B}\right) = 755\text{ mm} \times \frac{13.6\text{ g mL}^{-1}}{1.22\text{ g mL}^{-1}} = 8.42 \times 10^3\text{ mm}$$

11.50 Use Boyle's Law to solve for the second volume:

$$V_2 = \frac{P_1V_1}{P_2} = \frac{255\text{ mL}(725\text{ torr})}{365\text{ torr}} = 507\text{ mL}$$

11.52 Use Charles's Law to solve the second volume:

$$V_2 = \frac{V_1T_2}{T_1} = \frac{3.86\text{ L }(353\text{ K})}{318\text{ K}} = 4.28\text{ L}$$

11.54 Compare pressure change to temperature to solve for temperature change:

$$T_2 = \frac{P_2T_1}{P_1} = \frac{1700\text{ torr}(558\text{ K})}{850\text{ torr}} = 1120\text{ K} \qquad 1120\text{K} - 273\text{K} = 843\text{ °C}$$

11.56 In general the combined gas law equation is: $\frac{P_1V_1}{T_1} = \frac{P_2V_2}{T_2}$, and in particular, for this problem, we have:

$$P_2 = \frac{P_1V_1T_2}{T_1V_2} = \frac{(740\text{ torr})(2.58\text{ L})(348.2\text{ K})}{(297.2\text{ K})(2.81\text{ L})} = 796\text{ torr}$$

11.58 In general the combined gas law equation is $\frac{P_1V_1}{T_1}=\frac{P_2V_2}{T_2}$, and in particular, for this problem, we have:

$$V_2=\frac{P_1V_1T_2}{T_1P_2}=\frac{(745\text{ torr})(2.68\text{ L})(648.2\text{ K})}{(297.2\text{ K})(760\text{ torr})}=5.73\text{ L}$$

11.60 In general the combined gas law equation is: $\frac{P_1V_1}{T_1}=\frac{P_2V_2}{T_2}$, and in particular, for this problem, we have:

$$T_2=\frac{P_2V_2T_1}{P_1V_1}=\frac{(373\text{ torr})(9.45\text{ L})(293.2\text{ K})}{(761\text{ torr})(6.18\text{ L})}=220\text{ K}=-53^\circ\text{C}$$

11.62 $R=\left(0.0821\frac{\text{L atm}}{\text{mol K}}\right)\left(\frac{1000\text{ mL}}{1\text{ L}}\right)\left(\frac{760\text{ torr}}{1\text{ atm}}\right)=6.24\times10^4\frac{\text{mL torr}}{\text{mol K}}$

11.64

$$V=\frac{nRT}{P}=\frac{\left(0.136\text{g}\frac{1\text{ mol}}{32.0\text{ g}}\right)\left(0.0821\frac{\text{L atm}}{\text{mol K}}\right)(298\text{ K})}{\left(748\text{ torr}\frac{1\text{ atm}}{760\text{ torr}}\right)}=0.106\text{ L}$$

11.66

$$P=\frac{nRT}{V}=\frac{\left(10.0\text{g}\frac{1\text{ mol}}{32.0\text{ g}}\right)\left(0.0821\frac{\text{L atm}}{\text{mol K}}\right)(300\text{ K})}{(2.50\text{ L})}$$

$$=3.08\text{ atm}\frac{760\text{ torr}}{1\text{ atm}}=2{,}340\text{ torr}$$

11.68

$$n=\frac{PV}{RT}=\frac{(0.821\text{ atm})(0.0265\text{ L})}{\left(0.0821\frac{\text{L atm}}{\text{mol K}}\right)(293\text{ K})}$$

$$=\left(9.04\times10^{-4}\text{ mol}\right)\left(\frac{44\text{ g}}{1\text{ mol}}\right)=0.0398\text{ g}$$

11.70 Since PV = nRT, then;

$$P=\frac{nRT}{V}=\frac{(4.18\text{ mol})(0.0821\frac{\text{L atm}}{\text{mol K}})(291.2\text{ K})}{(24.0\text{ L})}=4.16\text{ atm}$$

11.72 a) $\text{density } C_2H_6 = \left(\frac{30.1\text{ g } C_2H_6}{1\text{ mol } C_2H_6}\right)\left(\frac{1\text{ mol}}{22.4\text{ L}}\right) = 1.34\text{ g L}^{-1}$

b) $\text{density } N_2 = \left(\frac{28.0\text{ g } N_2}{1\text{ mol } N_2}\right)\left(\frac{1\text{ mol}}{22.4\text{ L}}\right) = 1.25\text{ g L}^{-1}$

c) $\text{density } Cl_2 = \left(\frac{70.9\text{ g } Cl_2}{1\text{ mol } Cl_2}\right)\left(\frac{1\text{ mol}}{22.4\text{ L}}\right) = 3.17\text{ g L}^{-1}$

d) $\text{density Ar} = \left(\frac{39.9\text{ g Ar}}{1\text{ mol Ar}}\right)\left(\frac{1\text{ mol}}{22.4\text{ L}}\right) = 1.78\text{ g L}^{-1}$

11.74 In general PV = nRT, where n = mass ÷ formula mass. Thus

$$PV = \frac{\text{mass}}{(\text{formula mass})}RT$$

and we arrive at the formula for the density (mass divided by volume) of a gas:

$$d = \frac{P \times (\text{formula mass})}{RT}$$

$$d = \frac{(742\text{ torr})\left(\frac{1\text{ atm}}{760\text{ torr}}\right)(32.0\text{ g/mol})}{\left(0.0821\frac{\text{L atm}}{\text{mol K}}\right)(297.2\text{ K})}$$

$$d = 1.28\text{ g/L for } O_2$$

11.76 First determine the number of moles:

$$n = \frac{PV}{RT} = \frac{(10.0\text{ torr})\left(\frac{1\text{ atm}}{760\text{ torr}}\right)(255\text{ mL})\left(\frac{1\text{ L}}{1000\text{ mL}}\right)}{\left(0.0821\frac{\text{L atm}}{\text{mol K}}\right)(298.2\text{ K})} = 1.37 \times 10^{-4}\text{ mol}$$

Now calculate the molecular mass:

$$\text{molecular mass} = \frac{\text{mass}}{\#\text{ of moles}} = \frac{(12.1\text{ mg})\left(\frac{1\text{ g}}{1000\text{ mg}}\right)}{1.37 \times 10^{-4}\text{ mol}} = 88.3\text{ g/mol}$$

11.78

$$\text{molecular mass} = \frac{dRT}{P} = \frac{\text{mass RT}}{PV} = \frac{(1.13\text{ g/L})\left(0.0821\frac{\text{L atm}}{\text{mol K}}\right)(295\text{ K})}{(0.995\text{ atm})}$$

$$= 27.5\text{ g/mol}$$

11.80 When gases are held at the same temperature and pressure, and dispensed in this fashion during chemical reactions, then they react in a ratio of volumes that is equal to the ratio of the coefficients (moles) in the balanced chemical equation for the given reaction. We can, therefore, directly use the stoichiometry of the balanced chemical equation to determine the combining ratio of the gas volumes:

$$\# \text{L F}_2 = (4.00 \text{ L H}_2)\left(\frac{1 \text{ volume F}_2}{1 \text{ volume H}_2}\right) = 4.00 \text{ L F}_2$$

11.82 $\# \text{mL O}_2 = (175 \text{ mL C}_4\text{H}_{10})\left(\frac{13 \text{ mL O}_2}{2 \text{ mL C}_4\text{H}_{10}}\right) = 1.14\times10^3 \text{ mL}$

11.84

$$\# \text{mol C}_3\text{H}_6 = (18.0 \text{ g C}_3\text{H}_6)\left(\frac{1 \text{ mol C}_3\text{H}_6}{42.08 \text{ g C}_3\text{H}_6}\right) = 0.428 \text{ mol C}_3\text{H}_6$$

$$\# \text{mol H}_2 = (0.428 \text{ mol C}_3\text{H}_6)\left(\frac{1 \text{ mol H}_2}{1 \text{ mol C}_3\text{H}_6}\right) = 0.428 \text{ mol H}_2$$

$$V = \frac{nRT}{P} = \frac{(0.428 \text{ mol H}_2)(0.0821 \frac{\text{L atm}}{\text{mol K}})(297.2 \text{ K})}{(740 \text{ torr})(\frac{1 \text{ atm}}{760 \text{ torr}})} = 10.7 \text{ L H}_2$$

11.86 $CH_4 + 2O_2 \rightarrow CO_2 + 2H_2O$

$$n_{CH_4} = \frac{PV}{RT} = \frac{(725 \text{ torr})\left(\frac{1 \text{ atm}}{760 \text{ torr}}\right)(16.8 \times 10^{-3} \text{ L})}{\left(0.0821 \frac{\text{L atm}}{\text{mol K}}\right)(308 \text{ K})} = 6.34\times10^{-4} \text{ moles}$$

From the balanced equation, twice as many moles of oxygen are needed:

$\Rightarrow 1.27\times10^{-3}$ moles O_2

$$V_{O_2} = \frac{nRT}{P} = \frac{(1.27 \times 10^{-3} \text{ moles})\left(0.0821 \frac{\text{L atm}}{\text{mol K}}\right)(300 \text{ K})}{(654 \text{ torr})\left(\frac{1 \text{ atm}}{760 \text{ torr}}\right)} = 3.63\times10^{-2} \text{ L}$$

$\Rightarrow 36.3$ mL

11.88 $2CO + O_2 \rightarrow 2CO_2$

$$\# \text{moles CO} = \frac{(683 \text{ torr})\left(\frac{1 \text{ atm}}{760 \text{ torr}}\right)(0.300 \text{ L})}{\left(0.0821 \frac{\text{L atm}}{\text{mol K}}\right)(298 \text{ K})} = 1.10\times10^{-2} \text{ moles}$$

$$\#\,\text{moles O}_2 = \frac{(715\text{ torr})\left(\frac{1\text{ atm}}{760\text{ torr}}\right)(0.150\text{ L})}{\left(0.0821\frac{\text{L atm}}{\text{mol K}}\right)(398\text{ K})} = 4.32\times10^{-3}\text{ moles}$$

Therefore, oxygen is the limiting reagent and $2(4.32 \times 10^{-3})$ mol CO_2 will form, based on the balanced equation.

$$V_{CO_2} = \frac{nRT}{P} = \frac{(8.64\times10^{-3}\text{ moles})\left(0.0821\frac{\text{L atm}}{\text{mol K}}\right)(300\text{ K})}{(745\text{ torr})\left(\frac{1\text{ atm}}{760\text{ torr}}\right)} = 0.217\text{ L}$$

$$\Rightarrow 217\text{ mL}$$

11.90 P_{Tot} = 200 torr + 150 torr + 300 torr = 650 torr

11.92 Assume all gases behave ideally and recall that 1 mole of an ideal gas at 0°C and 1 atm occupies a volume of 22.4 L. So,

$$P_{N_2} = 0.30\text{ atm} = 228\text{ torr}$$
$$P_{O_2} = 0.20\text{ atm} = 152\text{ torr}$$
$$P_{He} = 0.40\text{ atm} = 304\text{ torr}$$
$$P_{CO_2} = 0.10\text{ atm} = 76\text{ torr}$$

11.94 $P_{total} = (P_{CO} + P_{H_2O})$

P_{H_2O} = 17.54 torr at 20 °C, from Table 10.2.

P_{CO} = 754 – 17.54 torr = **736 torr**

The temperature stays constant so, $P_1V_1 = P_2V_2$, and

$$V_2 = \frac{P_1V_1}{P_2} = \frac{(736\text{ torr})(268\text{ mL})}{(760\text{ torr})} = 260\text{ mL}$$

11.96 From Table 11.2, the vapor pressure of water at 20 °C is 17.54 torr. Thus only (742 – 17.54) = 724 torr is due to "dry" methane. In other words, the fraction of the wet methane sample that is pure methane is 724/742 = 0.976. The question can now be phrased: What volume of wet methane, when multiplied by 0.976, equals 244 mL?

Volume"wet" methane × 0.976 = 244 mL

Volume"wet" methane = 244 mL/0.976 = 250 mL

In other words, one must collect **250 mL** of "wet methane" gas in order to have collected the equivalent of 244 mL of pure methane.

11.98 Use equation 11.8 to convert all partial pressures to mole fraction. (P_{total} = 760 torr)

$$X_{N_2} = \frac{P_{N_2}}{P_{total}} = \frac{570\text{ torr}}{760\text{ torr}} = 0.750; 75.0\,\%$$

$$X_{O_2} = \frac{P_{O_2}}{P_{total}} = \frac{103\text{ torr}}{760\text{ torr}} = 0.136; 13.6\,\%$$

$$X_{CO_2} = \frac{P_{CO_2}}{P_{total}} = \frac{40\text{ torr}}{760\text{ torr}} = 0.053; 5.3\,\%$$

$$X_{H_2O} = \frac{P_{H_2O}}{P_{total}} = \frac{47\text{ torr}}{760\text{ torr}} = 0.062; 6.2\,\%$$

11.100 Effusion rates for gases are inversely proportional to the square root of the gas density, and the gas with the lower density ought to effuse more rapidly. Nitrogen in this problem has the higher effusion rate because it has the lower density:

$$\frac{\text{rate}(N_2)}{\text{rate}(CO_2)} = \sqrt{\frac{1.96\text{ g L}^{-1}}{1.25\text{ g L}^{-1}}} = 1.25$$

11.102 The relative rates are inversely proportional to the square roots of their molecular masses:

$$\frac{\text{rate}(^{235}UF_6)}{\text{rate}(^{238}UF_6)} = \sqrt{\frac{\text{molar mass }(^{238}UF_6)}{\text{molar mass }(^{235}UF_6)}} = \sqrt{\frac{352\text{ g mol}^{-1}}{349\text{ g mol}^{-1}}} = 1.0043$$

Additional Exercises

11.104 We found that 1 atm = 33.9 ft of water. This is equivalent to 33.9 ft × 12 in./ft = 407 in. of water, which in this problem is equal to the height of a water column that is uniformly 1.00 in.^{2} in diameter. Next, we convert the given density of water from the units g/mL to the units lb/in.^{3}:

$$\#\frac{\text{lb}}{\text{in.}^3} = \left(\frac{1.00\text{ g}}{1.00\text{ mL}}\right)\left(\frac{1\text{ lb}}{454\text{ g}}\right)\left(\frac{1\text{ mL}}{1\text{ cm}^3}\right)\left(\frac{2.54\text{ cm}}{1\text{ in.}}\right)^3 = 0.0361\frac{\text{lb}}{\text{in.}^3}$$

The area of the total column of water is now calculated: 1.00 in.^{2} × 407 in. = 407 in.^{3}, along with the mass of the total column of water: 407 in.^{3} × 0.0361 lb/in.^{3} = 14.7 lb. Finally, we can determine the pressure (force/unit area) that corresponds to one atm: 1 atm = 14.7 lb ÷ 1.00 in.^{2} = **14.7 lb/in^2**.

11.106

$$\text{Total weight} = (45.6 \text{ tons} + 8.3 \text{ tons})\left(\frac{2000 \text{ lb}}{1 \text{ ton}}\right) = 1.08\times10^5 \text{ lbs}$$

$$\text{Total pressure} = 85 \text{ psi} + 14.7 \text{ psi} = 99.7 \text{ psi/tire}$$

$$\# \text{ tires} = \frac{1.08 \text{ x } 10^5 \text{ lbs}}{(99.7 \text{ lbs in}^{-2}\text{/tire})(100 \text{ in}^2)} = 10.8 \text{ tires}$$

The minimal number of tires is 12 since tires are mounted in multiples of 2

*11.108

From the data we know that the pressure in flask 1 is greater than atmospheric pressure, and greater than the pressure in flask 2. The pressure in flask 1 can be determined from the manometer data. The pressure in flask 1 is:

$$(0.827 \text{ atm})\left(\frac{760 \text{ mm Hg}}{1 \text{ atm}}\right)\left(\frac{1 \text{ cm}}{10 \text{ mm}}\right) + 12.26 \text{ cm} = 75.11 \text{ cm Hg}$$

The pressure in flask 2 is lower than flask 1

$$P = 75.11 \text{ cm Hg} - (16.24 \text{ cm oil})\left(\frac{0.826 \text{ g mL}^{-1}}{13.6 \text{ g mL}^{-1}}\right) = 74.12 \text{ cm Hg} = 741.2 \text{ torr}$$

*11.110

First calculate the initial volume (V_1) and the final volume (V_2) of the cylinder, using the given geometrical data, noting that the radius is half the diameter (10.7/2 = 5.35 cm): $V_1 = \pi \times (5.35 \text{ cm})^2 \times 13.4 \text{ cm} = 1.20 \times 10^3 \text{ cm}^3$
$V_2 = \pi \times (5.35 \text{ cm})^2 \times (13.4 \text{ cm} - 12.7 \text{ cm}) = 62.9 \text{ cm}^3$

In general the combined gas law equation is: $\frac{P_1V_1}{T_1} = \frac{P_2V_2}{T_2}$, and in particular, for this problem, we have:

$$T_2 = \frac{P_2V_2T_1}{P_1V_1} = \frac{(34.0 \text{ atm})(62.9 \text{ cm}^3)(364 \text{ K})}{(1.00 \text{ atm})(1.20\times10^3 \text{ cm}^3)} = 649 \text{ K} = 376\ ^\circ\text{C}$$

11.112

The temperatures must first be converted to Kelvin:

$$^\circ\text{C} = \frac{5}{9}\times(^\circ\text{F} - 32) = \frac{5}{9}\times(-50 - 32) = -46\ ^\circ\text{C}$$

$$^\circ\text{C} = \frac{5}{9}\times(^\circ\text{F} - 32) = \frac{5}{9}\times(120 - 32) = 49\ ^\circ\text{C}$$

Next, the pressure calculation is done using the following equation:

$$P_2 = \frac{P_1T_2}{T_1} = \frac{(35 \text{ lb in.}^{-2})(322 \text{ K})}{(227 \text{ K})} = 50 \text{ lb in.}^{-2}$$

11.114

The relative rates are inversely proportional to the square roots of their molecular masses:

$$\frac{\text{rate}(NH_3)}{\text{rate(unknown gas)}} = \sqrt{\frac{\text{molar mass (unknown gas)}}{\text{molar mass }(NH_3)}} = 2.93$$

$$\frac{\text{molar mass (unknown gas)}}{\text{molar mass }(NH_3)} = (2.93)^2$$

$$\text{molar mass (unknown gas)} = \text{molar mass }(NH_3) \times (2.93)^2$$

$$\text{molar mass (unknown gas)} = 17.30\ \text{g/mol} \times 8.58 = 146\ \text{g/mol}$$

*11.116 $P_{total} = 740\ \text{torr} = P_{H_2} + P_{water}$

The vapor pressure of water at 25 °C is available in Table 11.2…23.76 torr. Hence:

$$P_{H_2} = (740 - 24)\ \text{torr} = 716\ \text{torr}$$

Next, we calculate the number of moles of hydrogen gas that this represents:

$$n = \frac{PV}{RT} = \frac{(716\ \text{torr})\left(\frac{1\ \text{atm}}{760\ \text{torr}}\right)(0.335\ \text{L})}{\left(0.0821\ \frac{\text{L atm}}{\text{mol K}}\right)(298.2\ \text{K})} = 0.0129\ \text{mol } H_2$$

The balanced chemical equation is: $Zn(s) + 2HCl(aq) \rightarrow H_2(g) + ZnCl_2(aq)$ and the quantities of the reagents that are needed are:

$$\#\text{g Zn} = (0.0129\ \text{mol } H_2)\left(\frac{1\ \text{mol Zn}}{1\ \text{mol } H_2}\right)\left(\frac{65.39\ \text{g Zn}}{1\ \text{mol Zn}}\right) = 0.844\ \text{g Zn}$$

$$\#\text{mL HCl} = (0.0129\ \text{mol } H_2)\left(\frac{2\ \text{mol HCl}}{1\ \text{mol } H_2}\right)\left(\frac{1000\ \text{mL HCl}}{6.00\ \text{mol HCl}}\right) = 4.30\ \text{mL HCl}$$

11.118

Using the ideal gas law, determine the number of moles of H_2 and O_2 gas initially present:

For hydrogen:

$$n = \frac{PV}{RT} = \frac{(1250\ \text{torr})\left(\frac{1\ \text{atm}}{760\ \text{torr}}\right)(400\ \text{mL})\left(\frac{1\ \text{L}}{1000\ \text{mL}}\right)}{\left(0.0821\ \frac{\text{L atm}}{\text{mol K}}\right)(318\ \text{K})} = 2.52\times10^{-2}\ \text{mol } H_2$$

For oxygen:

$$n = \frac{PV}{RT} = \frac{(740\ \text{torr})\left(\frac{1\ \text{atm}}{760\ \text{torr}}\right)(300\ \text{mL})\left(\frac{1\ \text{L}}{1000\ \text{mL}}\right)}{\left(0.0821\ \frac{\text{L atm}}{\text{mol K}}\right)(298\ \text{K})} = 1.19\times10^{-2}\ \text{mol } O_2$$

This problem is an example of a limiting reactant problem in that we know the amounts of H_2 and O_2 initially present. Since 1 mol of O_2 reacts completely with 2 mol of H_2, we can see, by inspection, that there is excess H_2 present. Using the amounts calculated above, we can make 2.38 x 10^{-2} mol of H_2O and have an excess of 1.4 x 10^{-3} mol of H_2. Thus, the total amount of gas present after complete reaction is 2.52 x 10^{-2} mol. Using this value for n, we can calculate the final pressure in the reaction vessel:

$$P = \frac{nRT}{V} = \frac{(2.52\times10^{-2}\text{ mol})(0.0821\,\frac{\text{L atm}}{\text{mol K}})(393\text{ K})}{(500\text{ mL})(\frac{1\text{ L}}{1000\text{ mL}})} = 1.63\text{ atm} = 1.24\times10^{3}\text{ torr}$$

11.120 $2H_2 + O_2 \rightarrow 2H_2O$

$$\#\text{ moles }H_2 = (12.7\text{ g }H_2)\left(\frac{1\text{ mol }H_2}{2.02\text{ g }H_2}\right) = 6.30\text{ moles }H_2$$

$$\#\text{ moles }O_2 = (87.5\text{ g }O_2)\left(\frac{1\text{ mol }O_2}{32.0\text{ g }O_2}\right) = 2.73\text{ moles }O_2$$

$\therefore O_2$ is the limiting reactant

$$\#\text{ moles }H_2O = (2.73\text{ mol }O_2)\left(\frac{2\text{ mol }H_2O}{1\text{ mol }O_2}\right) = 5.47\text{ moles }H_2O$$

$$\#\text{ moles }H_2\text{ needed} = (2.73\text{ mol }O_2)\left(\frac{2\text{ mol }H_2}{1\text{ mol }O_2}\right) = 5.47\text{ moles }H_2$$

$$P_{H_2O} = \frac{(5.47\text{ mol})\left(0.0821\frac{\text{L atm}}{\text{mol K}}\right)(433\text{ K})}{12.0\text{ L}} = 16.2\text{ atm}$$

$$P_{H_2} = \frac{(0.83\text{ mol})\left(0.0821\frac{\text{L atm}}{\text{mol K}}\right)(433\text{ K})}{12.0\text{ L}} = 2.5\text{ atm}$$

$$P_{Tot} = 16.2\text{ atm} + 2.5\text{ atm} = 18.7\text{ atm}$$

11.122 a) First determine the % by mass S and O in the sample:

% S = 1.448 g/3.620 g × 100 = 40.00 % S

% O = 2.172 g/3.620 g × 100 = 60.00 % O

b) Next, determine the number of moles of S and O in a sample of the material weighing 100 g exactly, in order to make the conversion from % by mass to grams straightforward: In 100 g of material there are 40.00 g S and 60.00 g O:

$$40.00 \text{ g S} \div 32.07 \text{ g/mol} = 1.247 \text{ mol S}$$
$$60.00 \text{ g O} \div 16.00 \text{ g/mol} = 3.750 \text{ mol O}$$

Dividing each of these mole amounts by the smaller of the two gives the relative mole amounts of S and O in the material: for S, 1.247 mol ÷ 1.247 mol = 1.000 relative moles, for O, 3.750 mol ÷ 1.247 mol= 3.007 relative moles, and the empirical formula is, therefore, SO_3.

c) We determine the formula mass of the material by use of the ideal gas law:

$$n = \frac{PV}{RT} = \frac{(750 \text{ torr})\left(\frac{1 \text{ atm}}{760 \text{ torr}}\right)(1.120 \text{ L})}{\left(0.0821 \frac{\text{L atm}}{\text{mol K}}\right)(298.2 \text{ K})} = \quad 0.0451 \text{ mol}$$

The formula mass is given by the mass in grams (given in the problem) divided by the moles determined here: formula mass = 3.620 g ÷ 0.0451 mol = 80.2 g mol^{-1}. Since this is equal to the formula mass of the empirical unit determined in step (b) above, namely SO_3, then the molecular formula is also SO_3.

11.124 a) The equation can be rearranged to give:

$$0.04489 \times \frac{V(P - P_{H_2O})}{273 + {}^\circ C} = \%N \times W$$

This means that the left side of the above equation should be obtainable simply from the ideal gas law, applied to the nitrogen case. If PV = nRT, then for nitrogen: PV = (mass nitrogen)/(28.01 g/mol) × RT, and the mass of nitrogen that is collected is given by: (mass nitrogen) = PV(28.01)/RT, where R = 82.1 mL atm/K mol × 760 torr/atm = 6.24×10^4 mL torr/K mol. Using this value for R in the above equation, we have the following result for the mass of nitrogen, remembering that the pressure of nitrogen is less than the total pressure, by an amount equal to the vapor pressure of water:

$$(\text{mass nitrogen}) = \frac{28.01 \times V \times (P_{total} - P_{H_2O})}{\left(6.24 \times 10^4 \frac{\text{mL torr}}{\text{mol K}}\right)(273 + {}^\circ C)}$$

Finally, it is only necessary to realize that the value

$$\frac{28.01}{6.24 \times 10^4} \times 100 = 0.04489$$

is exactly the value given in the problem.

$$\text{b)}\ \%N = \ 0.04489 \times \frac{(18.90 \text{ mL})(746 \text{ torr} - 22.1 \text{ torr})}{(0.2394 \text{ g})(273.15 + 23.80)} = 8.639\,\%$$

Chapter 12

Practice Exercises

12.1 Propylamine would have a substantially higher boiling point because of its ability to form hydrogen bonds (there are N-H bonds in propylamine, but not in trimethylamine.)

12.2 The number of molecules in the vapor will increase, and the number of molecules in the liquid will decrease, but the sum of the molecules in the vapor and the liquid remains the same.

12.3 We use the curve for water, and find that at 330 torr, the boiling point is approximately 75 °C.

12.4 Adding heat will shift the equilibrium to the right, producing more vapor. This increase in the amount of vapor causes a corresponding increase in the pressure, such that the vapor pressure generally increases with increasing temperature.

12.5 Refer to the phase diagram for water, Figure 12.28. We "move" along a horizontal line marked for a pressure of 2.15 torr. At –20 °C, the sample is a solid. If we bring the temperature from –20 °C to 50 °C, keeping the pressure constant at 2.15 torr, the sample becomes a gas. The process is thus solid → gas, i.e. sublimation.

12.6 As diagramed in Figure 12.28, this falls in the liquid region.

Review Problems

12.86 Diethyl ether has the faster rate of evaporization, since it does not have hydrogen bonds, as does butanol.

12.88 London forces are possible in them all. Where another intermolecular force can operate, it is generally stronger than London forces, and this other type of interaction overshadows the importance of the London force. The substances in the list that can have dipole–dipole attractions are those with permanent dipole moments: (a), (b), and (d). SF_6, (c), is a non–polar molecular substance. HF, (a), has hydrogen bonding.

12.90 Chloroform would be expected to display larger dipole-dipole attractions because it has a larger dipole moment than bromoform. (Chlorine has a higher electronegativity which results in each C-Cl bond having a larger dipole than each C-Br bond.) On the other hand, bromoform would be expected to show stronger London forces due to having larger electron clouds which are more polarizable than those of chlorine.

Since bromoform in fact has a higher boiling point that chloroform, we must conclude that it experiences stronger intermolecular attractions than chloroform, which can only be due to London forces. Therefore, London forces are more important in determining the boiling points of these two compounds.

12.92 Ethanol, because it has H-bonding.

12.94 ether < acetone < benzene < water < acetic acid

12.96 Relative humidity is determined by the formula given in Chapter 12, shown below. The equilibrium vapor pressure of water at 25 °C may be found in the table "Vapor Pressure of Water…" in the Appendix of the text.

$$\%\ \text{relative humidity} = \left(\frac{\text{actual } P_{H_2O}}{\text{equilibrium } P_{H_2O}}\right) \times 100$$

$$= \left(\frac{18.42 \text{ torr}}{23.8 \text{ torr}}\right) \times 100 = 77.4\%$$

12.98 $\#\text{kJ} = (125 \text{ g } H_2O)\left(\frac{1 \text{ mol } H_2O}{18.015 \text{ g } H_2O}\right)\left(\frac{43.9 \text{ kJ}}{1 \text{ mol } H_2O}\right) = 305 \text{ kJ}$

12.100 We can approach this problem by first asking either of two equivalent questions about the system: how much heat energy (q) is needed in order to melt the entire sample of solid water (105 g), or how much energy is lost when the liquid water (45.0 g) is cooled to the freezing point? Regardless, there is only one final temperature for the combined (150.0 g) sample, and we need to know if this temperature is at the melting point (0 °C, at which temperature some solid water remains in equilibrium with a certain amount of liquid water) or above the melting point (at which temperature all of the solid water will have melted).

Heat flow supposing that all of the solid water is melted:

$q = 6.01 \text{ kJ/mole} \times 105 \text{ g} \times 1 \text{ mol}/18.0 \text{ g} = 35.1 \text{ kJ}$

Heat flow on cooling the liquid water to the freezing point:

$q = 45.0 \text{ g} \times 4.18 \text{ J/g °C} \times 85 \text{ °C} = 1.60 \times 10^4 \text{ J} = 16.0 \text{ kJ}$

The lesser of these two values is the correct one, and we conclude that 16.0 kJ of heat energy will be transferred from the liquid to the solid, and that the final temperature of the mixture will be 0 °C. The system will be an equilibrium mixture weighing 150 g and having some solid and some liquid in equilibrium with one another. The amount of solid that must melt in order to decrease the temperature of 45.0 g of water from 85 °C to 0 °C is: 16.0 kJ ÷ 6.01 kJ/mol = 2.66 mol of solid water. 2.66 mol × 18.0 g/mol = 47.9 g of water must melt.

a) The final temperature will be 0 °C.
b) 47.9 g of water must melt.

12.102 Water boils at a higher temperature than ethanol. This reveals large intermolecular attractions in water, which would cause us to expect a larger molar heat of vaporization for water than ethanol.

12.104

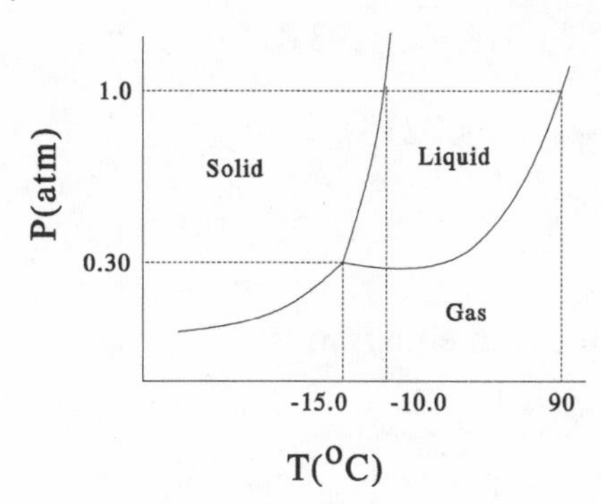

12.106 a) solid b) gas c) liquid d) solid, liquid, and gas

12.108 At –56 °C, the vapor is compressed until the liquid–vapor line is reached, at which point the vapor condenses to a liquid. As the pressure is increased further, the solid–liquid line is reached, and the liquid freezes.

At –58° C, the gas is compressed until the solid–vapor line is reached, at which point the vapor condenses directly to a solid.

Additional Exercises

12.110 Intermolecular (between different particles): The principal attractive forces are ion–ion forces and ion–dipole forces. These are of overwhelming strength compared to London forces, which do technically exist.

Intramolecular (within certain particles): The sulfite ion (SO_3^{2-}) is a polyatomic ion whose atoms are held together by *covalent bonds*. Although a covalent bond is not an "attraction" in itself, the attractive forces which make up these bonds are valence electron's attractions to the nuclei of neighboring atoms in the polyatomic ion.

12.112 Yes, because of the possibility of a weak hydrogen bond between the double-bonded oxygen of acetone and a hydrogen atom of water.

12.114 Using Hess's Law, sublimation may be considered equivalent to melting followed by vaporization.

$\Delta H_{sublimation} = \Delta H_{fusion} + \Delta H_{vaporization} = 10.8 \text{ kJ/mol} + 24.3 \text{ kJ/mol} = 35.1 \text{ kJ}$

*12.116 At the higher temperature, all of the carbon dioxide exists in the gas phase because the critical temperature has been exceeded.

12.118 The "two point" form of the Clausius-Clapeyron equation is:

$$\ln\left(\frac{P_1}{P_2}\right) = \frac{\Delta H_{vap}}{8.314\ \text{J mol}^{-1}\text{K}^{-1}}\left(\frac{1}{T_2} - \frac{1}{T_1}\right)$$

Inserting the given values from the problem, we have:

$$\ln\left(\frac{72.8\ \text{torr}}{186.2\ \text{torr}}\right) = \frac{\Delta H_{vap}}{8.314\ \text{J mol}^{-1}\text{K}^{-1}}\left(\frac{1}{313\ \text{K}} - \frac{1}{293\ \text{K}}\right)$$

$$-0.939 = \frac{\Delta H_{vap}}{8.314\ \text{J mol}^{-1}\text{K}^{-1}}(-0.000218)$$

$\Delta H_{vap} = 35{,}800\ \text{J/mol} = 35.8\ \text{kJ/mol}$

12.120 The "two point" form of the Clausius-Clapeyron equation is:

$$\ln\left(\frac{P_1}{P_2}\right) = \frac{\Delta H_{vap}}{8.314\ \text{J mol}^{-1}\text{K}^{-1}}\left(\frac{1}{T_2} - \frac{1}{T_1}\right)$$

Inserting the given values from the problem, we have:

$$\ln\left(\frac{28.1\ \text{torr}}{47.0\ \text{torr}}\right) = \frac{\Delta H_{vap}}{8.314\ \text{J mol}^{-1}\text{K}^{-1}}\left(\frac{1}{60\ \text{K}} - \frac{1}{57\ \text{K}}\right)$$

$$-0.514 = \frac{\Delta H_{vap}}{8.314\ \text{J mol}^{-1}\text{K}^{-1}}(-0.000877)$$

$\Delta H_{vap} = 4{,}870\ \text{J/mol} = 4.87\ \text{kJ/mol}$

The boiling point of liquid nitrogen is the temperature where its vapor pressure equals atmospheric pressure, or 760 torr:

$$\ln\left(\frac{P_1}{P_2}\right) = \frac{\Delta H_{vap}}{8.314\ \text{J mol}^{-1}\text{K}^{-1}}\left(\frac{1}{T_2} - \frac{1}{T_1}\right)$$

$$\ln\left(\frac{28.1}{760}\right) = \frac{4{,}870\ \text{J/mol}}{8.314\ \text{J mol}^{-1}\text{K}^{-1}}\left(\frac{1}{T_2} - \frac{1}{57\ \text{K}}\right)$$

$$-3.30 = 586\ \text{K} \times \left(\frac{1}{T_2} - \frac{1}{57\ \text{K}}\right)$$

$$-5.63 \times 10^{-3}\ \text{K}^{-1} = \frac{1}{T_2} - \frac{1}{57\ \text{K}}$$

$$\frac{1}{T_2} = 1.19 \times 10^{-2}\ \text{K}^{-1};\ T_2 = 84\ \text{K} = -189\ ^\circ\text{C}$$

12.122 The "two point" form of the Clausius-Clapeyron equation is:

$$\ln\left(\frac{P_1}{P_2}\right) = \frac{\Delta H_{vap}}{8.314\ \text{J mol}^{-1}\text{K}^{-1}}\left(\frac{1}{T_2} - \frac{1}{T_1}\right)$$

Inserting the given values from the problem, we have:

$$\ln\left(\frac{P_1}{P_2}\right) = \frac{\Delta H_{vap}}{8.314\ \text{J mol}^{-1}\text{K}^{-1}}\left(\frac{1}{T_2} - \frac{1}{T_1}\right)$$

$$\ln\left(\frac{20.0}{P_2}\right) = \frac{38{,}000\ \text{J/mol}}{8.314\ \text{J mol}^{-1}\text{K}^{-1}}\left(\frac{1}{298} - \frac{1}{291.6\ \text{K}}\right)$$

$$\ln\left(\frac{20.0}{P_2}\right) = 4{,}570\ \text{K} \times (-0.000{,}0737\text{K})$$

$$\ln\left(\frac{20.0}{P_2}\right) = -0.337$$

$$\frac{20.0}{P_2} = e^{-0.337}$$

$$\frac{20.0}{P_2} = 0.714$$

$$P_2 = 28.0\ \text{torr}$$

12.124 Here we need only solve for the temperature T_2 at which the vapor pressure P_2 has reached 760 torr:

$$\ln\left(\frac{P_1}{P_2}\right) = \frac{\Delta H_{vap}}{8.314\ \text{J mol}^{-1}\text{K}^{-1}}\left(\frac{1}{T_2} - \frac{1}{T_1}\right)$$

$$\ln\left(\frac{31.6}{760}\right) = \frac{42.09\times10^{3}\ \text{J/mol}}{8.314\ \text{J mol}^{-1}\text{K}^{-1}}\left(\frac{1}{T_2} - \frac{1}{293\ \text{K}}\right)$$

$$-3.18 = 5063\ \text{K} \times \left(\frac{1}{T_2} - \frac{1}{293\ \text{K}}\right)$$

$$-6.28\times10^{-4}\ \text{K}^{-1} = \frac{1}{T_2} - \frac{1}{293\ \text{K}}$$

$$\frac{1}{T_2} = 2.78\times10^{-3}\ \text{K}^{-1};\ T_2 = 359\ \text{K} = 86\ ^\circ\text{C}$$

*12.126 We have four unknowns and four equations.

For liquid A:

$$\ln\left(\frac{100\ \text{torr}}{P_{A2}}\right) = \frac{32{,}000\ \text{J mol}^{-1}}{8.314\ \text{J mol}^{-1}\text{K}^{-1}}\left(\frac{1}{T_{A2}} - \frac{1}{298\ \text{K}}\right)$$

For liquid B:

$$\ln\left(\frac{200\text{ torr}}{P_{B2}}\right)=\frac{18{,}000\text{ J mol}^{-1}}{8.314\text{ J mol}^{-1}\text{K}^{-1}}\left(\frac{1}{T_{B2}}-\frac{1}{298\text{ K}}\right)$$

(Note that the ΔH_{vap} values were converted to J to cancel with the unit in the gas constant in the denominator.)

The other two equations are:

$P_{A2}=P_{B2}$ (since we are looking for the point at which they have the same vapor pressure), and

$T_{A2}=T_{B2}$ (since we are looking for one temperature at which the vapor pressures are equal). Therefore, we can simply write P_2 and T_2 in the first two equations:

$$\ln\left(\frac{100\text{ torr}}{P_2}\right)=3848.93\text{ K}\left(\frac{1}{T_2}-\frac{1}{298\text{ K}}\right) \quad \text{and}$$

$$\ln\left(\frac{200\text{ torr}}{P_2}\right)=2165.02\text{ K}\left(\frac{1}{T_2}-\frac{1}{298\text{ K}}\right)$$

The right side of both equations may be distributed out,

$$\ln\left(\frac{100\text{ torr}}{P_2}\right)=\frac{3848.93\text{ K}}{T_2}-\frac{3848.93\text{ K}}{298\text{ K}} \quad \text{and}$$

$$\ln\left(\frac{200\text{ torr}}{P_2}\right)=\frac{2165.02\text{ K}}{T_2}-\frac{2165.02\text{ K}}{298\text{ K}}$$

and divided,

$$\ln\left(\frac{100\text{ torr}}{P_2}\right)=\frac{3848.93\text{ K}}{T_2}-12.9159 \quad \text{(Equation 1), and}$$

$$\ln\left(\frac{200\text{ torr}}{P_2}\right)=\frac{2165.02\text{ K}}{T_2}-7.26517 \quad \text{(Equation 2)}$$

Using Equation 1, we can rearrange for T_2:

$$\ln\left(\frac{100\text{ torr}}{P_2}\right)=\frac{3848.93\text{ K}}{T_2}-12.9159$$

$$\ln\left(\frac{100\text{ torr}}{P_2}\right)+12.9159=\frac{3848.93\text{ K}}{T_2}$$

$$T_2=\frac{3848.93\text{ K}}{\left[\ln\left(\frac{100\text{ torr}}{P_2}\right)+12.9159\right]}$$

Now we may substitute this into Equation 2:

$$\ln\left(\frac{200\text{ torr}}{P_2}\right)=\frac{2165.02\text{ K}}{T_2}-7.26517$$

$$\ln\left(\frac{200\text{ torr}}{P_2}\right)=\frac{2165.02\text{ K}}{\frac{3848.93\text{ K}}{\left[\ln\left(\frac{100\text{ torr}}{P_2}\right)+12.9159\right]}}-7.26517$$

$$\ln\left(\frac{200\text{ torr}}{P_2}\right)=2165.02\text{ K}\left[\frac{\ln\left(\frac{100\text{ torr}}{P_2}\right)+12.9159}{3848.93\text{ K}}\right]-7.26517$$

$$\ln\left(\frac{200\text{ torr}}{P_2}\right)=0.562499\left[\ln\left(\frac{100\text{ torr}}{P_2}\right)+12.9159\right]-7.26517$$

$$\ln\left(\frac{200\text{ torr}}{P_2}\right)=0.562499\ln\left(\frac{100\text{ torr}}{P_2}\right)+7.26517-7.26517$$

$$\ln\left(\frac{200\text{ torr}}{P_2}\right)=0.562499\ln\left(\frac{100\text{ torr}}{P_2}\right)$$

$$\frac{\ln\left(\frac{200\text{ torr}}{P_2}\right)}{\ln\left(\frac{100\text{ torr}}{P_2}\right)} = 0.562499$$

Therefore, $P_2 = 488$ torr.

Inserting this value back into either Equation 1 or Equation 2 gives:

$$\ln\left(\frac{100\text{ torr}}{488}\right) = \frac{3848.93\text{ K}}{T_2} - 12.9159$$

$$-1.59 = \frac{3848.93\text{ K}}{T_2} - 12.9159$$

$$11.33 = \frac{3848.93\text{ K}}{T_2}$$

$$T_2 = \frac{3848.93\text{ K}}{11.33} = 340\text{ K} = 67\ ^\circ\text{C}$$

Chapter 13

Practice Exercises

13.1 For cesium:

8 corners x 1/8 Cs^+ per corner = 1 Cs^+

For chloride:

1 Cl^- in center, Total: 1 Cl^-

Thus, the ratio is 1 to 1.

13.2 Because this is a high melting, hard material, it must be a covalent or network solid. Covalent bonds link the various atoms of the crystal.

13.3 Since the melt does not conduct electricity, it is not an ionic substance. The softness and the low melting point suggest that this is a molecular solid, and indeed the formula is most properly written S_8.

13.4 $-(CF_2\text{-}CF_2)\text{-}(CF_2\text{-}CF_2)\text{-}(CF_2\text{-}CF_2)-$

Review Problems

13.92 For zinc:

4 surrounding center = **4 Zn^{2+}**

For sulfide:

8 corners x 1/8 S^{2-} per corner = 1 S^{2-}

6 faces x 1/2 S^{2-} per face = 3 S^{2-}

Total = **4 S^{2-}**

13.94 From figure 13.6, we can see that the length of the diagonal of the cell = 4r, where r = radius of the atom. According to the Pythagorean theorem,

$$a^2 + b^2 = c^2$$

for a right triangle. Since a = b here, we may re-write this as

$$2l^2 = c^2,$$

where l = length of the edge of the unit cell. As mentioned above, the diagonal of the unit cell = 4r, so we may say that

$$2l^2 = (4r)^2$$
$$l^2 = (4r)^2/2$$
$$l^2 = 16r^2/2$$
$$l^2 = 8r^2$$
$$l = \sqrt{(8r^2)}$$

Finally, substituting the value provided for r in the problem, $l = \sqrt{[8(1.24\text{Å})^2]}$ = **3.51 Å**. Using the conversion factor 1pm = 100 Å, this is **351 pm.**

13.96 Each edge is composed of 2 × radius of the cation plus 2 × radius of the anion. The edge is therefore 2 × 133 + 2 × 195 = 656 pm.

13.98 Using the Bragg equation (eqn. 13.1), $n\lambda = 2d\sin\theta$:

a)

$$n(229\text{pm}) = 2(1{,}000)\sin\theta$$
$$0.1145n = \sin\theta$$
$$\boldsymbol{\theta = 6.57°}$$

b)

$$n(229\text{pm}) = 2(250)\sin\theta$$
$$0.458n = \sin\theta$$
$$\boldsymbol{\theta = 27.3°}$$

13.100 From figure 13.6, we can see that the length of the diagonal of the cell = 4r, where r = radius of the atom. According to the Pythagorean theorem,

$$a^2 + b^2 = c^2$$

for a right triangle. Since a = b here, we may re-write this as

$$2l^2 = c^2,$$

where l = length of the edge of the unit cell. As mentioned above, the diagonal of the unit cell = 4r, so we may say that

$$2l^2 = (4r)^2$$
$$l^2 = (4r)^2/2$$
$$l^2 = 16r^2/2$$
$$l^2 = 8r^2$$
$$l = \sqrt{(8r^2)}$$

Finally, substituting the value provided for r in the problem, $l = \sqrt{[8(1.43\text{Å})^2]}$ = **4.04 Å**.

13.102 First, let us convert the given density into units of amu/pm^{-3}:

$$\frac{\text{amu}}{\text{pm}^3} = \left(\frac{3.99\text{ g}}{1\text{ cm}^3}\right)\left(\frac{1\text{ amu}}{1.66\times10^{-24}\text{ g}}\right)\left(\frac{1\times10^{-10}\text{ cm}}{1\text{ pm}}\right)^3 = 2.40\times10^{-6}\text{ amu/pm}^3$$

Thus, the density of CsCl is $\underline{2.40 \times 10^{-6}\text{ amu/pm}^3}$.

If CsCl were *body-centered cubic* (see Figure 13.7)…
Its mass would be:

(2 Cs ions x 132.91) + (2 chloride ions x 35.45) = 336.72 amu

and its volume would be:

$(412.3\text{ pm})^3 = 7.009 \times 10^7\text{ pm}^3$

Therefore, its density would be:

$d = m/V = \underline{4.804 \times 10^{-6}\text{ amu/pm}^3}$

If CsCl were *face-centered cubic* (see Figure 13.10)…
Its mass would be:

(4 Cs ions x 132.91) + (4 chloride ions x 35.45) = 673.44 amu

and its volume would be:

$(412.3\text{ pm})^3 = 7.009 \times 10^7\text{ pm}^3$

Therefore, its density would be:

$$d = m/V = \underline{9.609 \times 10^{-6}\ amu/pm^3}$$

Thus, CsCl is neither face-centered cubic, nor body-centered cubic.

13.104 According to the Pythagorean theorem,

$$a^2 + b^2 = c^2$$

for a right triangle. First, we need to find the length of a diagonal on a face of the unit cell. Since a = b here, we may re-write this as

$$2l^2 = c^2,$$

where l = length of the edge of the unit cell and c = the diagonal length. Using the given 412.3 pm as the length of the edge, c = 583.1 pm. The diagonal length inside the cell from corner to opposite corner may now be found by the same theorem:

$$a^2 + b^2 = c^2$$
$$(412.3)^2 + (583.1)^2 = c^2$$
$$c = 714.1\ pm$$

This diagonal length inside the cell from corner to opposite corner is due to 1 Cs^+ ion and 1 Cl^- ion (see Figure 13.9). Therefore:

$$2r_{Cs+} + 2r_{Cl-} = 714.1pm$$
$$2r_{Cs+} + 2(181pm) = 714.1pm$$
$$2r_{Cs+} = 352\ pm$$
$$r_{Cs+} = \mathbf{176\ pm}$$

13.106 This must be a molecular solid, because if it were ionic it would be high–melting, and the molten material would conduct.

13.108 This is a metallic solid.

13.110 This is a metallic solid.

13.112 a) molecular d) metallic g) ionic
b) ionic e) covalent
c) ionic f) molecular

13.114

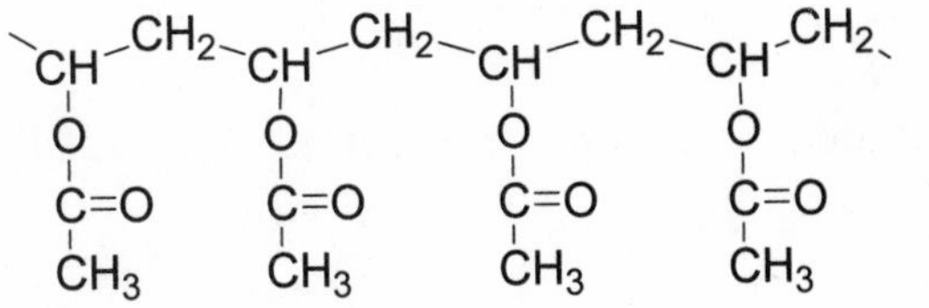

```
 \  CH2   CH2   CH2   CH2
 CH    CH    CH    CH    \
 |     |     |     |
 O     O     O     O
 |     |     |     |
 C=O   C=O   C=O   C=O
 |     |     |     |
 CH3   CH3   CH3   CH3
```

13.116

```
       O              O
       ||             ||
  —O—C—(benzene ring)—C—O—CH2—(cyclohexane ring)—CH2—
```

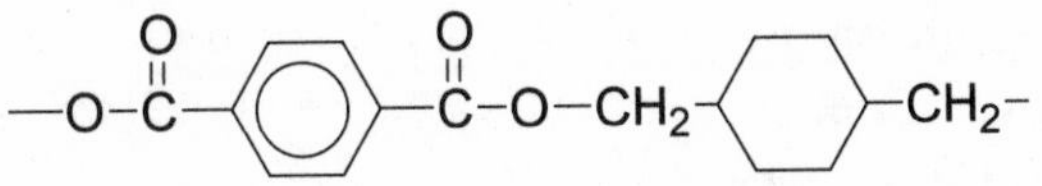

Additional Exercises

13.118

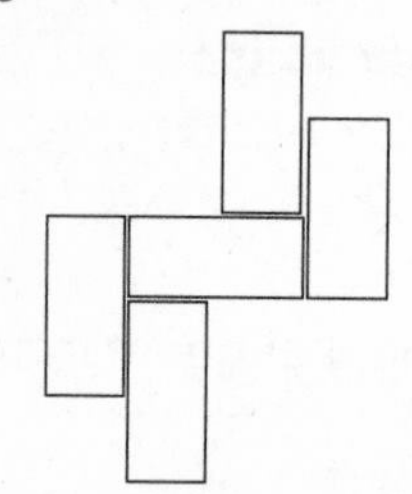

13.120 From figure 13.6, we can see that the length of the diagonal of the cell = 4r, where r = radius of the atom. According to the Pythagorean theorem,

$$a^2 + b^2 = c^2$$

for a right triangle. Since a = b here, we may re-write this as

$$2l^2 = c^2,$$

where l = length of the edge of the unit cell. As mentioned above, the diagonal of the unit cell = 4r, so we may say that

$$2l^2 = (4r)^2$$
$$2l^2 = 16r^2$$
$$(0.125)l^2 = r^2$$
$$\sqrt{[(0.125)l^2]} = r$$

Finally, substituting the value provided for l in the problem, r = $\sqrt{[(0.125)(407.86)^2]}$ = **144.20 pm.**

*13.122

From Figure 13.6, we can see that the length of the diagonal of the cell = 4r, where r = radius of the particle. In our case, we will consider r as the radius of the bromide ion. According to the Pythagorean theorem,

$$a^2 + b^2 = c^2$$

for a right triangle. Since a = b here, we may re-write this as

$$2l^2 = c^2,$$

where l = length of the edge of the unit cell. As mentioned above, the diagonal of the unit cell = 4r, so we may say that

$$2l^2 = (4r)^2$$
$$2l^2 = 16r^2$$
$$(0.125)l^2 = r^2$$
$$\sqrt{[(0.125)l^2]} = r$$

Finally, substituting the value provided for l in the problem, r = $\sqrt{[(0.125)(550)^2]}$ = **194 pm.**

By inspection of Figure 13.6, we see that the length of the unit cell is equal to 2r + d, where r = radius of bromide ion, and d = the space in-between, which corresponds to the diameter of the lithium ion:

$$550 = 2r + d$$
$$550 = 2(194 \text{ pm}) + d$$
$$d = 162 \text{ pm}$$

Therefore the radius of the lithium ion is half of this, or **81 pm**.

In this case, the value we have calculated is larger than that given in the table because the bromide ions are so large that they leave a bigger space to fill than the actual diameter of the lithium ion. (81 pm is simply the radius of that space.)

*13.124

Using the Bragg equation (eqn. 13.1), $n\lambda = 2d\sin\theta$:

$$154 = 2d\sin(6.97°)$$
$$77 = d(0.124)$$
$$d = 635 \text{ pm}$$

The mass of the unit cell would be:

(4 potassium ions × 39.10) + (4 chloride ions × 35.45) = 298.20 amu

and its volume would be:

$(635 \text{ pm})^3 = 2.56 \times 10^8 \text{ pm}^3$

Therefore, its density would be:

$d = m/V = \underline{1.16 \times 10^{-6} \text{ amu/pm}^3}$

Chapter 14

Practice Exercises

14.1 Since these solutions are saturated, the maximum amount of each gas is dissolved in the solution. Hence, 0.00430 g of O_2 and 0.00190 g of N_2 are dissolved in the water.

14.2 The total mass of the solution is to be 250 g. If the solution is to be 1.00 % (w/w) NaOH, then the mass of NaOH will be: 250 g × 1.00 g NaOH/100 g solution = 2.50 g NaOH. We therefore need 2.50 g of NaOH and (250 – 2.50) = 248 g H_2O. The volume of water that is needed is: 248 g ÷ 0.988 g/mL = 251 mL H_2O.

14.3 An HCl solution that is 37 % (w/w) has 37 grams of HCl for every 1.0 x 10^2 grams of solution.

$$\#\,\text{g solution} = (7.5\text{ g HCl})\left(\frac{1.0\,\text{X}\,10^2\text{ g solution}}{37\text{ g HCl}}\right) = 2.0\times10^1\text{ g solution}$$

14.4
$$\#\,\text{g CH}_3\text{OH} = (2000\text{ g H}_2\text{O})\left(\frac{0.250\text{ mol CH}_3\text{OH}}{1000\text{ g H}_2\text{O}}\right)\left(\frac{32.0\text{ g CH}_3\text{OH}}{1\text{ mol CH}_3\text{OH}}\right)$$
$$= 16.0\text{ g CH}_3\text{OH} \Rightarrow 20\text{ g CH}_3\text{OH (rounded to 1 sig. fig.)}$$

14.5 We need to know the number of moles of NaOH and the number of kg of water.
4.00 g NaOH ÷ 40.0 g/mol = 0.100 mol NaOH
250 g H_2O × 1 kg/1000 g = 0.250 kg H_2O

The molality is thus given by:
m = 0.100 mol/0.25 kg = 0.40 mol NaOH/kg H_2O = 0.40 m

14.6 If a solution is 37.0 % (w/w) HCl, then 37.0 % of the mass of any sample of such a solution is HCl and (100.0 – 37.0) = 63.0 % of the mass is water. In order to determine the molality of the solution, we can conveniently choose 100.0 g of the solution as a starting point. Then 37.0 g of this solution are HCl and 63.0 g are H_2O. For molality, we need to know the number of moles of HCl and the mass in kg of the solvent:

37.0 g HCl ÷ 36.46 g/mol = 1.01 mol HCl
63.0 g H_2O × 1 kg/1000 g = 0.0630 kg H_2O

molality = mol HCl/kg H_2O = 1.01 mol/0.0630 kg = 16.1 m

14.7 We first determine the mass of one L (1000 mL) of this solution, using the density:

1000 mL × 1.38 g/mL = 1.38 x 10^3 g

Next, we use the fact that 40.0 % of this total mass is due to HBr, and calculate the mass of HBr in the 1000 mL of solution:

$0.400 \times 1.38 \times 10^3 = 552$ g HBr

This is converted to the number of moles of HBr in 552 g:

552 g HBr ÷ 80.91 g/mol = 6.82 mol HBr

Last, the molarity is the number of moles of HBr per liter of solution

6.82 mol/1 L = 6.82 M

14.8 First determine the number of moles of each component of the solution:

For $C_{16}H_{22}O_4$, 20.0 g/278 g/mol = 0.0719 mol

For C_8H_{18}, 50.0 g/114 g/mol = 0.439 mol

The mole fraction of solvent is:

0.439 mol/(0.439 mol + 0.0719 mol) = 0.859

Using Raoult's Law, we next find the vapor pressure to expect for the solution, which arises only from the solvent (since the solute is known to be nonvolatile):

$P_{solvent} = X_{solvent} \times P°_{solvent} = 0.859 \times 10.5 \text{ torr} = 9.02 \text{ torr}$

14.9 $P_{cyclohexane} = X_{cyclohexane} \times P°_{cyclohexane} = 0.500 \times 66.9 \text{ torr} = 33.5 \text{ torr}$

$P_{toluene} = X_{toluene} \times P°_{toluene} = 0.500 \times 21.1 \text{ torr} = 10.6 \text{ torr}$

$P_{total} = P_{cyclohexane} + P_{toluene} = 33.5 \text{ torr} + 10.6 \text{ torr} = 44.1 \text{ torr}$

14.10 A 10 % solution contains 10 g sugar and 90 g water.

10 g $C_{12}H_{22}O_{11}$ ÷ 342 g/mol = 0.029 mol $C_{12}H_{22}O_{11}$

90 g $H_2O \times$ 1 kg/1000 g = 0.090 kg H_2O

m = 0.029 mol/0.090 kg = 0.32 mol/kg

$\Delta T_b = K_b \times m = 0.51\ °C\ m^{-1} \times 0.32\ m = 0.16\ °C$

$T_b = 100.16\ °C$

14.11 It is first necessary to obtain the values of the freezing point of pure benzene and the value of K_f for benzene from Table 12.4 of the text. We proceed to determine the number of moles of solute that are present and that have caused this depression in the freezing point: $\Delta T = K_f m$

$\therefore m = \Delta T/K_f = (5.45\ °C - 4.13\ °C)/(5.07\ °C\ \text{kg mol}^{-1}) = 0.260\ m$

Next, use this molality to determine the number of moles of solute that must be present:

0.260 mol solute/kg solvent × 0.0850 kg solvent = 0.0221 mol solute

Last, determine the formula mass of the solute:

3.46 g/0.0221 mol = 157 g/mol

14.12 We can use the equation $\Pi = MRT$:

$\Pi = (0.0115 \text{ mol/L})(0.0821 \text{ L atm/K mol})(310 \text{ K})$

$\Pi = 0.293 \text{ atm}$

Converting to torr, we have:

$$\# \text{ torr} = (0.293 \text{ atm})\left(\frac{760 \text{ torr}}{1 \text{ atm}}\right) = 223 \text{ torr}$$

14.13 We can use the equation $\Pi = MRT$, remembering to convert pressure to atm:

$$\#\text{atm} = (25.0\text{ torr})\left(\frac{1\text{ atm}}{760\text{ torr}}\right) = 0.0329\text{ atm}$$

$\Pi = 0.0329$ atm = M × (0.0821 L atm/K mol)(298 K)
M = 1.34 x 10^{-3} mol L^{-1}
# mol = 1.34 x 10^{-3} mol L^{-1} × 0.100 L = 1.34 x 10^{-4} mol

$$\text{formula mass} = \frac{72.4 \text{ X } 10^{-3}\text{ g}}{1.34 \text{ X } 10^{-4}\text{ mol}} = 5.38 \text{ X } 10^{2}\text{ g mol}^{-1}$$

14.14 For the solution as if the solute were 100 % dissociated:

$\Delta T = (1.86\ °C\ m^{-1})(2 \times 0.237\ m) = 0.882\ °C$ and the freezing point should be –0.882 °C.

For the solution as if the solute were 0 % dissociated:

$\Delta T = (1.86\ °C\ m^{-1})(1 \times 0.237\ m) = 0.441\ °C$ and the freezing point should be –0.441 °C.

Review Problems:

14.63 This is to be very much like that shown in Figures 14.8 and 14.9:

(a) $KCl(s) \rightarrow K^{+}(g) + Cl^{-}(g)$, $\Delta H = +690$ kJ mol^{-1}
(b) $K^{+}(g) + Cl^{-}(g) \rightarrow K^{+}(aq) + Cl^{-}(aq)$, $\Delta H = -686$ kJ mol^{-1}

$KCl(s) \rightarrow K^{+}(aq) + Cl^{-}(aq)$, $\Delta H = +4$ kJ mol^{-1}

14.65 lattice energy + hydration energy = enthalpy of solution

Note: It is sometimes conventional to list lattice energies as negative values, however it always *requires* energy to separate oppositely-charged ions, therefore the lattice energy should be a positive value in the equation below. (Energy is put in to the system.)

(+630 kJ/mol) + hydration energy = +14 kJ/mol
hydration energy = – 616 kJ/mol

14.67 Henry's Law says that the solubility of a gas in a liquid is proportional to the pressure above the liquid. Since the pressure goes up by 1.5, the solubility does as well:

Solubility = 0.025 g/L (1.5 atm/1.0 atm) = 0.038 g/L

14.69 Henry's Law says that the solubility of a gas in a liquid is proportional to the pressure above the liquid. Since the pressure doubles, the solubility does as well:

Solubility = 0.010 g/L(2) = 0.020 g/L

14.71 One liter of solution has a mass of:

$$\#\text{g solution} = 1\,\text{L solution}\left(\frac{1{,}000\text{ mL solution}}{1\text{ L solution}}\right)\left(\frac{1.07\text{ g solution}}{1\text{ mL solution}}\right) = 1{,}070\text{ g}$$

According to the given molarity, it contains 3.000 mol NaCl. This has a mass of:

$$\#\text{g NaCl} = 3.000\text{ mol NaCl}\left(\frac{58.45\text{ g NaCl}}{1\text{ mol NaCl}}\right) = 175.4\text{ g NaCl}$$

Thus, the mass of water in 1 L solution must be:
1,070 g − 175.4 g = 895 g water

$$m = \frac{\#\text{mol solute}}{\text{kg solvent}} = \left(\frac{3.000\text{ mol NaCl}}{0.895\text{ kg solvent}}\right) = 3.35\,m$$

14.73 24.0 g glucose ÷ 180 g/mol = 0.133 mol glucose
molality = 0.133 mol glucose/1.00 kg solvent = 0.133 m

mole fraction = moles glucose/total moles
moles glucose = 0.133

$$\text{moles } H_2O = (1.00 \times 10^3\text{ g } H_2O)\left(\frac{1\text{ mole } H_2O}{18\text{ g } H_2O}\right) = 55.5\text{ moles } H_2O$$

$$\text{mole fraction glucose} = \frac{1.33}{55.5 + 0.133} = 2.39 \times 10^{-3}$$

mass % = (mass glucose/total mass)100% = (24.0 g/1024 g)100% = 2.34%

14.75 Start by assuming a 1 kg water is present (although one could use any amount):

$$\frac{\#\text{g ethanol}}{\text{g solution}} = \left(\frac{1.25\text{ mol ethanol}}{1\text{ kg water}}\right)\left(\frac{46.08\text{ g ethanol}}{1\text{ mol ethanol}}\right) = 57.6\text{ g ethanol}$$

Mass % ethanol = (mass ethanol/(total solution mass)x100%
Mass % ethanol = (57.6 g ethanol/(1,000 g water + 57.5 g ethanol)x100%
= 5.45 %

14.77 If we assume 100g of solution we have 5 g NH_3 and 95 g H_2O.

$$\#\text{moles } NH_3 = (5.00\text{ g } NH_3)\left(\frac{1\text{ mole } NH_3}{17.03\text{ g } NH_3}\right) = 0.294\text{ moles } NH_3$$

$$\#\text{moles } H_2O = (95.0\text{ g})\left(\frac{1\text{ mol}}{18.02\text{ g}}\right) = 5.27\text{ moles water}$$

mol percent = 0.294 mol/(5.27 moles + 0.294 moles)x100% = 5.28%

$$\# \text{kg } H_2O = (95.0 \text{ g})\left(\frac{1 \text{ kg}}{1{,}000 \text{ g}}\right) = 0.0950 \text{ kg water}$$

$m = 0.294$ mol/0.095 kg = 3.09 m

14.79 If we choose, for convenience, an amount of solution that contains 1 kg of solvent, then it also contains 0.363 moles of $NaNO_3$. The number of moles of solvent is:

1.00×10^3 g ÷ 18.0 g/mol = 55.6 mol H_2O

Now, convert the number of moles to a number of grams: for $NaNO_3$, 0.363 mol × 85.0 g/mol = 30.9 g; for H_2O, 1000 g was assumed and the percent (w/w) values are:

% $NaNO_3$ = 30.9 g/1031 g × 100 = 3.00 %
% H_2O = 1000 g/1031 g × 100 = 97.0 %

Now determine the mass of 1.00 L of this solution, using the known density: 1000 mL × 1.0185 g/mL = 1018.5 g. Next, we use the fact that 3.00 % of any sample of this solution is $NaNO_3$, and calculate the mass of $NaNO_3$ that is contained in 1000 mL of the solution: 0.0300 × 1018.5 g = 30.6 g $NaNO_3$. The number of moles of $NaNO_3$ is given by: 30.6 g ÷ 85.0 g/mol = 0.359 mol $NaNO_3$. The molarity is the number of moles of $NaNO_3$ per liter of solution: 0.359 mol/1.00 L = 0.359 M. Once again assume1 kg of solvent so we have 0.363 moles $NaNO_3$.

$$\# \text{mol } H_2O = (1000 \text{ g})\left(\frac{1 \text{ mol}}{18.02 \text{ g}}\right) = 55.6 \text{ moles } H_2O$$

$$X_{NaNO_3} = \frac{0.363}{55.6 + 0.363} = 6.49 \times 10^{-3}$$

14.81 $P_{solution} = P°_{solvent} \times X_{solvent}$ We need to determine $X_{solvent}$:

$$\# \text{mol glucose} = (65.0 \text{ g})\left(\frac{1 \text{ mol}}{180.2 \text{ g}}\right) = 0.361 \text{ moles}$$

$$\# \text{mol } H_2O = (150 \text{ g } H_2O)\left(\frac{1 \text{ mol } H_2O}{18.02 \text{ g } H_2O}\right) = 8.32 \text{ mol } H_2O$$

The total number of moles is thus: 8.32 mol + 0.361 mol = 8.69 mol and the mole fraction of the solvent is: $X_{solvent} = \left(\frac{8.32 \text{ mol solvent}}{8.69 \text{ mol solution}}\right)$ = 0.957. Therefore,

$P_{solution}$ = 23.8 torr × 0.957 = 22.8 torr.

14.83 $P_{benzene} = X_{benzene} \times P°_{benzene}$
$P_{toluene} = X_{toluene} \times P°_{toluene}$
$P_{Tot} = P_{benzene} + P_{Toluene}$

$$\# \text{ mol benzene} = (60.0 \text{ g})\left(\frac{1 \text{ mol}}{78.11 \text{ g}}\right) = 0.768 \text{ mol benzene}$$

$$\# \text{ mol toluene} = (40.0 \text{ g})\left(\frac{1 \text{ mol}}{92.14 \text{ g}}\right) 0.434 \text{ mol toluene}$$

$$X_{\text{benzene}} = \frac{0.768}{0.768 + 0.434} = 0.639$$

$$X_{\text{toluene}} = \frac{0.434}{0.768 + 0.434} = 0.361$$

$$P_{\text{benzene}} = (0.639)(93.4 \text{ torr}) = 59.7 \text{ torr}$$

$$P_{\text{toluene}} = (0361)(26.9 \text{ torr}) = 9.71 \text{ torr}$$

$$P_{\text{Tot}} = 59.7 \text{ torr} + 9.71 \text{ torr} = 69.4 \text{ torr}$$

*14.85 The following relationships are to be established:

$$P_{\text{Total}} = 96 \text{ torr} = (P°_{\text{benzene}})(X_{\text{benzene}}) + (P°_{\text{Toluene}})(X_{\text{Toluene}})$$

The relationship between the two mole fractions is:

$$X_{\text{benzene}} = 1 - X_{\text{Toluene}},$$

since the sum of the two mole fractions is one. Substituting this expression for X_{benzene} into the first equation gives:

$$96 = P°_{\text{benzene}}(1 - X_{\text{Toluene}}) + (P°_{\text{Toluene}})(X_{\text{Toluene}})$$

$$96 = 180(1 - X_{\text{Toluene}}) + 60(X_{\text{Toluene}})$$

Solving for X_{Toluene} we get:

$$120(X_{\text{Toluene}}) = 84$$

$$X_{\text{Toluene}} = 0.70$$

$$X_{\text{benzene}} = 0.30$$

Therefore, the mole % values are to be 70 % toluene and 30 % benzene.

14.87 a) $X_{\text{solvent}} = P/P° = 511 \text{ torr}/526 \text{ torr} = 0.971$

$X_{\text{solute}} = 1 - X_{\text{solvent}} = 0.029$

b) We know 0.971 = 1 mol/1 mol + x moles, $x = 2.99 \times 10^{-2}$ moles

c) Molecular mass = 8.3 g/2.99 x 10^{-2} moles = 278 g/mol

14.89 $\Delta T = K_b m = (0.51 \text{ °C kg mol}^{-1})(2.0 \text{ mol kg}^{-1}) = 1.0 \text{ °C}$

$\therefore T_b = 100.0 + 1.0 = 101 \text{ °C}$

$\Delta T = K_f m = (1.86 \text{ °C kg mol}^{-1})(2.00 \text{ mol kg}^{-1}) = 3.72 \text{ °C}$

$\therefore T_f = 0.0 - 3.72 = -3.72 \text{ °C}$

14.91 $\Delta T_f = K_f m$

$m = \Delta T_f / K_f = 3.00 \text{ °C}/1.86 \text{ °C kg/mol} = 1.61 \text{ mol/kg}$

$$\# \text{kg} = (100 \text{ g})\left(\frac{1 \text{ kg}}{1000 \text{ g}}\right) = 0.1 \text{ kg}$$

$$\# \text{mol} = (1.61 \text{ mol/kg})(0.1 \text{ kg}) = 0.161 \text{ mol}$$

$$\# \text{g} = (0.161 \text{ mol})\left(\frac{342.3 \text{ g}}{1 \text{ mol}}\right) = 60 \text{ g}$$

14.93 $\Delta T = (5.45 - 3.45) = 2.00\ °C$

$$K_f \times m = 2.00\ °C$$

$$(5.07\ °C\ \text{kg mol}^{-1})(m) = 2.00\ °C$$

$$m = 0.394 \text{ mol solute/kg solvent}$$

(0.394 mol/kg benzene)(0.200 kg benzene) = 0.0788 mol solute

Molecular mass = g/mol = 12.00 g/0.0788 mol = 152 g/mol

14.95 $\Delta T_f = K_f m$

$$m = \Delta T_f / K_f$$

$$m = (0.307\ °C)/(5.07\ °C\ \text{kg/mol}) = 0.0606 \text{ mol/kg}$$

$$\# \text{mol} = (0.0606 \text{ mol/kg})(0.5 \text{ kg}) = 0.0303 \text{ mol}$$

$$\text{molar mass} = \frac{3.84 \text{ g}}{0.0303 \text{ mol}} = 127 \text{ g/mol}$$

The empirical formula has a mass of 64.1 g/mol. So the molecular formula is $C_8H_4N_2$

14.97 a) If the equation is correct, the units on both sides of the equation should be g/mol. The units on the right side of this equation are:

$$\frac{(\text{g}) \times (\text{L atm mol}^{-1}\ \text{K}^{-1}) \times (\text{K})}{\text{L} \times \text{atm}} = \text{g/mol}$$

which is correct.

b) $\Pi = MRT = (n/V)RT$, $n = \Pi V/RT$

This means that we can calculate the number of moles of solute in one L of solution, as follows:

$$n = \frac{(0.021 \text{ torr})(1 \text{ atm}/760 \text{ torr})(1.0 \text{ L})}{(0.0821 \text{ L atm mol}^{-1}\ \text{K}^{-1})(298 \text{ K})} = 1.1 \text{ X } 10^{-6} \text{ mol}$$

The molecular mass is the mass in 1 L divided by the number of moles in 1 L:

$$2.0 \text{ g}/1.1 \text{ x } 10^{-6} \text{ mol} = 1.8 \times 10^{6} \text{ g/mol}$$

14.99 The equation for the vapor pressure is:

$$P_{solution} = P°_{H_2O} \times X_{H_2O}$$

where $P°_{H_2O}$ is 17.5 torr. To calculate the vapor pressure we need to find the mole fraction of water first:

$$X_{H_2O} = \text{moles } H_2O/(\text{moles } H_2O + \text{moles NaCl})$$

Calculate the moles of NaCl in 10.0 g

$$\#\text{ mol NaCl} = (10.0\text{ g NaCl})\left(\frac{1\text{ mol NaCl}}{58.44\text{ g NaCl}}\right) = 0.171\text{ moles NaCl}$$

When NaCl dissolves in water, Na^+ and Cl^- are formed. So, for every mole of NaCl that dissolves, two moles of ions are formed. For this solution, the number of moles of ions is 0.342.

The number of moles of solvent (water) is:

$$\#\text{ mol } H_2O = (100\text{ g } H_2O)\left(\frac{1\text{ mol } H_2O}{18.02\text{ g } H_2O}\right) = 5.55\text{ moles } H_2O$$

Calculate the mole fraction as

$$X_{H_2O} = \frac{(\text{moles } H_2O)}{(\text{moles } H_2O + \text{moles NaCl})} = \frac{5.55\text{ mol}}{(5.55\text{ mol} + 0.342\text{ mol})} = 0.942$$

The vapor pressure is then $P_{solution} = P°_{H_2O} \times X_{H_2O} = 17.5\text{ torr} \times 0.942 = 16.5\text{ torr}$

14.101 $\Pi = MRT$

$$M = \frac{(2.0\text{ g NaCl})\left(\frac{1\text{ mol NaCl}}{58.45\text{ g NaCl}}\right)}{0.100\text{ L}} = 0.34\text{ M}$$

For every NaCl there are two ions produced so $M = 0.68$ M

$$\Pi = (0.68\text{ M})(0.0821\text{ Latm/molK})(298\text{ K})\left(\frac{760\text{ torr}}{1\text{ atm}}\right) = 1.3 \times 10^4\text{ torr}$$

14.103 $CaCl_2 \rightarrow Ca^{2+} + 2Cl^-$; van't Hoff factor, $i = 3$

$\Delta T_f = i \times k_f \times m = (3)(1.86\ °C\ m^{-1})(0.20\ m) = 1.1\ °C$

The freezing point is –1.1 °C.

14.105 The freezing point depression that is expected from this solution if HF behaves as a nonelectrolyte is:

$$\Delta T_f = 1.86\ °C/m \times 1.00\ m = 1.86\ °C.$$

The freezing point that is expected upon complete dissociation of HF is:

$$\Delta T_f = K_f \times (2 \times m) = 3.72\ °C.$$

The observed freezing point depression is 1.91 °C, and the apparent molality is:

$$m = \Delta T_f / K_f$$

$$m = (1.91\ °C)/(1.86\ °C/m) = 1.03\ m,$$

or 1.03 mol solute particles per kg of solvent. This represents an excess of 3 solute particles per mol of HF (1.03 m – 1.00 m = 0.03 m)., and we conclude that the percent ionization is 3 %.

14.107 Any electrolyte such as $NiSO_4$, that dissociated to give 2 ions, if fully dissociated should have a van't Hoff factor of 2.

14.109 $\Delta T_f = i \times k_f \times m$

$i = \Delta T_f/(k_f \times m) = 0.415°C/(1.86\ °C{\cdot}m^{-1})(0.118\ m) = 1.89$

Additional Exercises

*14.111

The partial pressure of N_2 in air is:

$$P_{N2} = 1.00atm(0.78\ mol\%) = 0.78\ atm$$

Therefore, according to Henry's Law, the amount of N_2 dissolved per liter of blood at 1.00 atm is:

$$(1\ L)(0.015\ g/L)(0.78/1.00) = 0.012\ g\ N_2$$

$$0.012\ g\ N_2\ (1\ mol\ N_2/28.0\ g\ N_2) = 0.00043\ mol\ N_2$$

The amount of N_2 dissolved per liter of blood at 4.00 atm would be four times that, or: 0.0017 mol N_2

The amount of nitrogen released per liter of blood upon quickly surfacing is the difference between the two, or (0.0017 mol − 0.00043 mol) = 0.0013 mol N_2. The volume of that gas at 1.00 atm and 37 °C would be given by the ideal gas law:

$$PV = nRT$$

$$V = nRT/P$$

$V = (0.0013\ mol\ N_2)(0.0821\ L{\cdot}atm/mol{\cdot}K)[(273+37)K]/1\ atm$

$V = 0.033\ L = 33\ mL\ N_2$ per liter of blood

14.113 Around 70 °C, water molecules begin to acquire enough energy on average so that the gaps created between them can accommodate a molecule of nitrogen.

*14.115

a) The formula masss are $Na_2Cr_2O_7{\cdot}2H_2O$: 298 g/mol, C_3H_8O: 60.1 g/mol, and C_3H_6O: 58.1 g/mol.

$$\#\,g\ Na_2Cr_2O_7 \bullet 2H_2O = (21.4\,g\ C_3H_8O)\left(\frac{1\,mol\ C_3H_8O}{60.1\,g\ C_3H_8O}\right)$$

$$\times\left(\frac{1\,mol\ Na_2Cr_2O_7 \bullet 2H_2O}{3\,mol\ C_3H_8O}\right)\left(\frac{298\,g\ Na_2Cr_2O_7 \bullet 2H_2O}{1\,mol\ Na_2Cr_2O_7 \bullet 2H_2O}\right)$$

$$= 35.4\,g\ Na_2Cr_2O_7 \bullet 2H_2O$$

b) The theoretical yield is:

$$\#\,g\ C_3H_6O = (21.4\,g\ C_3H_8O)\left(\frac{1\,mol\ C_3H_8O}{60.1\,g\ C_3H_8O}\right)\left(\frac{3\,mol\ C_3H_6O}{3\,mol\ C_3H_8O}\right)\left(\frac{58.1\,g\ C_3H_6O}{1\,mol\ C_3H_6O}\right)$$

$$= 20.7\,g\ C_3H_6O$$

The percent yield is therefore: $12.4/20.7 \times 100\ \% = 59.9\ \%$

c) First, we determine the number of grams of C, H, and O that are found in the products, and then the % by mass of C, H, and O that were present in the sample that was analyzed by combustion, i.e. the by–product:

$$\# \text{g C} = (22.368 \text{ X } 10^{-3} \text{ g } CO_2)\left(\frac{12.011 \text{ g C}}{44.010 \text{ g } CO_2}\right) = 6.1046 \text{ X } 10^{-3} \text{ g C}$$

and the % C is: 6.1046×10^{-3} g/8.654×10^{-3} g × 100 % = 70.54 % C

$$\# \text{g H} = (10.655 \text{ X } 10^{-3} \text{ g } H_2O)\left(\frac{2.0159 \text{ g H}}{18.015 \text{ g } H_2O}\right) = 1.1923 \text{ X } 10^{-3} \text{ g H}$$

and the % H is: 1.1923×10^{-3} g H/8.654×10^{-3} g × 100 % = 13.78 % H
For O, the mass is the total mass minus that of C and H in the sample that was analyzed:

8.654×10^{-3} g total – (6.1046×10^{-3} g C + 1.1923×10^{-3} g H)
= 1.357×10^{-3} g O

and the % O is: 1.357×10^{-3} g/8.654×10^{-3} g × 100 % = 15.68 % O.
Alternatively, we could have determined the amount of oxygen by using the mass % values, realizing that the sum of the mass percent values should be 100.

Next, we convert these mass amounts for C, H, and O into mole amounts by dividing the amount of each element by the atomic mass of each element:
For C, 6.1046×10^{-3} g C ÷ 12.011 g/mol = 0.50825×10^{-3} mol C
For H, 1.1923×10^{-3} g H ÷ 1.0079 g/mol = 1.1829×10^{-3} mol H
For O, 1.357×10^{-3} g O ÷ 16.00 g/mol = 0.08481×10^{-3} mol O
Lastly, these are converted to relative mole amounts by dividing each of the above mole amounts by the smallest of the three (We can ignore the 10^{-3} term since it is common to all three components):
For C, 0.50825 mol/0.08481 mol = 5.993
For H, 1.1829 mol/0.08481 mol = 13.95
For O, 0.08481 mol/0.08481 mol = 1.000
and the empirical formula is given by this ratio of relative mole amounts, namely $C_6H_{14}O$.

d) $\Delta T_f = K_f m$, (5.45 °C – 4.87 °C) = (5.07 °C/*m*) × *m*, *m* = 0.11 *m*, and there are 0.11 moles of solute dissolved in each kg of solvent. Thus, the number of moles of solute that have been used here is:

0.11 mol/kg × 0.1150 kg = 1.31×10^{-2} mol solute.

The formula mass is thus: 1.338 g/0.0131 mol = 102 g/mol. Since the empirical formula has this same mass, we conclude that the molecular formula is the same as the empirical formula, i.e. $C_6H_{14}O$.

*14.117

a) Since –40 °F is also equal to –40 °C, the following expression applies: $\Delta T = K_f m$, so 40 °C = (1.86 °C kg mol^{-1}) × *m*, *m* = 40/1.86 mol/kg = 22 *m*
Therefore, 22 moles must be added to 1 kg of water.

b)

$$\#\,\text{mL} = (22\ \text{moles})\left(\frac{62.1\ \text{g}}{1\ \text{mol}}\right)\left(\frac{1.00\ \text{mL}}{1.11\ \text{g}}\right) = 1.2\times10^3\ \text{mL}$$

c) There are 946 mL in one quart. Thus, for 1 qt of water we are to have 946 mL, and the required number of quarts of ethylene glycol is:

$$\frac{\#\,\text{qt}\ C_2H_6O_2}{1\ \text{qt}\ H_2O} = \left(\frac{1.2\times10^3\ \text{mL}\ C_2H_6O_2}{1000\ \text{g}\ H_2O}\right)\left(\frac{1\ \text{g}\ H_2O}{1\ \text{mL}\ H_2O}\right)$$

$$\times\left(\frac{946\ \text{mL}\ H_2O}{1\ \text{qt}\ H_2O}\right)\left(\frac{1\ \text{qt}\ C_2H_6O_2}{946\ \text{mL}\ C_2H_6O_2}\right)$$

$$= 1.2\ \text{qt}\ C_2H_6O_2$$

The proper ratio of ethylene glycol to water is 1.2 qt to 1 qt.

14.119 a) Since the molarity of the solution is 4.613 mol/L, then one L of this solution contains:

$$4.613\ \text{mol} \times 46.07\ \text{g/mol} = 212.5\ \text{g}\ C_2H_5OH$$

The mass of the total 1 L of solution is:

$$1000\ \text{mL} \times 0.9677\ \text{g/mL} = 967.7\ \text{g.}$$

The mass of water is thus 967.7 g – 212.5 g = 755.2 g H_2O, and the molality is:

$$4.613\ \text{mol}\ C_2H_5OH/0.7552\ \text{kg}\ H_2O = 6.108\ m$$

b) % (w/w) C_2H_5OH = 212.5 g/967.7 g × 100 % = 21.96 %

14.121

a) Assume 100 mol of solution, so there are 0.9159 mol solute and 100–0.9159 = 99.0841 mol of water. Converting this we get 1.785 kg water. So the molality is 0.5131 *m*.

b) $$\%(\text{w/w})\ KNO_3 = \frac{\text{mass}\ KNO_3}{(\text{mass}\ KNO_3 + \text{mass}\ H_2O)}\times100\%$$

$$\#\,\text{g}\ KNO_3 = (0.9159\ \text{mol}\ KNO_3)\left(\frac{101.1\ \text{g}\ KNO_3}{1\ \text{mol}\ KNO_3}\right) = 92.6\ \text{g}\ KNO_3$$

$$\%(\text{w/w})\ KNO_3 = \frac{92.6\ \text{g}\ KNO_3}{(92.6\ \text{g} + 1785\ \text{g})}\times100\% = 4.93\%$$

c) Add the mass of water and the mass of KNO_3 then divide by the density to determine a volume: 0.5117 M

Chapter 15

Practice Exercises

15.1 From the coefficients in the balanced equation we see that, for every two moles of SO_2 that is produced, 2 moles of H_2S are consumed, three moles of O_2 are consumed, and two moles of H_2O are produced.

Rate of disappearance of $O_2 = \left(\dfrac{3\text{ mol }O_2}{2\text{ mol}SO_2}\right)\left(\dfrac{0.30\text{ mol}}{\text{L s}}\right) = 0.45\text{ mol L}^{-1}\text{ s}^{-1}$

Rate of disappearance of $H_2S = \left(\dfrac{2\text{ mol }H_2S}{2\text{ mol }SO_2}\right)\left(\dfrac{0.30\text{ mol}}{\text{L s}}\right) = 0.30\text{ mol L}^{-1}\text{ s}^{-1}$

15.2 The rate of the reaction after 250 seconds have elapsed is equal to the slope of the tangent to the curve at 250 seconds. First draw the tangent, and then estimate its slope as follows, where A is taken to represent one point on the tangent, and B is taken to represent another point on the tangent:

$$\text{rate} = \left(\frac{\text{A (mol/L)} - \text{B (mol/L)}}{\text{A (s)} - \text{B (s)}}\right) = \frac{\text{change in concentration}}{\text{change in time}}$$

A value near $1 \times 10^{-4}\text{ mol L}^{-1}\text{ s}^{-1}$ is correct.

15.3 a) First use the given data in the rate law:

Rate = $k[\text{HI}]^2$
$2.5 \times 10^{-4}\text{ mol L}^{-1}\text{ s}^{-1} = k[5.58 \times 10^{-2}\text{ mol/L}]^2$
$k = 8.0 \times 10^{-2}\text{ L mol}^{-1}\text{ s}^{-1}$

b) $\text{L mol}^{-1}\text{ s}^{-1}$

15.4 The order of the reaction with respect to a given substance is the exponent to which that substance is raised in the rate law:
order of the reaction with respect to $[BrO_3^-] = 1$
order of the reaction with respect to $[SO_3^{2-}] = 1$
overall order of the reaction = 1 + 1 = 2

15.5 In each case, $k = \text{rate}/[A][B]^2$, and the units of k are $\text{L}^2\text{ mol}^{-2}\text{ s}^{-1}$.

Each calculation is performed as follows, using the second data set as the example:

$$k = \frac{0.40\text{ mol L}^{-1}\text{ s}^{-1}}{(0.20\text{ mol L}^{-1})(0.10\text{ mol L}^{-1})^2} = 2.0\times10^2\text{ L}^2\text{ mol}^{-2}\text{ s}^{-1}$$

Each of the other data sets also gives the same value:
$k = 2.0 \times 10^2 \ L^2 \ mol^{-2} \ s^{-1}$.

15.6 When the concentration of sucrose is doubled, the rate doubles.
When the concentration of sucrose is raised five times, the rate goes up five times.
The concentration and rate are directly proportional, therefore the reaction is first order with respect to sucrose.

15.7 The rate law will likely take the form rate = $k[A]^n[B]^{n'}$, where n and n' are the order of the reaction with respect to A and B, respectively. On comparing the first two lines of data, in which the concentration of B is held constant, we note that increasing the concentration of A by a factor of 2 (from 0.40 to 0.80) causes an increase in the rate by a factor of 4 (from 1.0×10^{-4} to 4.0×10^{-4}). Thus, we have a rate increase by 2^2, caused by a concentration increase by a factor of 2. This corresponds to the case in Table 15.3 for which n = 2, and we conclude that the reaction is second–order with respect to A.

On comparing the second and third lines of data (wherein the concentration of A is held constant), we note that increasing the concentration of B by a factor of 2 (from 0.30 to 0.60) causes an increase in the rate by a factor of 4 (from 4.0×10^{-4} to 16.0×10^{-4}). This is an increase in rate by a factor of 2^2, and it forces us to conclude, using the information of Table 15.3, that the value of n' is 2. Thus the reaction is also second–order with respect to B. The rate law is then written:

Rate = $k[A]^2[B]^2$

15.8 a) We substitute into equation 15.3, first converting the time to seconds:
$t = 2 \text{ hr} \times 3600 \text{ s/hr} = 7200 \text{ s}$

$$\ln \frac{[A]_0}{[A]_t} = kt$$

$$\text{antiln}\left[\ln \frac{[A]_0}{[A]_t}\right] = \text{antiln}\,[kt] = \text{antiln}\,[(6.2 \text{ X } 10^{-5}\ s^{-1})(7200\ s)]$$

$$\frac{[A]_0}{[A]_t} = 1.56 = \frac{0.40 \text{ M}}{[A]_t}$$

$$[A]_t = \frac{0.40 \text{ M}}{1.56} = 0.26 \text{ M}$$

b) Again, we use equation 15.3, this time solving for time:

$$\ln \frac{[A]_0}{[A]_t} = kt$$

$$t = \frac{1}{k} \times \ln \frac{[A]_0}{[A]_t} = \frac{1}{6.2 \times 10^{-5}\ s^{-1}} \times \ln \frac{0.40\ M}{0.30\ M} = 4600\ s$$

4.6×10^3 s × 1 min/60 s = 77 min

15.9 This is a second–order reaction, and we use equation 15.4:

$$\frac{1}{[NOCl]_t} - \frac{1}{[NOCl]_0} = kt$$

$$\frac{1}{[0.010\ M]} - \frac{1}{[0.040\ M]} = (0.020\ L\ mol^{-1}\ s^{-1}) \times t$$

$t = 3.8 \times 10^3$ s
$t = 3.8 \times 10^3$ s × 1 min/60 s = 63 min

15.10 For a first–order reaction:

$$t_{1/2} = \frac{0.693}{k} = \frac{0.693}{6.17\ X\ 10^{-4}\ s^{-1}} = 1.12 \times 10^3\ s$$

$$t_{1/2} = 1.12\ X\ 10^3\ s \times \frac{1\ min}{60\ s}$$
$$= 18.7\ min$$

If we refer to the chart given in the text in example 15.10, we see that two half lives will have passed if there is to be only one quarter of the original amount of material remaining. This corresponds to:

18.7 min per half–life × 2 half lives = 37.4 min

15.11 The reaction is first–order. A second–order reaction should have a half–life that depends on the initial concentration according to equation 15.6.

15.12 a) Use equation 15.9:

$$\ln \frac{k_2}{k_1} = \frac{-E_a}{R} \left[\frac{1}{T_2} - \frac{1}{T_1} \right]$$

$$\ln \left[\frac{23\ L\ mol^{-1}\ s^{-1}}{3.2\ L\ mol^{-1}\ s^{-1}} \right] = \frac{-E_a}{8.314\ J\ mol^{-1}\ K^{-1}} \left[\frac{1}{673\ K} - \frac{1}{623\ K} \right]$$

Solving for E_a gives 1.4×10^5 J/mol = 1.4×10^2 kJ/mol

b) We again use equation 15.9, substituting the values:

$k_1 = 3.2 \text{ L mol}^{-1} \text{ s}^{-1}$ at $T_1 = 623$ K

$k_2 = ?$ at $T_2 = 573$ K

$$\ln\frac{k_2}{k_1} = \frac{-E_a}{R}\left[\frac{1}{T_2} - \frac{1}{T_1}\right]$$

$$\ln\left[\frac{k_2}{3.2 \text{ L mol}^{-1} \text{ s}^{-1}}\right] = \frac{-1.4\times10^5 \text{ J mol}^{-1}}{8.314 \text{ J mol}^{-1} \text{ K}^{-1}}\left[\frac{1}{573 \text{ K}} - \frac{1}{623 \text{ K}}\right]$$

Solving for k_2 gives 0.30 L mol^{-1} s^{-1}.

15.13 If the reaction occurs in a single step, one molecule of each reactant must be involved, according to the balanced equation. Therefore, the rate law is expected to be: Rate = $k[NO][O_3]$.

15.14 The slow step (second step) of the mechanism determines the rate law:

$$\text{Rate} = k[NO_2Cl]^1[Cl]^1$$

However, Cl is an intermediate and cannot be part of the rate law expression. We need to solve for the concentration of Cl by using the first step of the mechanism. Assuming that the first step is an equilibrium, the rates of the forward and reverse reactions are equal:

$$\text{Rate} = k_{forward}[NO_2Cl] = k_{reverse}[Cl][NO_2]$$

Solving for [Cl] we get

$$[Cl] = \frac{k_f}{k_r}\frac{[NO_2Cl]}{[NO_2]}$$

Substituting into the rate law expression for the second step yields:

Rate = $\dfrac{k[NO_2Cl]^2}{[NO_2]}$, where all the constants have been combined into one new constant.

Review Problems

15.64

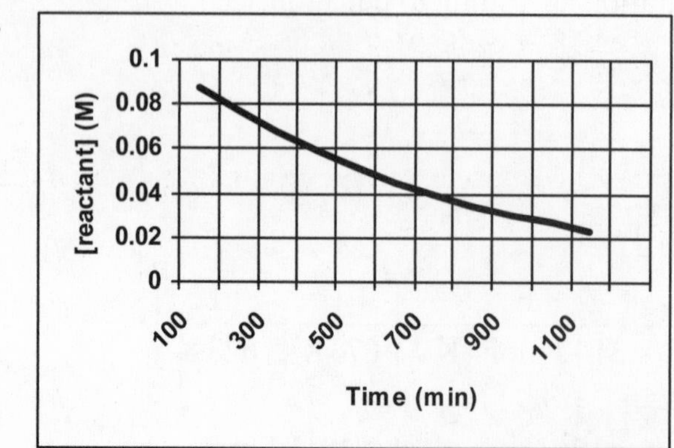

Since they are in a 1-to-1 mol ratio, the rate of formation of SO_2 is *equal* and *opposite* to the rate of consumption of SO_2Cl_2. This is equal to the slope of the curve at any point on the graph (see below). At 200 min, we obtain a value of about 1 x 10^{-4} M/s. At 600 minutes, this has decreased to about 7 x 10^{-5} M/s.

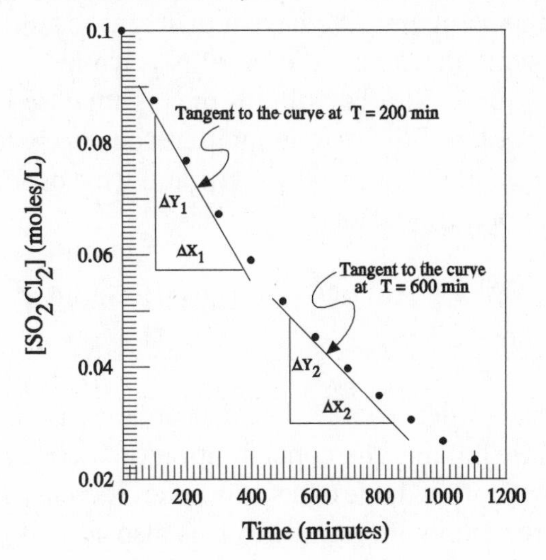

15.66 This is determined by the coefficients of the balanced chemical equation. For every mole of N_2 that reacts, 3 mol of H_2 will react. Thus the rate of disappearance of hydrogen is three times the rate of disappearance of nitrogen. Similarly, the rate of disappearance of N_2 is half the rate of appearance of NH_3, or NH_3 appears twice as fast as N_2 disappears.

15.68 a) rate for $O_2 = -1.20$ mol L^{-1} $s^{-1} \times 19/2 = -11.4$ mol L^{-1} s^{-1}

By convention, this is reported as a positive number: 11.4 mol L^{-1} s^{-1}

b) rate for $CO_2 = +1.20$ mol L^{-1} $s^{-1} \times 12/2 = 7.20$ mol L^{-1} s^{-1}

c) rate for $H_2O = +1.20$ mol L^{-1} $s^{-1} \times 14/2 = 8.40$ mol L^{-1} s^{-1}

15.70 The rate can be found by simply inserting the given concentration values:

Rate = $(5.0 \times 10^5\ L^5\ mol^{-5}\ s^{-1})[H_2SeO_3][I^-]^3[H^+]^2$

Rate = $(5.0 \times 10^5\ L^5\ mol^{-5}\ s^{-1})[2.0 \times 10^{-2}\ M][2.0 \times 10^{-3}\ M]^3[1.0 \times 10^{-3}\ M]^2$

Rate = 8.0×10^{-11} mol L^{-1} s^{-1}

15.72 rate = $(7.1 \times 10^9\ L^2\ mol^{-2}\ s^{-1})(1.0 \times 10^{-3}\ mol/L)^2(3.4 \times 10^{-2}\ mol/L)$

rate = 2.4×10^2 mol L^{-1} s^{-1}

15.74 In each case, the order with respect to a reactant is the exponent to which that reactant's concentration is raised in the rate law.

a) For $HCrO_4^-$, the order is 1.
For HSO_3^-, the order is 2.
For H^+, the order is 1.

b) The overall order is 1 + 2 + 1 = 4.

15.76 On comparing the data of the first and second experiments, we find that, whereas the concentration of N is unchanged, the concentration of M has been doubled, causing a doubling of the rate. This corresponds to the fourth case in Table 15.3, and we conclude that the order of the reaction with respect to M is 1. In the second and third experiments, we have a different result. When the concentration of M is held constant, the concentration of N is tripled, causing an increase in the rate by a factor of nine. This constitutes the eighth case in Table 15.3, and we conclude that the order of the reaction with respect to N is 2. This means that the overall rate expression is: rate = $k[M][N]^2$ and we can solve for the value of k by substituting the appropriate data:

5.0×10^{-3} mol L^{-1} s^{-1} = $k \times$ [0.020 mol/L][0.010 mol/L]^2
$k = 2.5 \times 10^3$ L^2 mol^{-2} s^{-1}

15.78 The reaction is first–order in OCl^-, because an increase in concentration by a factor of two, while holding the concentration of I^- constant (compare the first and second experiments of the table), has caused an increase in rate by a factor of 2^1 = 2. The order of reaction with respect to I^- is also 1, as is demonstrated by a comparison of the first and third experiments: Rate = $k[OCl^-][I^-]$

Using the last data set:
3.5×10^4 mol L^{-1} s^{-1} = k[1.7×10^{-3} mol/L][3.4×10^{-3} mol/L]
$k = 6.1 \times 10^9$ L mol^{-1} s^{-1}

15.80 Compare the first and second experiments. On doubling the ICl concentration, the rate is found to increase by a factor of $2 = 2^1$, and the order of the reaction with respect to ICl is 1 (case number four in Table 15.3). In the first and third experiments, the concentration of ICl is constant, whereas the concentration of H_2 in the first experiment is twice that in the third. This causes a change in the rate by a factor of 2 also, and the rate law is found to be: Rate = $k[ICl][H_2]$. Using the data of the first experiment:
1.5×10^{-3} mol L^{-1} s^{-1} = k[0.10 mol L^{-1}][0.10 mol L^{-1}]
$k = 1.5 \times 10^{-1}$ L mol^{-1} s^{-1}

15.82 A graph of ln $[SO_2Cl_2]_t$ versus t will yield a straight line if the data obeys a first–order rate law.

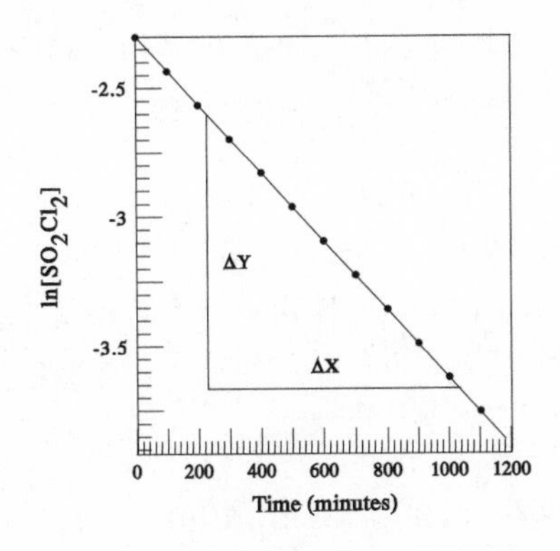

These data do yield a straight line when ln $[SO_2Cl_2]_t$ is plotted against the time, t. The slope of this line equals $-k$. Plotting the data provided and using linear regression to fit the data to a straight line yields a value of 1.32×10^{-3} min^{-1} for k.

15.84 a) The time involved must be converted to a value in seconds:

1 hr × 3600 s/hr = 3.6×10^3 s, and then we make use of equation 13.5, where x is taken to represent the desired SO_2Cl_2 concentration:

$$\ln \frac{0.0040\ M}{x} = (2.2 \times 10^{-5}\ s^{-1})(3.6 \times 10^3\ s)$$

$x = 3.7 \times 10^{-3}$ M

b) The time is converted to a value having the units seconds
24 hr × 3600 s/hr = 8.64×10^4 s, and then we use equation 13.5, where x is taken to represent the desired SO_2Cl_2 concentration:

$$\ln \frac{0.0040\ M}{x} = (2.2 \times 10^{-5}\ s^{-1})(8.64 \times 10^4\ s)$$

$x = 6.0 \times 10^{-4}$ M

15.86 Any consistent set of units for expressing concentration may be used in equation 15.3, where we let A represent the drug that is involved:

$$\ln \frac{[A]_0}{[A]_t} = kt$$

$$\ln \frac{25.0\ ^{mg}/_{kg}}{15.0\ ^{mg}/_{kg}} = k(120\ min)$$

Solving for k we get 4.26×10^{-3} min^{-1}.

15.88 We use the equation:

$$\frac{1}{[\text{HI}]_t} - \frac{1}{[\text{HI}]_0} = kt$$

$$\frac{1}{[8.0 \text{ X } 10^{-4} \text{ M}]} - \frac{1}{[3.4 \text{ X } 10^{-2} \text{ M}]} = (1.6 \text{ x } 10^{-3} \text{ L mol}^{-1} \text{ s}^{-1}) \times t$$

Solving for t gives:
$t = 7.6 \times 10^5$ s or $t = 7.6 \times 10^5 \text{ s} \times 1 \text{ min}/60 \text{ s} = 1.3 \times 10^4$ min

15.90 $\#\text{ half lives} = (2.0 \text{ hrs})\left(\frac{60 \text{ min}}{1 \text{ hr}}\right)\left(\frac{1 \text{ half life}}{15 \text{ min}}\right) = 8.0 \text{ half lives}$

Eight half lives correspond to the following fraction of original material remaining:

Number of half lives	Fraction remaining
1	1/2
2	1/4
3	1/8
4	1/16
5	1/32
6	1/64
7	1/128
8	1/256

15.92 It requires approximately 500 min (as determined from the graph) for the concentration of SO_2Cl_2 to decrease from 0.100 M to 0.050 M, i.e. to decrease to half its initial concentration. Likewise, in another 500 minutes, the concentration decreases by half again, i.e. from 0.050 M to 0.025 M. This means that the half–life of the reaction is independent of the initial concentration, and we conclude that the reaction is first–order in SO_2Cl_2.

15.94 $t = 0.693/k = 0.693/1.6 \times 10^{-3} \text{ s}^{-1} = 4.3 \times 10^2$ seconds

15.96 The graph is prepared exactly as in example 15.12 of the text. The slope is found using linear regression, to be: -9.5×10^3 K. Thus $-9.5 \times 10^3 \text{ K} = -E_a/R$
$E_a = -(-9.5 \times 10^3 \text{ K})(8.314 \text{ J K}^{-1} \text{ mol}^{-1}) = 7.9 \times 10^4 \text{ J/mol} = 79 \text{ kJ/mol}$

Using the equation, we proceed as follows:

$$\ln\frac{k_2}{k_1}=\frac{-E_a}{R}\left[\frac{1}{T_2}-\frac{1}{T_1}\right]$$

$$\ln\left[\frac{1.94\times10^{-3}\ \text{L mol}^{-1}\ \text{s}^{-1}}{2.88\times10^{-4}\ \text{L mol}^{-1}\ \text{s}^{-1}}\right]=\frac{-E_a}{8.314\ \text{J mol}^{-1}\ \text{K}^{-1}}\left[\frac{1}{673\ \text{K}}-\frac{1}{593\ \text{K}}\right]$$

$$1.907=\frac{2.00\times10^{-4}\ \text{K}^{-1}}{8.314\ \text{J mol}^{-1}\ \text{K}^{-1}}\times E_a$$

$E_a = 7.93 \times 10^4$ J/mol = 79.3 kJ/mol

15.98 Using the equation we have:

$$\ln\frac{k_2}{k_1}=\frac{-E_a}{R}\left[\frac{1}{T_2}-\frac{1}{T_1}\right]$$

$$\ln\left[\frac{1.0\times10^{-3}\ \text{L mol}^{-1}\ \text{s}^{-1}}{9.3\times10^{-5}\ \text{L mol}^{-1}\ \text{s}^{-1}}\right]=\frac{-E_a}{8.314\ \text{J mol}^{-1}\ \text{K}^{-1}}\left[\frac{1}{403\ \text{K}}-\frac{1}{373\ \text{K}}\right]$$

$$2.37=\frac{2.00\times10^{-4}\ \text{K}^{-1}}{8.314\ \text{J mol}^{-1}\ \text{K}^{-1}}\times E_a$$

$E_a = 9.89 \times 10^4$ J/mol = 99 kJ/mol

Equation states $k = A\exp\left(\frac{-E_a}{RT}\right)$

$$A=\frac{k}{\exp\left(\frac{-E_a}{RT}\right)}$$

$$=\frac{9.3\times10^{-5}\ \text{L mol}^{-1}\ \text{s}^{-1}}{\exp\left(\frac{-9.89\times10^{4}\ \text{J/mol}}{(8.314\ \text{J/mol K})(373\ \text{K})}\right)}$$

$= 6.6 \times 10^9$ L mol^{-1} s^{-1}

15.100 Substituting into the equation:

$$\ln\frac{k_2}{k_1}=\frac{-E_a}{R}\left[\frac{1}{T_2}-\frac{1}{T_1}\right]$$

$$\ln\left[\frac{3.75\times10^{-2}\ \text{s}^{-1}}{2.1\times10^{-3}\ \text{s}^{-1}}\right]=\frac{-E_a}{8.314\ \text{J mol}^{-1}\ \text{K}^{-1}}\left[\frac{1}{298\ \text{K}}-\frac{1}{273\ \text{K}}\right]$$

$(3.70 \times 10^{-5}$ mol/J$)(E_a) = 2.88$
$E_a = 7.8 \times 10^4$ J/mol = 78 kJ/mol

15.102 We can use the equation:

a)

$$k = A \exp\left(\frac{-E_a}{RT}\right)$$

$$= \left(4.3 \text{ x } 10^{13} \text{ s}^{-1}\right)\exp\left(\frac{-103\times10^3 \text{ J mol}^{-1}}{\left(8.314 \ ^{J}/_{\text{mol K}}\right)(293 \text{ K})}\right)$$

$$= 1.9 \text{ x } 10^{-5} \text{ s}^{-1}$$

b)

$$k = A \exp\left(\frac{-E_a}{RT}\right)$$

$$= \left(4.3\times10^{13} \text{ s}^{-1}\right)\exp\left(\frac{-103\times10^3 \text{ J mol}^{-1}}{\left(8.314 \ ^{J}/_{\text{mol K}}\right)(373 \text{ K})}\right)$$

$$= 1.6\times10^{-1} \text{ s}^{-1}$$

Additional Exercises

15.104 The concentration of CH_4 at any time is equal to the starting concentration of CH_3CHO (0.200 M) minus the remaining concentration of CH_3CHO, because for every mole of CH_3CHO that disappears, one mole of CH_4 appears. The rates of reaction at t = 40 s and at t = 100 s are given by the slopes of tangents to the curve at t = 40 s and t = 100 s respectively.

$\text{rate}_{40} = 1.2 \times 10^{-3} \text{ mol L}^{-1} \text{ s}^{-1}$
$\text{rate}_{100} = 5.1 \times 10^{-4} \text{ mol L}^{-1} \text{ s}^{-1}$

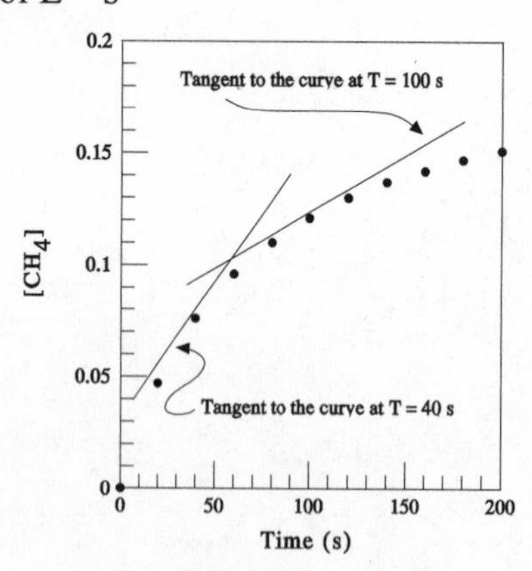

15.106 From Section 15.5, the fraction remaining after n half-lives is $1/2^n$. Therefore, we have:

$$0.810 = 1/2^n$$

Inverting both sides of the equation gives:

$$1.23 = 2^n$$

We can solve this for now using trial and error: We know that $2^1 = 2$, and that $2^0 = 1$. Therefore n must be between 0 and 1. We start by trying n = 0.50:

$$2^{0.50} = 1.41$$

This is too high, so we might try 0.30:

$$2^{0.30} = 1.23$$

This is just what we were looking for, therefore n = 0.30. Since one half life of C-14 is 5730 years, one might estimate the age of the mummy as (0.30)(5730) = about 1,700 years old.

*15.108

a) rate = $k_1[A]^2$

b) rate = $k_{-1}[A_2]^1$

c) rate = $k_2[A_2]^1[E]^1$

d) $2A + E \rightarrow B + C$

e) The rates for the forward and reverse directions of step one are set equal to each other in order to arrive at an expression for the intermediate $[A_2]$ in terms of the reactant [A]: $k_1[A]^2 = k_{-1}[A_2]$

$$[A_2] = \frac{k_1}{k_{-1}}[A]^2$$

This is substituted into the rate law for question (c) above, giving a rate expression that is written using only observable reactants:

$$\text{Rate} = k_2\frac{k_1}{k_{-1}}[A]^2[E]^1$$

15.110 First, determine a value for E_a using the equation:

$$\ln\frac{k_2}{k_1} = \frac{-E_a}{R}\left[\frac{1}{T_2} - \frac{1}{T_1}\right]$$

$$\ln\left[\frac{2.25 \times 10^{-5}\ \text{min}^{-1}}{5.84 \times 10^{-6}\ \text{min}^{-1}}\right] = \frac{-E_a}{8.314\ \text{J mol}^{-1}\ \text{K}^{-1}}\left[\frac{1}{343\ \text{K}} - \frac{1}{333\ \text{K}}\right]$$

$$1.35 = \frac{8.76 \times 10^{-5}\ \text{K}^{-1}}{8.314\ \text{J mol}^{-1}\ \text{K}^{-1}} \times E_a$$

$E_a = 1.28 \times 10^5$ J/mol = 128 kJ/mol

Next, use this value of E_a and the data at 70 °C to calculate a rate constant at 80 °C:

$$\ln\frac{k_2}{k_1} = \frac{-E_a}{R}\left[\frac{1}{T_2} - \frac{1}{T_1}\right]$$

$$\ln\left[\frac{k_2}{2.25 \times 10^{-5}\ \text{min}^{-1}}\right] = \frac{-1.28 \times 10^5\ \text{J mol}^{-1}}{8.314\ \text{J mol}^{-1}\ \text{K}^{-1}}\left[\frac{1}{353\ \text{K}} - \frac{1}{343\ \text{K}}\right]$$

$$\ln\left[\frac{k_2}{2.25 \times 10^{-5}\ \text{min}^{-1}}\right] = 1.27$$

$k_2 = 2.25 \times 10^{-5}\ \text{min}^{-1} \times \exp(1.27) = 8.02 \times 10^{-5}\ \text{min}^{-1}$.

Finally, use the first–order rate expression to determine time:

$$\ln \frac{0.0020\ \text{M}}{0.0012\ \text{M}} = 8.02\ \text{X}\ 10^{-5}\ \text{min}^{-1} \times \text{t}$$

Solving for t we get 6.34×10^3 min.

15.112 Molecular oxygen exists highly in the diradical state. That is, the two electrons in the highest, unoccupied molecular orbitals (HOMOs) exist in separate, equal energy orbitals (see discussion in section 10.7). This means that the "double bond" of O_2 does not have the stability of most other double bonds. Therefore diatomic oxygen is at a higher energy level relative to the transition state energy for its reactions, and the activation energy is smaller.

Molecular nitrogen, on the other hand, contains two π-bonds. Neither of these are diradical in character and lend exceptional stability to the molecule. Therefore diatomic nitrogen is at a lower energy level relative to the transition state energy for its reactions, and the activation energy is greater.

15.114 To solve this problem, plot the data provided as 1/T vs 1/t where T is the absolute temperature and 1/t is proportional to the rate constant.

t (min)	T (K)	1/T	ln(1/t)
10	291	0.003436	-2.302585
9	293	0.003412	-2.197224
8	294	0.003401	-2.079441
7	295	0.003389	-1.945910
6	297	0.003367	-1.791759
13.6	288	0.003472	-2.608

(See graph, next page.)

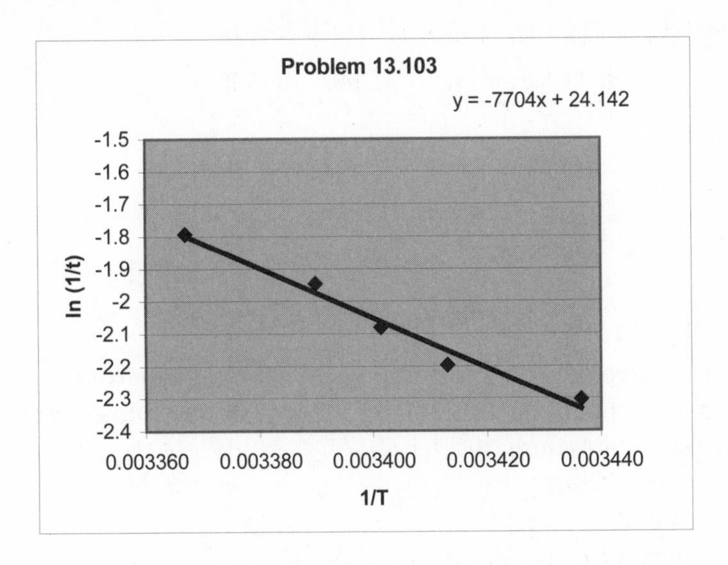

The slope of the graph is equal to $-E_a/R$, therefore:

$$-7{,}704 = -E_a/R$$
$$7{,}704R = E_a$$
$$7{,}704\ \text{K}(8.314\ \text{J/mol·K}) = E_a$$
$$E_a = 64{,}050\ \text{J/mol}$$
$$E_a = 64\ \text{kJ/mol}$$

From the straight–line equation, we can determine the time needed to develop the film at 15 °C is 14 min.

*15.116

Taking note of the inverse relationship between the reaction rate constant, *k*, and the cooking time, t, we set up equation 15.9 in the following manner:

$$\ln\frac{k_2}{k_1} = \frac{-E_a}{R}\left[\frac{1}{T_2} - \frac{1}{T_1}\right]$$

$$\ln\left[\frac{1/t_2}{1/t_1}\right] = \frac{-E_a}{R}\left[\frac{1}{T_2} - \frac{1}{T_1}\right]$$

We are provided with some subtle but key information about the physical conditions. For instance, the 3–minute traditional egg provides a cooking time of 3 minutes at the normal boiling point of water, 100 °C or 373 K. We are also given the atmospheric pressure (355 torr) on Mt. McKinley where the cooking is to be carried out at a lower temperature. At 355 torr, H_2O will boil when its vapor pressure equals 355 torr. The temperature corresponding to this pressure is 80 °C or 323 K (See Appendix). Thus, with the given value of the activation energy, i.e., $E_a = 418$ kJ/mol, we can proceed with the calculation to obtain t_2:

$$\ln\left[\frac{1/t_2}{1/3\,\text{min}}\right] = \frac{-418\times10^3\ \text{J/mol}}{8.314\ \text{J/mol K}}\left[\frac{1}{353\,\text{K}} - \frac{1}{373\,\text{K}}\right]$$

$$\ln\left[\frac{3\ \text{min}}{t_2}\right] = -7.64$$

$$\frac{3\ \text{min}}{t_2} = \exp(-7.64)$$

$$t_2 = 3\ \text{min}/\exp(-7.64) = 6.2\times10^3\ \text{min} = 104\ \text{hrs}$$

Thus, to get the same degree of protein denaturation, it would take roughly 4 days to cook the egg at an atmospheric pressure of 355 torr as opposed to cooking the egg at normal atmospheric pressure.

*15.118

a) The first step, in which a free radical is produced, is the initiation step.

b) Both the second and third steps are propagating steps since HBr, the desired product, and an additional free radical are produced.

c) The final step in which two bromine free radicals recombine to give a bromine molecule is the termination step.

The presence of the additional reaction step serves to decrease the concentration of HBr.

Chapter 16

Practice Exercises

16.1 a) $\frac{[H_2O]^2}{[H_2]^2[O_2]} = K_c$ b) $\frac{[CO_2][H_2O]^2}{[CH_4][O_2]^2} = K_c$

16.2 Since the starting equation has been reversed and divided by two, we must invert the equilibrium constant, and then take the square root: $K_c = 1.2 \times 10^{-13}$

16.3 If we divide both equations by 2 and reverse the second we get:

$CO(g) + 1/2O_2(g) \rightarrow CO_2(g)$ $K_c = 5.7 \times 10^{45}$
$H_2O(g) \rightarrow H_2(g) + 1/2O_2(g)$ $K_c = 3.3 \times 10^{-41}$

Note that when we divide the equation by two, we need to take the square root of the rate constant. When we reverse the reaction, we need to take the inverse.

Adding these equations we get the desired equation so we need simply multiply the values for K_c in order to obtain the new value: $K_c = 1.9 \times 10^5$

16.4 $K_c = \frac{(P_{HI})^2}{(P_{H_2})(P_{I_2})}$

16.5 Reaction (b) will proceed farthest to completion since it has the largest value for K_c.

16.6 We would expect K_P to be smaller than K_c since Δn_g is negative.

Use the equation:

$$K_p = K_c(RT)^{\Delta n_g}$$
$$K_c = \frac{K_p}{(RT)^{\Delta n_g}}$$

In this case, $\Delta n_g = (1 - 3) = -2$, and we have:

$$K_c = \frac{K_p}{(RT)^{\Delta n_g}} = \frac{3.8 \times 10^{-2}}{((0.0821 \frac{\text{L atm}}{\text{mol K}})(473\text{ K}))^{-2}} = 57$$

16.7 Use the equation $K_p = K_c(RT)^{\Delta n_g}$. In this reaction, $\Delta n_g = 3 - 2 = 1$, so

$$K_p = K_c(RT)^{\Delta n_g} = (7.3\times10^{34})\left((0.0821 \tfrac{\text{L atm}}{\text{mol K}})(298\text{ K})\right)^1 = 1.8\times10^{36}$$

16.8 a) $K_c = \dfrac{1}{[Cl_2\ (g)]}$

b) $K_c = \dfrac{1}{[NH_3\ (g)][HCl(g)]}$

c) $K_c = [Na^+\ (aq)][OH^-\ (aq)][H_2\ (g)]$

d) $K_c = [Ag^+]^2[CrO_4^{\ 2-}]$

e) $K_c = \dfrac{[Ca^{+2}(aq)][HCO_3^-(aq)]^2}{[CO_2(aq)]}$

16.9 a) The equilibrium will shift to the right, decreasing the concentration of Cl_2 at equilibrium, and consuming some of the added PCl_3. The value of K_p will be unchanged.

b) The equilibrium will shift to the left, consuming some of the added PCl_5 and increasing the amount of Cl_2 at equilibrium. The value of K_p will be unchanged.

c) For any exothermic equilibrium, an increase in temperature causes the equilibrium to shift to the left, in order to remove energy in response to the stress. This equilibrium is shifted to the left, making more Cl_2 and more PCl_3 at the new equilibrium. The value of K_p is given by the following:

$$K_p = \frac{P_{PCl_5}}{P_{PCl_3} \times P_{Cl_2}}$$

In this system, an increase in temperature (which causes an increase in the equilibrium concentrations of both PCl_3 and Cl_2 and a decrease in the equilibrium concentration of PCl_5) causes an increase in the denominator of the above expression as well as a decrease in the numerator of the above expression. Both of these changes serve to decrease the value of K_p.

d) Decreasing the container volume for a gaseous system will produce an increase in partial pressures for all gaseous reactants and products. In order to lower the increase in partial pressures, the equilibrium will shift so as to favor the reaction side having the smaller number of gaseous molecules, in this case to the right. This shift will decrease the amount of Cl_2 and PCl_3 at equilibrium, and it will increase the amount of PCl_5 at equilibrium. This increases the size of the numerator and decreases the size of the denominator in the above expression for K_p, causing the value of K_p to increase.

16.10 $K_c = \frac{[CO_2][H_2]}{[CO][H_2O]} = \frac{(0.150)(0.200)}{(0.180)(0.0411)} = 4.06$

16.11

$$2CO(g) + O_2(g) \rightarrow 2CO_2(g)$$

Using the stoichiometry of the reaction we can see that for every mol of O_2 that is used, twice as much CO will react and twice as much CO_2 will be produced. Consequently, if the $[O_2]$ decreases by 0.030 mol/L, the [CO] decreases by 0.060 mol/L and $[CO_2]$ increases by 0.060 mol/L.

16.12 a) The initial concentrations were:
$[PCl_3]$ = 0.200 mol/1.00 L = 0.20 M
$[Cl_2]$ = 0.100 mol/1.00 L = 0.100 M
$[PCl_5]$ = 0.00 mol/1.00 L = 0.000 M

b) The change in concentration of PCl_3 was (0.200 – 0.120) M = 0.080 mol/L. The other materials must have undergone changes in concentration that are dictated by the coefficients of the balanced chemical equation, namely: $PCl_3 + Cl_2 \rightarrow PCl_5$ or both PCl_3 and Cl_2 have decreased by 0.080 M and PCl_5 has increased by 0.080 M.

c) As stated in the problem, the equilibrium concentration of PCl_3 is 0.120 M. The equilibrium concentration of PCl_5 is 0.080 M since initially there was no PCl_5. The equilibrium concentration of Cl_2 equals the initial concentration minus the amount that reacted, 0.100 M - 0.080 M = 0.020 M.

d) $K_c = \frac{[PCl_5]}{[PCl_3][Cl_2]} = \frac{(0.080)}{(0.120)(0.020)} = 33$

16.13 $K_c = \frac{[CH_3CO_2C_2H_5][H_2O]}{[CH_3CO_2H][C_2H_5OH]} = \frac{(0.910)(0.00850)}{(0.210)[C_2H_5OH]} = 4.10$

$[C_2H_5OH] = 8.98 \times 10^{-3}$ M

16.14 Initially we have $[H_2] = [I_2]$ = 0.200 M.

	$[H_2]$	$[I_2]$	[HI]
I	0.200	0.200	–
C	–x	–x	+2x
E	0.200–x	0.200–x	+2x

Substituting the above values for equilibrium concentrations into the mass action expression gives:

$$K_c = \frac{[HI]^2}{[H_2][I_2]} = \frac{(2x)^2}{(0.200 - x)(0.200 - x)} = 49.5$$

Take the square root of both sides of this equation to get; $\frac{2x}{(0.200 - x)} = 7.04$.

This equation is easily solved giving x = 0.156. The substances then have the following concentrations at equilibrium: $[H_2] = [I_2] = 0.200 - 0.156 = 0.044$ M, $[HI] = 2(0.156) = 0.312$ M.

16.15 $$N_2(g) + O_2(g) \rightleftharpoons 2NO(g)$$

	$[N_2]$	$[O_2]$	[NO]
I	0.033	0.00810	–
C	–x	–x	+2x
E	0.033–x	0.00810–x	+2x

Substituting the above values for equilibrium concentrations into the mass action expression gives:

$$K_c = \frac{[NO]^2}{[N_2][O_2]} = \frac{(2x)^2}{(0.033 - x)(0.00810 - x)} = 4.8 \times 10^{-31}$$

If we assume that x << 0.033 and x << 0.00810, we can simplify this equation. (Because the value of K_c is so low, this assumption should be valid.) The equation simplifies as:

$$K_c = \frac{(2x)^2}{(0.033)(0.00810)} = 4.8 \times 10^{-31}$$

This equation is easily solved to give $x = 5.7 \times 10^{-18}$ M. The equilibrium concentration of NO is 2x according to the ICE table so, $[NO] = 1.1 \times 10^{-17}$ M.

Review Problems

16.19 (a) $K_c = \frac{[POCl_3]^2}{[PCl_3]^2[O_2]}$

(b) $K_c = \frac{[SO_2]^2[O_2]}{[SO_3]^2}$

(c) $K_c = \frac{[NO]^2[H_2O]^2}{[N_2H_4][O_2]^2}$

(d) $K_c = \frac{[NO_2]^2[H_2O]^8}{[N_2H_4][H_2O_2]^6}$

(e) $K_c = \frac{[SO_2][HCl]^2}{[SOCl_2][H_2O]}$

16.21 (a) $K_p = \frac{(P_{POCl_3})^2}{(P_{PCl_3})^2(P_{O_2})}$

(b) $K_p = \frac{(P_{SO_2})^2(P_{O_2})}{(P_{SO_3})^2}$

(c) $K_p = \frac{(P_{NO})^2(P_{H_2O})^2}{(P_{N_2H_4})(P_{O_2})^2}$

(d) $K_p = \frac{(P_{NO_2})^2(P_{H_2O})^8}{(P_{N_2H_4})(P_{H_2O_2})^6}$

(e) $K_p = \frac{(P_{SO_2})(P_{HCl})^2}{(P_{SOCl_2})(P_{H_2O})}$

16.23 (a) $K_c = \frac{[Ag(NH_3)_2^+]}{[Ag^+][NH_3]^2}$

(b) $K_c = \frac{[Cd(SCN)_4^{2-}]}{[Cd^{2+}][SCN^-]^4}$

16.25 The first equation has been reversed in making the second equation. We therefore take the inverse of the value of the first equilibrium constant in order to determine a value for the second equilibrium constant: $K = 1 \times 10^{85}$

16.27 (a) $K_c = \frac{[HCl]^2}{[H_2][Cl_2]}$

(b) $K_c = \frac{[HCl]}{[H_2]^{1/2}[Cl_2]^{1/2}}$

K_c for reaction (b) is the square root of K_c for reaction (a).

16.29 M = P/RT

$$M = \frac{(745\text{ torr})\left(\frac{1\text{ atm}}{760\text{ torr}}\right)}{(0.0821\ \frac{\text{L atm}}{\text{mol K}})(318\text{ K})} = 0.0375\text{ M}$$

16.31 (b), because the number of moles of gas do not change

16.33 $K_p = K_c \times (RT)^{\Delta n_g}$
$6.3 \times 10^{-3} = K_c[(0.0821\text{ L atm K}^{-1}\text{ mol}^{-1})(498\text{ K})]^{-2} = 5.98 \times 10^{-4} \times K_c$
$K_c = 11$

16.35 $K_p = K_c \times (RT)^{\Delta n_g}$
$K_p = 4.2 \times 10^{-4}[(0.0821\text{ L atm K}^{-1}\text{ mol}^{-1})(773\text{ K})]^1 = 2.7 \times 10^{-2}$

16.37 $K_p = K_c \times (RT)^{\Delta n_g}$
$K_p = (0.40)[(0.0821\text{ L atm K}^{-1}\text{ mol}^{-1})(1046\text{ K})]^{-2} = 5.4 \times 10^{-5}$

16.39 In each case we get approximately 55.5 M:

(a)

$$\#\,\text{mol H}_2\text{O} = (18.0\ \text{mL H}_2\text{O})\left(\frac{1\,\text{g}}{1\,\text{mL}}\right)\left(\frac{1\,\text{mol H}_2\text{O}}{18.02\ \text{g H}_2\text{O}}\right) = 0.999\ \text{mol H}_2\text{O}$$

$$\text{M} = \left(\frac{0.999\ \text{mol H}_2\text{O}}{18.0\ \text{mL H}_2\text{O}}\right)\left(\frac{1000\ \text{mL}}{1\,\text{L}}\right) = 55.5\ \text{M}$$

(b)

$$\#\,\text{mol H}_2\text{O} = (100.0\ \text{mL H}_2\text{O})\left(\frac{1\,\text{g}}{1\,\text{mL}}\right)\left(\frac{1\,\text{mol H}_2\text{O}}{18.02\ \text{g H}_2\text{O}}\right) = 5.549\ \text{mol H}_2\text{O}$$

$$\text{M} = \left(\frac{5.549\ \text{mol H}_2\text{O}}{100.0\ \text{mL H}_2\text{O}}\right)\left(\frac{1000\ \text{mL}}{1\,\text{L}}\right) = 55.49\ \text{M}$$

(c)

$$\#\,\text{mol H}_2\text{O} = (1.00\ \text{L H}_2\text{O})\left(\frac{1000\ \text{mL}}{1\,\text{L}}\right)\left(\frac{1\,\text{g}}{1\,\text{mL}}\right)\left(\frac{1\,\text{mol H}_2\text{O}}{18.02\ \text{g H}_2\text{O}}\right)$$

$$= 55.5\ \text{mol H}_2\text{O}$$

$$\text{M} = \left(\frac{55.5\ \text{mol H}_2\text{O}}{1.00\ \text{L H}_2\text{O}}\right) = 55.5\ \text{M}$$

16.41 (a) $K_c = \dfrac{[CO]^2}{[O_2]}$

(b) $K_c = [H_2O][SO_2]$

(c) $K_c = \dfrac{[CH_4][CO_2]}{[H_2O]^2}$

(d) $K_c = \dfrac{[H_2O][CO_2]}{[HF]^2}$

(e) $K_c = [H_2O]^5$

16.43

	[HCl]	[HI]	[Cl_2]
I	0.100	–	–
C	–2x	+2x	+x
E	0.100–2x	+2x	+x

Note: Since the I_2(s) has a constant concentration, it may be neglected.

$$K_c = \frac{[HI]^2[Cl_2]}{[HCl]^2} = 1.6 \times 10^{-34}$$

$$K_c = \frac{(2x)^2(x)}{(0.100 - 2x)^2} = 1.6 \times 10^{-34}$$

Because the value of K_c is so small, we make the simplifying assumption that $(0.100 - 2x) \approx 0.100$, and the above equation becomes:

$$K_c = \frac{[HI]^2[Cl_2]}{[HCl]^2} = 1.6 \times 10^{-34}$$

$$K_c = \frac{(2x)^2(x)}{(0.100)^2} = 1.6 \times 10^{-34}$$

$4x^3 = 1.6 \times 10^{-36}$; $\therefore$ $x = 7.37 \times 10^{-13}$, and the above assumption is seen to have been valid.

$[HI] = 2x = 1.47 \times 10^{-12}$ M
$[Cl_2] = x = 7.37 \times 10^{-13}$ M

$([HCl] = (0.100 - 2x) \approx 0.100$ M$)$

16.45 The mass action expression for this equilibrium is:

$$K_c = \frac{[PCl_5]}{[PCl_3][Cl_2]} = 0.18$$

The value for the reaction quotient for this system is:

$$Q = \frac{(0.00500)}{(0.0420)(0.0240)} = 4.96$$

(a) No. This is not the value of the equilibrium constant, and we conclude that the system is not at equilibrium.

(b) Since the value of the reaction quotient for this system is larger than that of the equilibrium constant, the system must shift to the left to reach equilibrium.

16.47 $$K_c = \frac{[CH_3OH]}{[CO][H_2]^2} = \frac{[CH_3OH]}{(0.180)(0.220)^2} = 0.500$$

$[CH_3OH] = 4.36 \times 10^{-3}$ M.

16.49 $K_c = \frac{[CH_3OH]}{[CO][H_2]^2} = \frac{(0.00261)}{(0.105)(0.250)^2} = 0.398$

16.51

	[HBr]	$[H_2]$	$[Br_2]$
I	0.500	–	–
C	–2x	+x	+x
E	0.500–2x	+x	+x

The problem tell us that $[Br_2] = 0.0955$ M = x at equilibrium. Using the ICE table as a guide we see that the equilibrium concentrations are; $[H_2] = [Br_2] = 0.0955$ M and [HBr] = 0.500–2(0.0955) = 0.309 M.

$$K_c = \frac{[H_2][Br_2]}{[HBr]^2} = \frac{(0.0955)(0.0955)}{(0.309)^2} = 0.0955$$

16.53 According to the problem, the concentration of NO_2 increases in the course of this reaction. This means our ICE table will look like the following:

	$[NO_2]$	[NO]	$[N_2O]$	$[O_2]$
I	0.0560	0.294	0.184	0.377
C	+x	+x	–x	–x
E	0.0560+x	0.294+x	0.184–x	0.377–x

The problem tell us that $[NO_2]$ = 0.118 M = 0.0560+x at equilibrium. Solving we get; x = 0.062 M. Using the ICE table as a guide we see that the equilibrium concentrations are; [NO] = 0.356 M, $[N_2O]$ = 0.122 M and $[O_2]$ = 0.315 M.

$$K_c = \frac{[N_2O][O_2]}{[NO_2][NO]} = \frac{(0.122)(0.315)}{(0.118)(0.356)} = 0.915$$

16.55 $2BrCl \rightleftharpoons Br_2 + Cl_2$

	[BrCl]	$[Br_2]$	$[Cl_2]$
I	0.050	–	–
C	–2x	+x	+x
E	0.050–2x	+x	+x

Substituting the above values for equilibrium concentrations into the mass action expression gives:

$$K_c = \frac{[Br_2][Cl_2]}{[BrCl]^2} = \frac{(x)(x)}{(0.050-2x)^2} = 0.145$$

Take the square root of both sides to get

$$K_c = \frac{x}{0.050 - 2x} = 0.381$$

Solving for x gives: $x = 0.011\ M = [Br_2] = [Cl_2]$

16.57 The initial concentrations are each 0.240 mol/2.00 L = 0.120 M.

	$[SO_3]$	$[NO]$	$[NO_2]$	$[SO_2]$
I	0.120	0.120	–	–
C	–x	–x	+x	+x
E	0.120–x	0.120–x	+x	+x

Substituting the above values for equilibrium concentrations into the mass action expression gives:

$$K_c = \frac{[NO_2][SO_2]}{[SO_3][NO]} = \frac{(x)(x)}{(0.120 - x)(0.120 - x)} = 0.500$$

Taking the square root of both sides of this equation gives: $0.707 = x/(0.120 - x)$
Solving for x we have: $1.707(x) = 0.0848$

$x = 0.0497\ mol/L = [NO_2] = [SO_2]$
$[NO] = [SO_3] = 0.120 - x = 0.0703\ mol/L$

16.59 The initial concentrations are all 1.00 mol/100 L = 0.0100 M. Since the initial concentrations are all the same, the reaction quotient is equal to 1.0, and we conclude that the system must shift to the left to reach equilibrium since $Q > K_c$.

	$[CO]$	$[H_2O]$	$[CO_2]$	$[H_2]$
I	0.0100	0.0100	0.0100	0.0100
C	+x	+x	–x	–x
E	0.0100+x	0.0100+x	0.0100–x	0.0100–x

Substituting the above values for equilibrium concentrations into the mass action expression gives:

$$K_c = \frac{[CO_2][H_2]}{[CO][H_2O]} = \frac{(0.0100 - x)(0.0100 - x)}{(0.0100 + x)(0.0100 + x)} = 0.400$$

We take the square root of both sides of the above equation:

$$\frac{(0.0100 - x)}{(0.0100 + x)} = 0.632$$

$$(0.632)(0.0100 + x) = 0.0100 - x$$
$$(1.632)x = 3.68 \times 10^{-3}$$
$$x = 2.25 \times 10^{-3} \text{ mol/L}$$

The equilibrium concentrations are then:
$[H_2] = [CO_2] = (0.0100 - 2.25 \times 10^{-3}) = 7.7 \times 10^{-3}$ M
$[CO] = [H_2O] = (0.0100 + 2.25 \times 10^{-3}) = 0.0123$ M

16.61

	[HCl]	$[H_2]$	$[Cl_2]$
I	0.0500	–	–
C	–2x	+x	+x
E	0.0500–2x	+x	+x

Substituting the above values for equilibrium concentrations into the mass action expression gives:

$$K_c = \frac{[H_2][Cl_2]}{[HCl]^2} = \frac{(x)(x)}{(0.0500 - 2x)^2} = 3.2 \times 10^{-34}$$

Because K_c is so exceedingly small, we can make the simplifying assumption that x is also small enough to make $(0.0500 - 2x) \approx 0.0500$. Thus we have:

$$3.2 \text{ x } 10^{-34} = (x)^2/(0.0500)^2$$

Taking the square root of both sides, and solving for the value of x gives:

$$x = 8.9 \text{ x } 10^{-19} \text{ M} = [H_2] = [Cl_2]$$

$[HCl] = (0.0500 - x) \approx 0.0500$ mol/L

16.63 $$K_c = \frac{[CO]^2[O_2]}{[CO_2]^2} = 6.4 \times 10^{-7}$$

	$[CO_2]$	[CO]	$[O_2]$
I	$1.0 \text{ x } 10^{-2}$	–	–
C	–2x	+2x	+x
E	$1.0 \text{ x } 10^{-2}$ – 2x	+2x	+x

$$K_c = \frac{[2x]^2[x]}{[1.0 \times 10^{-2} - 2x]^2} = 6.4 \times 10^{-7}$$

Assume x << $1.0 \text{ x } 10^{-2}$...

$$\frac{4x^3}{(1.0\times10^{-2})^2} = 6.4\times10^{-7} \quad x = 2.5 \times 10^{-4}$$

$[CO] = 2x = 5.0 \times 10^{-4}$ M

*16.65 We first approach the problem in the normal fashion with an initial concentration of PCl_5 = 0.013 M.

	$[PCl_3]$	$[Cl_2]$	$[PCl_5]$
I	–	–	0.013
C	+x	+x	–x
E	+x	+x	0.013–x

Substituting the above values for equilibrium concentrations into the mass action expression gives:

$$K_c = \frac{[PCl_5]}{[PCl_3][Cl_2]} = \frac{(0.013 - x)}{(x)(x)} = 0.18$$

rearranging; $0.18x^2 + x - 0.013 = 0$

We next attempt to use the quadratic equation to solve for the value of x, setting a = 0.18; b = 1; c = –0.013.

However, we find that unless we carry one more significant figure than is allowed, the quadratic formula for this problem gives us a concentration of zero for PCl_5. A better solution is obtained by "allowing" the initial equilibrium to shift completely to the left, giving us a new initial situation from which to work:

	$[PCl_3]$	$[Cl_2]$	$[PCl_5]$
I	0.013	0.013	–
C	–x	–x	+x
E	0.013–x	0.013–x	+x

Substituting the above values for equilibrium concentrations into the mass action expression gives:

$$K_c = \frac{[PCl_5]}{[PCl_3][Cl_2]} = \frac{(+x)}{(0.013 - x)(0.013 - x)} = 0.18$$

Now, we may assume that x << 0.013. The equation is simplified and we solve for x:

$x = [PCl_5] = 3.0 \times 10^{-5}$ M.

*16.67

	$[SO_3]$	[NO]	$[NO_2]$	$[SO_2]$
I	0.0500	0.100	–	–
C	–x	–x	+x	+x
E	0.0500–x	0.100–x	+x	+x

Substituting the above values for equilibrium concentrations into the mass action expression gives:

$$K_c = \frac{[NO_2][SO_2]}{[SO_3][NO]} = \frac{(x)(x)}{(0.0500 - x)(0.100 - x)} = 0.500$$

Since the equilibrium constant is not much larger than either of the values 0.0500 or 0.100, we cannot neglect the size of x in the above expression. A simplifying assumption is not therefore possible, and we must solve for the value of x using the quadratic equation. Multiplying out the above denominator, collecting like terms, and putting the result into the standard quadratic form gives:

$$0.500x^2 + (7.50 \text{ x } 10^{-2})x - (2.50 \text{ x } 10^{-3}) = 0$$

$$x = \frac{-7.50\times10^{-2} \pm \sqrt{(-7.50\times10^{-2})^2 - 4(0.500)(-2.50\times10^{-3})}}{2(0.500)} = 0.0281\,M,$$

using the (+) root. So, $[NO_2]$ = $[SO_2]$ = 0.0281 M

*16.69

$$Kc = \frac{[CO][H_2O]}{[HCHO_2]^2} = 4.3\times10^5$$

Since Kc is large, start by assuming all of the $HCHO_2$ decomposes to give CO and H_2O

	$[HCHO_2]$	[CO]	$[O_2]$
I	–	0.200	0.200
C	+x	–x	–x
E	+x	0.200 –x	0.200 –x

$$Kc = \frac{[0.200][0.200]}{[x]} = 4.3 \text{ x } 10^5$$

$$x = 9.3 \text{ x } 10^{-8}$$

So, at equilibrium: $[CO] = [H_2O] = 0.200 - x = 0.200$ M

Additional Exercises

16.71 a) The mass action expression is:

$$K_p = \frac{(P_{NO_2})^2}{(P_{N_2O_4})} = 0.140 \text{ atm}$$

Solving the above expression for the partial pressure of NO_2, we get:

$$P_{NO_2} = \sqrt{P_{N_2O_4} \times K_p} = \sqrt{(0.250 \text{ atm})(0.140 \text{ atm})} = 0.187 \text{ atm}$$

b) $P_{total} = P_{NO_2} + P_{N_2O_4} = 0.187 + 0.250 = 0.437$ atm

*16.73 The initial concentrations are:

$[NO_2] = (0.200 \text{ mol}/4.00 \text{ L}) = 0.0500$ M
$[NO] = (0.300 \text{ mol}/4.00 \text{ L}) = 0.0750$ M
$[N_2O] = (0.150 \text{ mol}/4.00 \text{ L}) = 0.0375$ M
$[O_2] = (0.250 \text{ mol}/4.00 \text{ L}) = 0.0625$ M

Substituting these values into the mass action expression we determine:

$$Q = \frac{[N_2O][O_2]}{[NO_2][NO]} = \frac{(0.0375)(0.0625)}{(0.0500)(0.0750)} = 0.625$$

Since $Q < K_c$, this reaction will proceed from left to right as written. The ICE table becomes:

	$[NO_2]$	$[NO]$	$[N_2O]$	$[O_2]$
I	0.0500	0.0750	0.0375	0.0625
C	–x	–x	+x	+x
E	0.0500–x	0.0750–x	0.0375+x	0.0625+x

Substituting the above values for equilibrium concentrations into the mass action expression gives:

$$K_c = \frac{[N_2O][O_2]}{[NO_2][NO]} = \frac{(0.0375 + x)(0.0625 + x)}{(0.0500 - x)(0.0750 - x)} = 0.914$$

To solve we need to use the quadratic equation. Expanding the above calculation we get:

$$0.086x^2 + 0.214x - 0.00109 = 0$$
$$x = 0.00508.$$

Therefore,

$[NO_2] = 0.0500 - x = 0.0449$M
$[NO] = 0.0750 - x = 0.0699$ M
$[N_2O] = 0.0375 + x = 0.0426$ M

$[O_2] = 0.0625 + x = 0.0676$ M

*16.75 First, calculate a value for K_c using a rearranged form of equation 16.4, and setting the value for Δn to –1:

$$K_c = \frac{K_p}{(RT)^{\Delta n_g}} = \frac{1.5\times10^{18}}{\left[(0.0821\ \frac{\text{L atm}}{\text{mol K}})(300\ \text{K})\right]^{-1}} = 3.7\times10^{19}$$

The value for the equilibrium constant is very large, indicating that the equilibrium lies far to the right. We therefore anticipate that the initial conditions are unrealistic. The system is "allowed" to come to a more realistic new initial set of concentrations, by reaction of all of the starting amount of NO, to give a stoichiometric amount of N_2O and NO_2. Only then can we solve for the equilibrium concentrations in the usual manner:

	[NO]	$[N_2O]$	$[NO_2]$
I	—	0.010	0.010
C	+3x	–x	–x
E	+3x	0.010–x	0.010–x

Substituting the above values for equilibrium concentrations into the mass action expression gives:

$$K_c = \frac{[N_2O][NO_2]}{[NO]^3} = \frac{(0.010 - x)(0.010 - x)}{(3x)^3} = 3.7\times10^{19}$$

The simplifying assumption can be made that the value for x is much smaller than the number 0.010. Upon solving for x we get:

$$27x^3 = 2.7 \times 10^{-24}$$
$$x = 4.6 \text{ x } 10^{-9}\ \text{M}$$

$[NO] = 3x = 1.4 \text{ x } 10^{-8}$ M
$[N_2O] = [NO_2] = 0.010 - x = 0.010$ M.

*16.77 First, calculate the number of moles of CO used in the experiment, using the ideal gas law:

$$n = \frac{PV}{RT} = \frac{(0.177\ \text{atm})(2.00\ \text{L})}{(0.0821\ \frac{\text{L atm}}{\text{mol K}})(298\ \text{K})} = 0.0145\ \text{mol}$$

Next, calculate the partial pressure of CO at 400 °C:

$$P = \frac{nRT}{V} = \frac{(0.0145\ \text{mol})(0.0821\ \frac{\text{L atm}}{\text{mol K}})(673\ \text{K})}{(2.00\ \text{L})} = 0.401\ \text{atm}$$

Next, we calculate the number of moles of water that are supplied to the reaction, and convert to partial pressure for water, using the ideal gas equation:

$$\#\,\text{mol}\ H_2O = (0.391\ \text{g}\ H_2O)\left(\frac{1\ \text{mol}\ H_2O}{18.12\ \text{g}\ H_2O}\right) = 0.0217\ \text{mol}\ H_2O$$

$$P = \frac{nRT}{V} = \frac{(0.0217\ \text{mol})(0.0821\ \frac{\text{L atm}}{\text{mol K}})(673\ \text{K})}{(2.00\ \text{L})} = 0.600\ \text{atm}$$

Finally, we solve for the equilibrium partial pressure in the usual manner:

	P_{HCHO_2}	P_{CO}	P_{H_2O}
I	–	0.401	0.600
C	+x	–x	–x
E	+x	0.401–x	0.600–x

Substituting the above values for equilibrium partial pressures into the mass action expression gives:

$$K_p = \frac{(P_{CO})(P_{H_2O})}{(P_{HCHO_2})} = \frac{(0.401 - x)(0.600 - x)}{x} = 1.6 \times 10^6$$

Because K_p is so large, we assume x << 0.401 and x << 0.600. We then solve for $P_{HCHO_2} = x = 1.5 \times 10^{-7}$ atm.

*16.79

$$Kc = \frac{[N_2O][O_2]}{[NO_2][NO]} = 0.914$$

Let z = the initial concentration.

	$[NO_2]$	$[NO]$	$[N_2O]$	$[O_2]$
I	z	z	–	–
C	–x	–x	+x	+x
E	z – 2x	z –x	+x	+x

$$K_c = \frac{[x][x]}{[z - x][z - x]} = 0.914$$

Take the square roots of both sides to get:

$$\frac{x}{z - x} = 0.956$$

x = 0.050 from the data in the problem, so:

$$\frac{0.500}{z - 0.500} = 0.956$$

$$z = 0.10$$

$$\#\text{ moles NO} = \#\text{ moles NO}_2 = \left(\frac{0.10\text{ mol}}{\text{L}}\right)(5.00\text{ L}) = 0.51\text{ moles}$$

Chapter 17

Practice Exercises

17.1 In each case the conjugate base is obtained by removing a proton from the acid:

(a) OH^- (b) I^- (c) NO_2^- (d) $H_2PO_4^-$
(e) HPO_4^{2-} (f) PO_4^{3-} (g) H^- (h) NH_3

17.2 In each case the conjugate acid is obtained by adding a proton to the base:

(a) H_2O_2 (b) HSO_4^- (c) HCO_3^- (d) HCN
(e) NH_3 (f) NH_4^+ (g) H_3PO_4 (h) $H_2PO_4^-$

17.3 The Brønsted acids are $H_2PO_4^-(aq)$ and $H_2CO_3(aq)$
The Brønsted bases are $HCO_3^-(aq)$ and $HPO_4^{2-}(aq)$

conjugate pair

$$PO_4^{3-}(aq) + HC_2H_3O_2(aq) \rightleftharpoons HPO_4^{2-}(aq) + C_2H_3O_2^-(aq)$$

base acid acid base

conjugate pair

17.4 conjugate pair

$$PO_4^{3-}(aq) + HC_2H_3O_2(aq) \rightleftharpoons HPO_4^{2-}(aq) + C_2H_3O_2^-(aq)$$

base acid acid base

conjugate pair

17.5 $HPO_4^{2-}(aq) + OH^-(aq) \rightarrow PO_4^{3-}(aq) + H_2O$; HPO_4^{2-} acting as an acid
$HPO_4^{2-}(aq) + H_3O^+(aq) \rightarrow H_2PO_4^- + H_2O$; HPO_4^{2-} acting as a base

17.6 The substances on the right because they are the weaker acid and base.

17.7 a) HBr is the stronger acid since binary acid strength increases from left to right within a period.
b) H_2Te is the stronger acid since binary acid strength increases from top to bottom within a group.
c) CH_3SH since acid strength increases from top to bottom within a group.

17.8 $HClO_3$ because Cl is more electronegative than Br.

17.9 a) HIO_4 b) H_2TeO_4 c) H_3AsO_4

17.10 a) Fluoride ions have a filled octet of electrons and are likely to behave as Lewis bases, i.e., electron pair donors.

b) $BeCl_2$ is a likely Lewis acid since it has an incomplete shell. The Be atom has only two valence electrons and it can easily accept a pair of electrons.

c) It could reasonably be considered a potential Lewis base since it contains three oxygens, each with lone pairs and partial negative charges. However, it is more effective as a Lewis acid, since the central sulfur bears a significant positive charge.

17.11 $K_w = 1.0 \times 10^{-14} = [H^+][OH^-]$

$$[H^+] = \frac{1.0\times10^{-14}}{[OH^-]} = \frac{1.0\times10^{-14}}{7.8\times10^{-6}} = 1.3 \times 10^{-9}\ M$$

Since $[OH^-] > [H^+]$, the solution is basic.

17.12 $pH = -\log[H^+] = -\log[3.67 \times 10^{-4}] = 3.44$
$pOH = 14.00 - pH = 14.00 - 3.44 = 10.56$

The solution is acidic since pH is below 7.0.

17.13 $pOH = -\log[OH^-] = -\log[1.47 \times 10^{-9}] = 8.83$
$pH = 14.00 - pOH = 14.00 - 8.83 = 5.17$

17.14 In general, we have the following relationships between pH, $[H^+]$, and $[OH^-]$:

$$[H^+] = 10^{-pH}$$
$$[H^+][OH^-] = 1.00 \times 10^{-14}$$

(a) $[H^+] = 10^{-2.90} = 1.3 \times 10^{-3}\ M$
$[OH^-] = 1.00 \times 10^{-14}/1.3 \times 10^{-3}\ M = 7.7 \times 10^{-12}\ M$
The solution is acidic.

(b) $[H^+] = 10^{-3.85} = 1.4 \times 10^{-4}\ M$
$[OH^-] = 1.00 \times 10^{-14}/1.4 \times 10^{-4}\ M = 7.1 \times 10^{-11}\ M$
The solution is acidic.

(c) $[H^+] = 10^{-10.81} = 1.5 \times 10^{-11}\ M$
$[OH^-] = 1.00 \times 10^{-14}/1.5 \times 10^{-11}\ M = 6.7 \times 10^{-4}\ M$
The solution is basic.

(d) $[H^+] = 10^{-4.11} = 7.8 \times 10^{-5}\ M$
$[OH^-] = 1.00 \times 10^{-14}/7.8 \times 10^{-5}\ M = 1.3 \times 10^{-10}\ M$
The solution is acidic.

(e) $[H^+] = 10^{-11.61} = 2.5 \times 10^{-12}\ M$
$[OH^-] = 1.00 \times 10^{-14}/2.5 \times 10^{-12}\ M = 4.0 \times 10^{-3}\ M$
The solution is basic.

17.15 $[OH^-] = 0.0050\ M$
$pOH = -\log[OH^-] = -\log[0.0050] = 2.30$
$pH = 14.0 - pOH = 14.00 - 2.30 = 11.70$

$[H^+] = 10^{-11.70} = 2.0 \times 10^{-12}$ M

17.16 $[H^+] = 10^{-5.5} = 3.2 \times 10^{-6}$ M

Review Problems

17.40 (a) HF (b) $N_2H_5^+$ (c) $C_5H_5NH^+$
(d) HO_2^- (e) H_2CrO_4

17.42 (a)

conjugate pair (HNO_3 / NO_3^-)

$$HNO_3 + N_2H_4 \rightleftharpoons N_2H_5^+ + NO_3^-$$

acid base acid base

conjugate pair (N_2H_4 / $N_2H_5^+$)

(b)

conjugate pair ($N_2H_5^+$ / N_2H_4)

$$N_2H_5^+ + NH_3 \rightleftharpoons NH_4^+ + N_2H_4$$

acid base acid base

conjugate pair (NH_3 / NH_4^+)

(c)

conjugate pair ($H_2PO_4^-$ / HPO_4^{2-})

$$H_2PO_4^- + CO_3^{2-} \rightleftharpoons HCO_3^- + HPO_4^{2-}$$

acid base acid base

conjugate pair (CO_3^{2-} / HCO_3^-)

(d)

conjugate pair (HIO_3 / IO_3^-)

$$HIO_3 + HC_2O_4^- \rightleftharpoons H_2C_2O_4 + IO_3^-$$

acid base acid base

conjugate pair ($HC_2O_4^-$ / $H_2C_2O_4$)

17.44 a) H_2Se, larger central atom
b) HI, more electronegative atom
c) PH_3, larger central atom

17.46 a) HIO_4, because it has more oxygen atoms
b) H_3AsO_4, because it has more oxygen atoms

17.48 a) H_3PO_4, since P is more electronegative
b) HNO_3, because N is more electronegative (HNO_3 is a strong acid)

c) $HClO_4$, because Cl is more electronegative

17.50

$$[H-\ddot{N}-H]^- + H^+ \longrightarrow NH_3$$

17.52

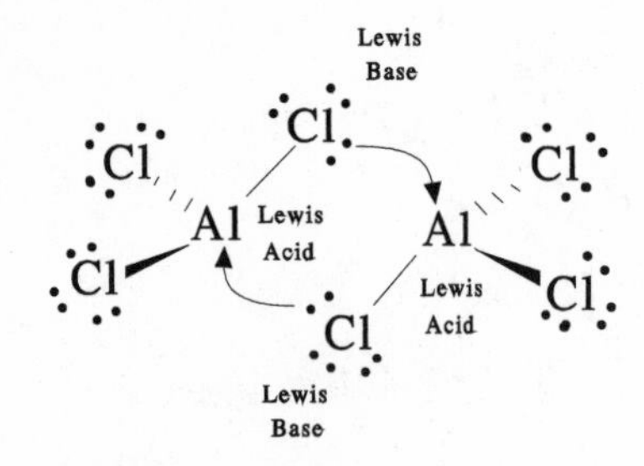

17.54 Lewis base Lewis acid

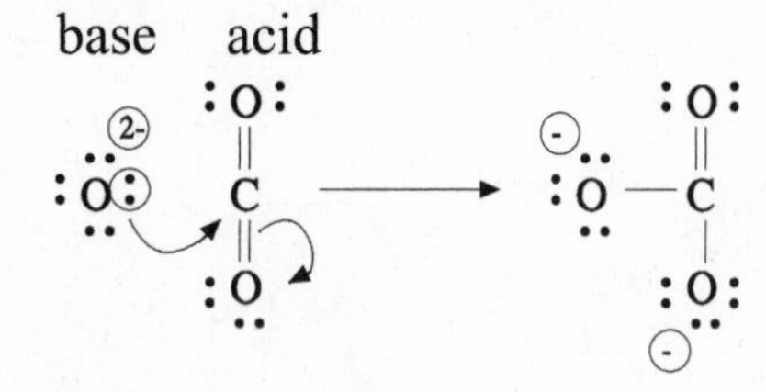

17.56 Lewis base Lewis acid

$$H_2N^- + H-OH \longrightarrow H-N(H)-H + {}^-OH$$

17.58 It should be stated from the outset that water at this temperature is neutral by definition, since $[H^+] = [OH^-]$. In other words, the self–ionization of water still occurs on a one–to–one mole basis:

$$H_2O \rightleftharpoons H^+ + OH^-$$

$$K_w = 2.4 \times 10^{-14} = [H^+][OH^-]$$

Since $[H^+] = [OH^-]$, we can rewrite the above relationship:

$$2.4 \times 10^{-14} = ([H^+])^2$$

$\therefore [H^+] = [OH^-] = 1.5 \times 10^{-7}$ M

$pH = -\log[H^+] = -\log(1.5 \times 10^{-7}) = 6.82$
$pOH = -\log[OH^-] = -\log(1.5 \times 10^{-7}) = 6.82$
$pK_w = pH + pOH = 6.82 + 6.82 = 13.64$

Alternatively, for the last calculation we can write:
$pK_w = -\log(K_w) = -\log(2.4 \text{ x } 10^{-14}) = 13.62$

Water is neutral at this temperature because the concentration of the hydrogen ion is the same as the concentration of the hydroxide ion.

17.60 At 25 °C: $K_w = 1.0 \times 10^{-14} = [H^+][OH^-]$

Let $x = [H^+]$, for each of the following:

(a) $x(0.0024) = 1.0 \times 10^{-14}$
$[H^+] = (1.0 \text{ x } 10^{-14}) \div (0.0024) = 4.2 \times 10^{-12}$ M

(b) $x(1.4 \times 10^{-5}) = 1.0 \times 10^{-14}$
$[H^+] = (1.0 \times 10^{-14}) \div (1.4 \times 10^{-5}) = 7.1 \times 10^{-10}$ M

(c) $x(5.6 \times 10^{-9}) = 1.0 \times 10^{-14}$
$[H^+] = (1.0 \times 10^{-14}) \div (5.6 \times 10^{-9}) = 1.8 \times 10^{-6}$ M

(d) $x(4.2 \times 10^{-13}) = 1.0 \times 10^{-14}$
$[H^+] = (1.0 \times 10^{-14}) \div (4.2 \times 10^{-13}) = 2.4 \times 10^{-2}$ M

17.62 $pH = -\log[H^+]$

$[H^+] = 4.2 \text{ x } 10^{-12}$ M	pH = 11.38
$[H^+] = 7.1 \text{ x } 10^{-10}$ M	pH = 9.15
$[H^+] = 1.8 \text{ x } 10^{-6}$ M	pH = 5.74
$[H^+] = 2.4 \text{ x } 10^{-2}$ M	pH = 1.62

17.64 $pH = -\log[H^+] = -\log(1.9 \times 10^{-5}) = 4.72$

17.66 $[H^+] = 10^{-pH}$ and $[OH^-] = 10^{-pOH}$

At 25 °C: pH + pOH = 14.00

(a) $[H^+] = 10^{-pH} = 10^{-3.14} = 7.2 \times 10^{-4}$ M
pOH = 14.00 – pH = 14.00 – 3.14 = 10.86
$[OH^-] = 10^{-pOH} = 10^{-10.86} = 1.4 \times 10^{-11}$ M

(b) $[H^+] = 10^{-pH} = 10^{-2.78} = 1.7 \times\times 10^{-3}$ M
pOH = 14.00 – pH = 14.00 – 2.78 = 11.22
$[OH^-] = 10^{-pOH} = 10^{-11.22} = 6.0 \times 10^{-12}$ M

(c) $[H^+] = 10^{-pH} = 10^{-9.25} = 5.6 \times 10^{-10}$ M
pOH = 14.00 – pH = 14.00 – 9.25 = 4.75
$[OH^-] = 10^{-pOH} = 10^{-4.75} = 1.8 \times 10^{-5}$ M

(d) $[H^+] = 10^{-pH} = 10^{-13.24} = 5.8 \times 10^{-14}$ M
pOH = 14.00 – pH = 14.00 – 13.24 = 0.76

$[OH^-] = 10^{-pOH} = 10^{-0.76} = 1.7 \times 10^{-1}$ M

(e) $[H^+] = 10^{-pH} = 10^{-5.70} = 2.0 \times 10^{-6}$ M
$pOH = 14.00 - pH = 14.00 - 5.70 = 8.30$
$[OH^-] = 10^{-pOH} = 10^{-8.30} = 5.0 \times 10^{-9}$ M

17.68 $[H^+] = 10^{-pH} = 10^{-5.6} = 2.5 \times 10^{-6}$ M
$pOH = 14.00 - pH = 14.00 - 5.6 = 8.4$
$[OH^-] = 10^{-pOH} = 10^{-8.4} = 4.0 \times 10^{-9}$ M

17.70 Since HNO_3 is a strong acid, $[H^+] = [HNO_3]$:

$[H^+] = 0.030$ M
$pH = -\log [0.030] = 1.5$
$pOH = 14.00 - pH = 14.00 - 1.5 = 12.5$
$[OH^-] = 10^{-pOH} = 10^{-12.5} = 3.2 \text{ x } 10^{-13}$ M

17.72 $$\text{M OH}^- = \frac{\text{\# moles OH}^-}{\text{\# L solution}} = \left(\frac{6.0 \text{ g NaOH}}{1.00 \text{ L solution}}\right)\left(\frac{1 \text{ mole NaOH}}{40.0 \text{ g NaOH}}\right)\left(\frac{1 \text{ mole OH}^-}{1 \text{ mole NaOH}}\right)$$
$= 0.15 \text{ M OH}^-$

$pOH = -\log[OH^-] = -\log(0.15) = 0.82$
$pH = 14.00 - pOH = 14.00 - 0.82 = 13.18$
$[H^+] = 10^{-pH} = 10^{-13.18} = 6.6 \times 10^{-14}$ M

17.74 $pOH = 14.00 - pH = 14.00 - 11.60 = 2.40$
$[OH^-] = 10^{-pOH} = 10^{-2.40} = 4.0 \times 10^{-3}$ M
$$[Ca(OH)_2] = \left(\frac{4.0\times10^{-3} \text{ mol OH}^-}{1 \text{ L solution}}\right)\left(\frac{1 \text{ mol Ca(OH)}_2}{2 \text{ mol OH}^-}\right)$$
$$= 2.0\times10^{-3} \text{ M Ca(OH)}_2$$
$= 2.0 \times 10^{-3}$ M $Ca(OH)_2$

17.76 First, we must find the concentration of the HCl solution. Since HCl is a strong acid, we know that $[HCl] = [H^+]$. We can find $[H^+]$ from the pH:

$[H^+] = 10^{-pH} = 10^{-2.25} = 5.62 \times 10^{-3}$ M $= [HCl]$

Now we can solve the problem using the given conversion factors:

$$\text{mL KOH soln.} = 300 \text{ mL HCl soln}\left(\frac{1 \text{ L HCl soln}}{1{,}000 \text{ mL HCl soln}}\right)\left(\frac{5.62\times10^{-3} \text{mol HCl}}{1 \text{ L HCl soln}}\right)$$
$$\times\left(\frac{1 \text{ mol KOH}}{1 \text{ mol HCl}}\right)\left(\frac{1 \text{ L KOH soln}}{0.100 \text{ mol KOH}}\right)\left(\frac{1{,}000 \text{ ml KOH soln}}{1 \text{ L KOH soln}}\right)$$

$= 16.9$ mL KOH solution

17.78 Since NaOH is a strong base, $[NaOH] = [OH^-] = 0.0020$ M OH^-. (Next, we make the simplifying assumption that the amount of hydroxide ion formed from the dissociation of water is so small that we can neglect it in calculating pOH for the solution.)

$pOH = -\log[OH^-] = -\log(0.0020) = 2.7$
$pH = 14.00 - pOH = 14.00 - 2.7 = 11.3$
$[H^+] = 10^{-pH} = 10^{-11.3} = 5.01 \times 10^{-12}$ M

Recall that the amount of H^+ formed from the dissociation of water must be equal to the amount of OH^- formed from the dissociation of water. Therefore,

$[OH^-] = [H^+] = 5.01 \times 10^{-12}$ M

Additional Exercises

17.80 Conjugate acid: $(CH_3)_2NH_2^+$; conjugate base: $(CH_3)_2N^-$

17.82 From problem 17.81, we find there are 3.95×10^{-4} mol HCl present. We need to know how many mol NaOH are in the NaOH solution to react with the HCl. We can find $[OH^-]$ from the given pH:

$pOH = 14.00 - pH = 14.00 - 10.50 = 3.50$
$[OH^-] = 10^{-pOH} = 10^{-3.50} = 3.16 \times 10^{-4}$ M

$$\text{mol NaOH} = 200 \text{ mL NaOH soln}\left(\frac{1 \text{ L NaOH soln}}{1{,}000 \text{ mL NaOH soln}}\right)\left(\frac{3.16\times10^{-4} \text{mol NaOH}}{1 \text{ L NaOH soln}}\right)$$

$= 6.32 \times 10^{-5}$ mol NaOH

Since HCl and NaOH react in a 1-to-1 ratio, 6.32×10^{-5} mol of each will neutralize one another. This leaves (3.95×10^{-4} mol HCl $- 6.32 \times 10^{-5}$ mol HCl) $= 3.32 \times 10^{-4}$ mol HCl remaining.

The concentration of H^+ is now:

$[H^+] =$ mol H^+/L solution
$= 3.32 \times 10^{-4}$ mol/0.200 L solution
$= 1.66 \times 10^{-3}$ M

Therefore $pH = -\log [H^+] = -\log [1.66 \times 10^{-3}] = 2.78$

17.84 a) In H_2O_2, the extra oxygen atom helps to stabilize the HO_2^- ion making it easier for H_2O_2 to lose a proton than can H_2O.
b) Acidic

17.86 The equilibrium lies to the right since reactions favor the weaker acid and base.

17.88 NH_3OH^+ is the strongest acid.

*17.90 The total $[H^+]$ is from the HCl and from the dissociation of H_2O. Since HCl is a strong acid, it will contribute 1.0×10^{-7} mol of H^+ per liter of solution. We need to use the equilibrium expression to determine the amount of H^+ contributed by the water.

$$K_w = [H^+][OH^-] = (1.0 \times 10^{-7} + x)(x) = 1.0 \times 10^{-14}$$

Solving a quadratic equation we see that $x = 6 \times 10^{-8} = [OH^-] = [H^+]$ from water dissociation.

So, $[H^+]_{total} = 1.0 \times 10^{-7} + 6 \times 10^{-8} = 1.6 \times 10^{-7}$ and pH = 6.80.

*17.92 A solution with pH = 3.00 would have:

$$\begin{aligned} [H^+] = 10^{-pH} = 10^{-3.00} &= 1.00 \times 10^{-3}\ M \\ &= 1.00 \times 10^{-3}\ \text{mol } [H^+]/\text{L solution} \\ &= 0.00100\ \text{mol } [H^+]/\text{L solution} \end{aligned}$$

We begin with:

$$\text{mol HCl} = 200\ \text{mL HCl soln}\left(\frac{1\ \text{L HCl soln}}{1{,}000\ \text{mL HCl soln}}\right)\left(\frac{0.010\ \text{mol HCl}}{1\ \text{L HCl soln}}\right)$$

$$= 0.002\ \text{mol HCl}$$
$$= 0.002\ \text{mol } H^+$$

Our final molarity will be: M = mol $[H^+]$ /L solution. This can be written as:

$$M = (0.002 - n_{NaOH})/(0.200 + x),$$

where n_{NaOH} = moles of NaOH added and, x = volume of 0.10 M NaOH added.

But n_{NaOH} is simply $M_{NaOH} \cdot V_{NaOH}$, so

$$\begin{gathered} M = (0.002 - 0.10x)/(0.200 + x) \\ 0.00100 = (0.002 - 0.10x)/(0.200 + x) \\ (0.200 + x)0.00100 = (0.002 - 0.10x) \\ 0.000200 + 0.00100\,x = 0.002 - 0.10x \\ 0.1001\,x = 0.0018 \\ x = 0.018, \text{ or } 18\ \text{mL} \end{gathered}$$

*17.94 $[H^+] = 10^{-pH} = 4.27 \times 10^{-3}$ M

$$\% \text{ ionization} = \frac{4.27 \times 10^{-3}}{1.0} \times 100\% = 0.43\%$$

Chapter 18

Practice Exercises

18.1 a) $HCHO_2 + H_2O \rightleftharpoons H_3O^+ + CHO_2^-$

$$K_a = \frac{[H_3O^+][CHO_2^-]}{[HCHO_2]}$$

b) $(CH_3)_2NH_2^+ + H_2O \rightleftharpoons H_3O^+ + (CH_3)_2NH$

$$K_a = \frac{[H_3O^+][CH_3NH]}{[(CH_3)_2NH_2^+]}$$

c) $H_2PO_4^- + H_2O \rightleftharpoons H_3O^+ + HPO_4^{2-}$

$$K_a = \frac{[H_3O^+][HPO_4^{2-}]}{[H_2PO_4^-]}$$

18.2 The acid with the smaller pKa (H_A) is the strongest acid.
Since pKa = – log Ka, Ka = 10^{-pKa}
For H_A: Ka = $10^{-3.16} = 6.9 \times 10^{-4}$
For H_B: Ka = $10^{-4.14} = 7.2 \times 10^{-5}$

18.3 a) $(CH_3)_3N + H_2O \rightleftharpoons (CH_3)_3NH^+ + OH^-$

$$K_b = \frac{[(CH_3)_3NH^+][OH^-]}{[(CH_3)_3N]}$$

b) $SO_3^{2-} + H_2O \rightleftharpoons HSO_3^- + OH^-$

$$K_b = \frac{[HSO_3^-][OH^-]}{[SO_3^{2-}]}$$

c) $NH_2OH + H_2O \rightleftharpoons NH_3OH^+ + OH^-$

$$K_b = \frac{[NH_3OH^+][OH^-]}{[NH_2OH]}$$

18.4 For conjugate acid base pairs, $K_a \times K_b = K_w$:
$K_b = K_w \div K_a = 1.0 \times 10^{-14} \div 1.8 \times 10^{-4} = 5.6 \times 10^{-11}$

18.5 $HBu \rightleftharpoons H^+ + Bu^-$

$$K_a = \frac{[H^+][Bu^-]}{[HBu]}$$

	[HBu]	$[H^+]$	$[Bu^-]$
I	0.01000	–	–
C	–x	+x	+x
E	0.01000–x	+x	+x

We know that the acid is 4.0% ionized so x = 0.01000 M × 0.040 = 0.00040 M. Therefore, our equilibrium concentrations are $[H^+] = [Bu^-]$ = 0.00040 M, and [HBu] = 0.01000 M – 0.00040 M = 0.00960 M.

Substituting these values into the mass action expression gives:

$$K_a = \frac{(0.00040)(0.00040)}{0.00960} = 1.7 \times 10^{-5}$$

$pK_a = -\log(K_a) = -\log(1.7 \times 10^{-5}) = 4.78$

18.6 We will use the symbol Mor and $HMor^+$ for the base and its conjugate acid respectively: $Mor + H_2O \rightleftharpoons HMor^+ + OH^-$

$$K_b = \frac{[HMor^+][OH^-]}{[Mor]}$$

	[Mor]	$[HMor^+]$	$[OH^-]$
I	0.010	–	–
C	–x	+x	+x
E	0.010–x	+x	+x

At equilibrium, the pH = 10.10 and the pOH = 14.00 – 10.10 = 3.90. The $[OH^-] = 10^{-pOH} = 10^{-3.90} = 1.3 \times 10^{-4}$ M = x.

Substituting these values into the mass action expression gives:

$$K_b = \frac{(1.3\times10^{-4})(1.3\times10^{-4})}{0.010 - 1.3\times10^{-4}} = 1.7 \times 10^{-6}$$

$pK_b = -\log(K_b) = -\log(1.7 \times 10^{-6}) = 5.77$

18.7 $HC_2H_6NO_2 \rightleftharpoons H^+ + C_2H_6NO_2^-$

$$K_a = \frac{[H^+][C_2H_6NO_2^-]}{[HC_2H_6NO_2]} = 1.4\times10^{-5}$$

	$[HC_2H_6NO_2]$	$[H^+]$	$[C_2H_6NO_2^-]$
I	0.050	–	–
C	–x	+x	+x
E	0.050–x	+x	+x

Assume that x << 0.050 and substitue the equilibrium values into the mass action expression to get:

$$K_a = \frac{[x][x]}{[0.050]} = 1.4 \times 10^{-5}$$

Solving for x we determine that x = 8.4×10^{-4} M = $[H^+]$.
pH = $-\log[H^+] = -\log(8.4 \times 10^{-4}) = 3.08$

18.8 $C_5H_5N + H_2O \rightleftharpoons C_5H_5NH^+ + OH^-$

$$K_b = \frac{[C_5H_5NH^+][OH^-]}{[C_5H_5N]} = 1.5 \times 10^{-9}$$

	$[C_5H_5N]$	$[C_5H_5NH^+]$	$[OH^-]$
I	0.010	–	–
C	–x	+x	+x
E	0.010–x	+x	+x

Assume that x << 0.010 and substitute the equilibrium values into the mass action expression to get:

$$K_b = \frac{[x][x]}{[0.010]} = 1.7 \times 10^{-9}$$

Solving for x we determine that x = 4.1×10^{-6} M = $[OH^-]$.
pOH = $-\log[OH^-] = -\log(4.1 \times 10^{-6}) = 5.38$
pH = 14.00 – pOH = 8.62

18.9 We will use the notation Hphenol and phenol for the acid and its conjugate base:
Hphenol $\rightleftharpoons$ H^+ + phenol

$$K_a = \frac{[H^+][\text{phenol}]}{[\text{Hphenol}]} = 1.3 \times 10^{-10}$$

	[Hphenol]	$[H^+]$	[phenol]
I	0.15	–	–
C	–x	+x	+x
E	0.15–x	+x	+x

If we assume that x << 0.15, a good assumption based upon the size of K_a, we can substitute the equilibrium values in to the mass action expression to get:

$$K_a = \frac{(x)(x)}{0.15} = 1.3 \times 10^{-10}$$

Solving gives $x = 4.4 \times 10^{-6}$ M = $[H^+]$
pH = $-\log[H^+] = -\log(4.4 \times 10^{-6}) = 5.36$

18.10 We examine the ions in solution one at a time: The cation for acidity, and the anion for basicity. Na^+ is an ion of a Group IA metal and is *not* acidic. NO_3^- is the conjugate base of HNO_3 (a strong acid), therefore it is *not* a strong base. This solution should be neutral.

18.11 We examine the ions in solution one at a time: The cation for acidity, and the anion for basicity. K^+ is an ion of a Group IA metal and is *not* acidic. Cl^- is the conjugate base of HCl (a strong acid), therefore it is *not* a strong base. This solution should be neutral.

18.12 We examine the ions in solution one at a time: The cation for acidity, and the anion for basicity. NH_4^+ is the conjugate acid of NH_3, a weak, molecular base. Therefore it is slightly *acidic.* Br^- is the conjugate base of HBr (a strong acid), therefore it is *not* a strong base. This solution should be acidic.

18.13 The sodium ion is neutral since it is the salt of the strong base, NaOH. The nitrite ion is basic since it is the salt of nitrous acid, HNO_2, a weak acid. The equilibrium we are interested in for this problem is: $NO_2^- + H_2O \rightleftharpoons HNO_2 + OH^-$.

$$K_b = \frac{[HNO_2][OH^-]}{[NO_2^-]}$$

In order to determine the value for K_b recall that $K_a \times K_b = K_w$. We can look for the value of K_a for HNO_2 $K_b = 1.0 \times 10^{-14} \div 7.1 \times 10^{-4} = 1.4 \times 10^{-11}$.

	$[NO_2^-]$	$[HNO_2]$	$[OH^-]$
I	0.10	–	–
C	–x	+x	+x
E	0.10–x	+x	+x

Assume that x << 0.10 and substitue the equilibrium values into the mass action expression to get:

$$K_b = \frac{[x][x]}{[0.10]} = 1.4 \times 10^{-11}$$

Solving we determine that $x = 1.2 \times 10^{-6}$ M = $[OH^-]$.
pOH = $-\log[OH^-] = -\log(1.2 \times 10^{-6}) = 5.93$
pH = 14.00 – pOH = 8.07

18.14 As previously determined, a solution of NH_4Br will be acidic since NH_4^+ is the salt of a weak base and Br^- is the salt of a strong acid. As in the previous Practice Exercise, we need to determine the value for the dissociation constant using the relationship $K_a \times K_b = K_w$ and the value of K_b for NH_3 as listed in Table 16.5. $K_a = 1.0 \times 10^{-14} \div 1.8 \times 10^{-5} = 5.6 \times 10^{-10}$. The equilibrium reaction is:
$NH_4^+ \rightleftharpoons NH_3 + H^+$

	$[NH_4^+]$	$[NH_3]$	$[H^+]$
I	0.10	–	–
C	–x	+x	+x
E	0.10–x	+x	+x

Assume that x << 0.10 and substitute the equilibrium values into the mass action expression to get:

$$K_a = \frac{[x][x]}{[0.10]} = 5.6 \times 10^{-10}$$

Solving we determine that $x = 7.5 \times 10^{-6}$ M = $[H^+]$.
$pH = -\log[H^+] = -\log(7.5 \times 10^{-6}) = 5.13$

18.15 Since the ammonium ion is the salt of a weak base, NH_3, it is acidic. The cyanide ion is the salt of a weak acid, HCN, so it is basic. In order to determine if the solution is acidic or basic, we need to determine the relative strength of the two components. Use the relationship $K_a \times K_b = K_w$ in order to determine the dissociation constants for the cyanide ion and the ammonium ion.

$$K_a(NH_4^+) = K_w \div K_b(NH_3) = 1.0 \times 10^{-14} \div 1.8 \times 10^{-5} = 5.6 \times 10^{-10}$$
$$K_b(CN^-) = K_w \div K_a(HCN) = 1.0 \times 10^{-14} \div 6.2 \times 10^{-14} = 1.6 \times 10^{-5}$$

Since the $K_b(CN^-)$ is larger than the $K_a(NH_4^+)$ the NH_4CN solution will be basic.

18.16 $(CH_3)_2NH + H_2O \rightleftharpoons (CH_3)_2NH_2^+ + OH^-$

$$K_b = \frac{[(CH_3)_2NH_2^+][OH^-]}{[(CH_3)_2NH]} = 9.6 \times 10^{-4}$$

	$[(CH_3)_2NH]$	$[(CH_3)_2NH_2^+]$	$[OH^-]$
I	0.0010	–	–
C	–x	+x	+x
E	0.0010–x	+x	+x

We cannot neglect x in this calculation due to the large size of the dissociation constant. Consequently, we will have to solve the quadratic equation. Substituting the equilibrium values into the mass action expression gives:

$$K_b = \frac{[x][x]}{[0.0010 - x]} = 9.6\times10^{-4}$$

Rearranging and collecting terms on one side of the equal sign gives:

$$x^2 + (9.6 \times 10^{-4})x - (9.6 \times 10^{-7}) = 0$$

Using the quadratic equation

$x = 6.1 \times 10^{-4}$ M = $[OH^-]$, pOH = 3.21 and the pH = 10.79

18.17 The equation is: $C_2H_3O_2^- + H_2O \rightleftharpoons HC_2H_3O_2 + OH^-$

Start by determining Kb for acetate ion using $K_w = K_aK_b$

$$K_b = K_w/K_a = 1.0 \times 10^{-14}/1.8 \times 10^{-5} = 5.6 \times 10^{-10}$$

$$K_b = \frac{[HC_2H_3O_2][OH^-]}{[C_2H_3O_2^-]}$$

	$[C_2H_3O_2^-]$	$[HC_2H_3O_2]$	$[OH^-]$
I	0.11	0.090	–
C	–x	+x	+x
E	0.11 – x	0.090 +x	+x

$$K_b = \frac{[x][0.090 + x]}{[0.11 - x]}$$

assume x << 0.090 and solve for x

$x = [OH^-] = 6.8\times10^{-10}$

pOH = 9.16

pH = 14.00 – 9.16 = 4.84

18.18 We will use the relationship $K_a \times K_b = K_w$.

$$K_a(NH_4^+) = K_w \div K_b(NH_3) = 1.0 \times 10^{-14} \div 1.8 \times 10^{-5} = 5.6 \times 10^{-10}$$

The equation for the acid dissociation is $NH_4^+ \rightleftharpoons H^+ + NH_3$. The dissociation constant is written:

$$K_a = \frac{[H^+][NH_3]}{[NH_4^+]} = 5.6\times10^{-10} = \frac{[H^+](0.12)}{(0.095)}$$

Solving gives $[H^+] = 4.4 \times 10^{-10}$ M

$pH = -\log[H^+] = -\log(4.4 \times 10^{-10}) = 9.36$

18.19 Yes, formic acid and sodium formate would make a good buffer solution since pK_a = 3.74 and the desired pH is within one pH unit of this value.

Using Equation 18.12; $[H^+] = K_a \times \frac{\text{mol } HCHO_{2\,\text{initial}}}{\text{mol } CHO_2^-{}_{\text{initial}}}$

Rearranging and substituting the known values we get;

$$\frac{\text{mol HCHO}_{2\text{initial}}}{\text{mol CHO}_2{}^-{}_{\text{initial}}} = \frac{[H^+]}{K_a} = \frac{1.3\times10^{-4}}{1.8\times10^{-4}} = 0.72 \text{ mol } HCHO_2 \text{ for every mol } CHO_2^-.$$

For 0.10 mol $HCHO_2$, 0.072 mol $NaCHO_2$ would be needed. Converting to grams, this is (0.072 mol $NaCHO_2$)(68.02 g/1 mol) = 4.9 g $NaCHO_2$.

18.20 Using the data provided in Example 18.17, the initial pH of this buffer is 4.74. When OH^- is added to the solution, it will react with the H^+ present from the dissociation of the acetic acid. The acetic acid in solution dissociates further to maintain the equilibrium and any unreacted hydroxide will react with H^+ as it is produced. Eventually, the hydroxide will be completely reacted. The net change that occurs is a reduction in the amount of acetic acid in solution and an equivalent increase in the amount of acetate ion in solution. We started with 1 mol of acetic acid and 1 mol of acetate ion. The addition of 0.11 mol OH^- will reduce the amount of acetic acid to 0.89 mol and increase the amount of acetate ion to 1.11 mol. Substituting these amounts into the mass action expression gives:

$$K_a = \frac{[H^+](1.11)}{(0.89)} = 1.8\times10^{-5}$$

$$[H^+] = 1.4\times10^{-5} \text{ M}$$

$pH = -\log[H^+] = 4.84$

18.21 $[H^+]$ is determined by the first protic equilibrium:
$H_2C_6H_6O_6 \rightleftharpoons H^+ + HC_6H_6O_6^-$

The mass action expression is: $K_{a_1} = 6.8 \times 10^{-5} = x^2/0.10$
$x = [H^+] = 2.6 \text{ x } 10^{-3}$ M
$pH = -\log(2.6 \text{ x } 10^{-3}) = 2.6$

The concentration of the anion, $[HC_6H_6O_6^-]$, is given almost entirely by the second ionization equilibrium: $HC_6H_6O_6^- \rightleftharpoons H^+ + C_6H_6O_6^{2-}$ for which the mass action expression is:

$$K_{a2} = \frac{[H^+][C_6H_6O_6^{2-}]}{[HC_6H_6O_6^-]} = 2.7\times10^{-12}$$

We have used the value for K_{a_2} from Table 18.3. Using the value of x from the first step above gives:

$$2.7\times10^{-12} = \frac{(2.6\times10^{-3})[C_6H_6O_6^{2-}]}{(2.6\times10^{-3})}$$

$$[HC_6H_6O_6^-] = 2.7 \times 10^{-12}$$

18.22 The equilibrium we are interested in for this problem is:

$$SO_3^{2-}(aq) + H_2O(\ell) \rightleftharpoons HSO_3^-(aq) + OH^-(aq)$$

$$K_b = K_w/K_{a_2} = 1.0 \times 10^{-14} / 6.6 \times 10^{-8} = 1.5 \times 10^{-7}$$

$$K_b = 1.5\times10^{-7} = \frac{[HSO_3^-][OH^-]}{[SO_3^{2-}]}$$

	$[SO_3^{2-}]$	$[HSO_3^-]$	$[OH^-]$
I	0.20	–	–
C	–x	+x	+x
E	0.20–x	+x	+x

Substituting these values into the mass action expression gives:

$$K_b = 1.5\times10^{-7} = \frac{(x)(x)}{0.20 - x}$$

Assume that $x << 0.20$ and solving gives $x = 1.7 \times 10^{-4}$.
$x = [OH^-] = 1.7 \times 10^{-4}$ M
$pOH = -\log(1.7 \times 10^{-4}) = 3.76$
$pH = 14.00 - pOH = 14.00 - 3.76 = 10.24$

18.23 For weak polyprotic acids, the concentration of the polyvalent ions is equal to the volume of K_{an} where n is the valency. By analogy, the concentration of H_2SO_3 in 0.010 M Na_2SO_3 will be equal to K_{b2} for SO_3^{2-}.

18.24 $HCHO_2 + H_2O \rightleftharpoons H_3O^+ + CHO_2^-$

$$Ka = \frac{[H_3O^+][CHO_2^-]}{[HCHO_2]} = 1.8\times10^{-4}$$

a)

	$[HCHO_2]$	$[H_3O^+]$	$[CHO_2^-]$
I	0.100	–	–
C	–x	+x	+x
E	0.100–x	+x	+x

Assume $x << 0.100$. Solving we get $x = [H_3O^+] = 4.2 \times 10^{-3}$. The pH is 2.37.

b) $[HCHO_2] = [CHO_2^-]$ so $[H_3O^+] = Ka = 1.8 \times 10^{-4}$ and the pH = 3.74.

c)

$$\#\text{ moles base added} = (15.0\text{ mL})\left(\frac{0.100\text{ mol}}{1000\text{ mL}}\right) = 1.50\times10^{-3}$$

$$\#\text{ moles acid initially} = (20.0\text{ mL})\left(\frac{0.10\text{ mol}}{1000\text{ mL}}\right) = 2.00\times10^{-3}$$

$$\text{excess acid} = 2.00\times10^{-3} - 1.50\times10^{-3} = 0.50\times10^{-3}\text{ moles acid}$$

$$[\text{acid}] = \frac{0.50\times10^{-3}\text{ moles}}{(35\text{ mL})\left(\frac{1\text{ L}}{1000\text{ mL}}\right)} = 1.43\times10^{-2}\text{ M}$$

$$[\text{base}] = \frac{1.50\times10^{-3}\text{ moles}}{(35\text{ mL})\left(\frac{1\text{ L}}{1000\text{ mL}}\right)} = 4.29\times10^{-2}\text{ M}$$

Substituting into the equilibrium expression and solving we get $[H_3O^+] = 6.00 \times 10^{-5}$ and the pH = 4.22.

d) We now have a solution of formate ion with a concentration of 0.0500 M. We need Kb for formate ion: $K_b = K_w/K_a = 5.6 \times 10^{-11}$. If we set up the equilibrium problem and solve we get: $[OH^-] = 1.7 \times 10^{-6}$.

The pOH = 5.78 and the pH = 8.22.

18.25

$$\#\text{ moles base added} = (30.0\text{ mL})\left(\frac{0.15\text{ mol}}{1000\text{ mL}}\right) = 4.50\times10^{-3}$$

$$\#\text{ moles acid initially} = (50.0\text{ mL})\left(\frac{0.20\text{ mol}}{1000\text{ mL}}\right) = 1.00\times10^{-2}$$

$$\text{excess acid} = 1.00\times10^{-2} - 4.50\times10^{-3} = 5.50\times10^{-3}\text{ moles acid}$$

$$[\text{acid}] = \frac{5.50\times10^{-3}\text{ moles}}{(80\text{ mL})\left(\frac{1\text{ L}}{1000\text{ mL}}\right)} = 6.88\times10^{-2}\text{ M}$$

$$[\text{base}] = \frac{4.50\times10^{-3}\text{ moles}}{(80\text{ mL})\left(\frac{1\text{ L}}{1000\text{ mL}}\right)} = 5.63\times10^{-2}\text{ M}$$

$$HCHO_2 + H_2O \rightleftharpoons H_3O^+ + CHO_2^-$$

$$K_a = \frac{[H_3O^+][CHO_2^-]}{[HCHO_2]}$$

	$[HCHO_2]$	$[H_3O^+]$	$[CHO_2^-]$
I	6.88×10^{-2}	–	5.63×10^{-2}
C	$-x$	$+x$	$+x$
E	$6.88 \times 10^{-2} - x$	$+x$	$5.63 \times 10^{-2} + x$

Assume $x << 5.63 \times 10^{-2}$

$$Ka = \frac{[x][5.63\times10^{-2}]}{[6.88\times10^{-2}]}$$

$$x = 2.20\times10^{-2} = [H_3O^+]$$

pH = 3.66

Review Problems

18.42 At 25 °C, $K_a \times K_b = K_w$
$K_b = K_w/K_a = 1.0 \times 10^{-14} \div 6.8 \times 10^{-4} = 1.5 \times 10^{-11}$

18.44 At 25 °C, $K_a \times K_b = K_w$
$K_b = K_w/K_a = 1.0 \times 10^{-14} \div 1.8 \times 10^{-12} = 5.6 \times 10^{-3}$

18.46 At 25 °C, $K_a \times K_b = K_w$
$K_b = K_w/K_a = 1.0 \times 10^{-14} \div 1.4 \times 10^{-4} = 7.1 \times 10^{-11}$

18.48 $[H+] = 10^{-pH} = 10^{-3.22} = 6.03 \times 10^{-4}$ M

% ionization = $([H^+]/[\text{total acid}])$ x 100 % = $(6.03 \times 10^{-4}/0.20)$x100 = 0.30 %

18.50

	[HA]	$[H^+]$	$[A^-]$
I	0.20	–	–
C	-6.03×10^{-4}	$+6.03 \times 10^{-4}$	$+6.03 \times 10^{-4}$
E	~0.20	6.03×10^{-4}	6.03×10^{-4}

Since the percent ionization is so low, we can simplify the equilibrium concentration of HA to be 0.20 (as above):

$$K_a = \frac{[H^+][A^-]}{[HA]} = \frac{[6.03\times10^{-4}][6.03\times10^{-4}]}{[0.20]}$$

$$= 1.8 \times 10^{-6}$$

18.52 $HIO_4 \rightleftharpoons H^+ + IO_4^-$

$$K_a = \frac{[H^+][IO_4^-]}{[HIO_4]}$$

	$[HIO_4]$	$[H^+]$	$[IO_4^-]$
I	0.10	–	–
C	–x	+x	+x
E	0.10–x	+x	+x

We know that at equilibrium $[H^+] = 0.038$ M = x. The equilibrium concentrations of the other components of the mixture are:

$[HIO_4] = 0.10 - x = 0.06$ M and $[IO_4^-] = x = 0.038$ M.

Substituting the above values for equilibrium concentrations into the mass action expression gives:

$$K_a = \frac{(0.038)(0.038)}{0.06} = 2 \times 10^{-2}$$

$pK_a = -\log(K_a) = -\log(2 \times 10^{-2}) = 1.7$

18.54 pOH = 14.00 – pH = 14.00 – 11.86 = 2.14

$[OH^-] = 10^{-pOH} = 10^{-2.14} = 7.2 \times 10^{-3}$ M

$$CH_3CH_2NH_2 + H_2O \rightleftharpoons CH_3CH_2NH_3^+ + OH^-$$

$$K_b = \frac{[CH_3CH_2NH_3^+][OH^-]}{[CH_3CH_2NH_2]}$$

	$[CH_3CH_2NH_2]$	$[CH_3CH_2NH_3^+]$	$[OH^-]$
I	0.10	–	–
C	–x	+x	+x
E	0.10–x	+x	+x

In the equilibrium analysis, the value of x is 7.2×10^{-3} M. Therefore, our equilibrium concentrations are $[CH_3CH_2NH_3^+] = [OH^-] = 7.2 \times 10^{-3}$ M, and $[CH_3CH_2NH_2] = 0.10$ M – 7.2×10^{-3} M = 0.09 M.

Substituting these values into the mass action expression gives:

$$K_b = \frac{(7.2\times10^{-3})(7.2\times10^{-3})}{0.09} = 6 \times 10^{-4}$$

$pK_b = -\log(K_b) = -\log(6 \times 10^{-4}) = 3.2$

% ionization = ([OH^-]/[total base]) x 100 % = ($7.2 \times 10^{-3}/0.10$) × 100 = 7.2 %

18.56 $HC_3H_5O_2 + H_2O \rightleftharpoons C_3H_5O_2^- + H_3O^+$

	$[HC_3H_5O_2]$	$[C_3H_5O_2^-]$	$[H_3O^+]$
I	0.150	—	—
C	−x	+x	+x
E	0.150−x	+x	+x

$$K_a = \frac{[C_3H_5O_2^-][H_3O^+]}{[HC_3H_5O_2]}$$

$$1.4\times10^{-4} = \frac{[x][x]}{[0.150 - x]}$$

Since the percent ionization is low (small K_a), we can assume $0.150 - x \approx 0.150$:

$$1.4\times10^{-4} = \frac{[x][x]}{[0.150]}$$

$$x^2 = 2.1 \times 10^{-5}$$

$$x = 0.0046 \text{ M}$$

Looking at the values for E in the table above, and inserting x, we have:

$[HC_3H_5O_2] = 0.150 - 0.0046 = 0.145$ M
$[C_3H_5O_2^-] = [H_3O^+] = 0.0046$ M

pH = − log [H^+] = 2.3

18.58 $HN_3 + H_2O \rightleftharpoons H_3O^+ + N_3^-$

$$K_a = \frac{[H_3O^+][N_3^-]}{[HN_3]} = 1.8\times10^{-5}$$

	$[HN_3]$	$[H_3O^+]$	$[N_3^-]$
I	0.15	—	—
C	−x	+x	+x
E	0.15 −x	+x	+x

Assume x << 0.15

$$K_a = \frac{[x][x]}{[0.15]} = 1.8\times10^{-5} \quad x = 1.6\times10^{-3} = [H_3O^+]$$

pH = 2.78

% ionization = ([H$^+$]/[total acid]) x 100 % = $(1.6 \times 10^{-3}/0.15) \times 100 = 1.1$ %

18.60 $K_b = 10^{-pKb} = 10^{-5.79} = 1.6 \times 10^{-6}$

$Cod + H_2O \rightleftharpoons HCod^+ + OH^-$

$$K_b = \frac{[HCod^+][OH^-]}{[Cod]} = 1.6 \times 10^{-6}$$

	[Cod]	[HCod$^+$]	[OH$^-$]
I	0.020	–	–
C	–x	+x	+x
E	0.020–x	+x	+x

Substituting these values into the mass action expression gives:

$$K_b = \frac{(x)(x)}{0.020 - x} = 1.6 \times 10^{-6}$$

If we assume that x << 0.020 we get:

$x^2 = 3.2 \times 10^{-8}$

$x = 1.8 \times 10^{-4}$ M = [OH$^-$]

pOH = –log[OH$^-$] = $-\log(1.8 \times 10^{-4}) = 3.74$

pH = 14.00 – pOH = 14.00 – 3.74 = 10.26

18.62 [H$^+$] = $10^{-pH} = 10^{-2.54} = 2.9 \times 10^{-3}$ M

$HC_2H_3O_2 \rightleftharpoons H^+ + C_2H_3O_2^-$

$$K_a = \frac{[H^+][C_2H_3O_2^-]}{[HC_2H_3O_2]} = 1.8 \times 10^{-5}$$

	[$HC_2H_3O_2$]	[H$^+$]	[$C_2H_3O_2^-$]
I	Z	–	–
C	–x	+x	+x
E	Z–x	+x	+x

Substituting these values into the mass action expression gives:

$$K_a = \frac{(x)(x)}{Z - x} = 1.8 \times 10^{-5}$$

Assuming x << Z and knowing that x = 2.9×10^{-3} M, we can solve for Z and find Z = 0.47. The initial concentration of $HC_2H_3O_2$ is 0.47 M.

18.64 NaCN will be basic in solution since CN^- is a basic ion and Na^+ is a neutral ion.

$$CN^- + H_2O \rightleftharpoons HCN + OH^-$$

For HCN, $K_a = 6.2 \times 10^{-10}$, we need K_b for CN^-
$K_b = K_w/K_a = (1.0 \times 10^{-14}) \div (6.2 \times 10^{-10}) = 1.6 \times 10^{-5}$

$$K_b = \frac{[HCN][OH^-]}{[CN^-]} = 1.6 \times 10^{-5}$$

	$[CN^-]$	$[HCN]$	$[OH^-]$
I	0.20	–	–
C	–x	+x	+x
E	0.20–x	+x	+x

Substituting these values into the mass action expression gives:

$$K_b = \frac{(x)(x)}{0.20 - x} = 1.6 \times 10^{-5}$$

Assuming that x << 0.20 we can solve for x an determine;
$x = 1.8 \times 10^{-3}$ M = $[OH^-]$
$pOH = -\log[OH^-] = -\log(1.8 \times 10^{-3}) = 2.74$
$pH = 14.00 - pOH = 14.00 - 2.74 = 11.26$

Concentration of HCN is equal to that of hydroxide ion: 1.8×10^{-3} M

18.66 A solution of CH_3NH_3Cl will be acidic since the Cl^- ion is neutral and the $CH_3NH_3^+$ ion is acidic.

$$CH_3NH_3^+ \rightleftharpoons H^+ + CH_3NH_2$$

For CH_3NH_2, $K_b = 4.4 \times 10^{-4}$. We need K_a for $CH_3NH_3^+$
$K_a = K_w/K_b = (1.0 \times 10^{-14}) \div (4.4 \times 10^{-4}) = 2.3 \times 10^{-11}$

$$K_a = \frac{[H^+][CH_3NH_2]}{[CH_3NH_3^+]} = 2.3 \times 10^{-11}$$

	$[CH_3NH_3^+]$	$[H^+]$	$[CH_3NH_2]$
I	0.15	–	–
C	–x	+x	+x
E	0.15–x	+x	+x

Substituting these values into the mass action expression gives:

$$K_a = \frac{(x)(x)}{0.15 - x} = 2.3\times10^{-11}$$

Assuming that x << 0.15 we can solve for x an determine;
$x = 1.9 \times 10^{-6}$ M = $[H_3O^+]$
pH = $-\log[H_3O^+] = -\log(1.9 \times 10^{-6}) = 5.72$

18.68 The reaction for this problem is: $H\text{–}Mor^+ \rightleftharpoons H^+ + Mor$
We know that $pK_a + pK_b = pK_w = 14.00$. So,
$pK_a = 14.00 - pK_b = 14.00 - 6.13 = 7.87$, and $K_a = 10^{-pKa} = 1.3 \times 10^{-8}$

$$K_a = \frac{[H^+][Mor]}{[H - Mor^+]} = 1.3\times10^{-8}$$

	$[H\text{–}Mor^+]$	$[H^+]$	[Mor]
I	0.20	–	–
C	–x	+x	+x
E	0.20–x	+x	+x

Substituting these values into the mass action expression gives:

$$K_a = \frac{(x)(x)}{0.20 - x} = 1.3\times10^{-8}$$

Assuming that x << 0.20 we can solve for x an determine;
$x = 5.1 \times 10^{-5}$ M = $[H^+]$
pH = $-\log[H^+] = -\log(5.1 \times 10^{-5}) = 4.29$

18.70 Let HNic symbolize the nicotinic acid: $Nic^- + H_2O \rightleftharpoons HNic + OH^-$

$$K_b = \frac{[HNic][OH^-]}{[Nic^-]}$$

	$[Nic^-]$	[HNic]	$[OH^-]$
I	0.18	–	–
C	–x	+x	+x
E	0.18–x	+x	+x

Assume x << 0.18

$$K_b = \frac{(x)(x)}{0.18} \qquad \text{we know } x = [OH^-]$$

$$\text{we know pH} = 9.05 \text{ so pOH} = 4.95$$

$$\text{and } [OH^-] = 10^{-pOH} = 1.12 \times 10^{-5} = x$$

$$\text{Now } K_b = \frac{(1.12 \times 10^{-5})^2}{0.18} = 7.0 \times 10^{-10}$$

For a conjugate acid-base pair:

$$K_a \cdot K_b = 1.00 \times 10^{-14}$$
$$K_a \cdot 7.0 \times 10^{-10} = 1.00 \times 10^{-14}$$
$$K_a = 1.4 \times 10^{-5}$$

18.72 $C_5H_5NH^+$ is the conjugate acid of pyridine, C_5H_5N, which has a K_b listed in Table 18.2; $K_b = 1.7 \times 10^{-9}$. For a conjugate acid-base pair:

$$K_a \cdot K_b = 1.00 \times 10^{-14}$$
$$K_a \cdot 1.7 \times 10^{-9} = 1.00 \times 10^{-14}$$
$$K_a = 5.9 \times 10^{-6}$$

$$C_5H_5NH^+ + H_2O \rightleftharpoons H_3O^+ + C_5H_5N$$

$$K_a = \frac{[H_3O^+][C_5H_5N]}{[C_5H_5N_3H^+]} = 5.9 \times 10^{-6}$$

	$[C_5H_5NH^+]$	$[H_3O^+]$	$[C_5H_5N]$
I	0.10	–	–
C	–x	+x	+x
E	0.10 –x	+x	+x

Assume x << 0.10

$$K_a = \frac{[x][x]}{[0.10]} = 5.9 \times 10^{-6} \qquad x = 7.7 \times 10^{-4} = [C_5H_5N]$$

% reacting = $([C_5H_5N]/[\text{total acid}]) \times 100\ \% = (7.7 \times 10^{-4}/0.10) \times 100 = 0.77\ \%$

*18.74 $OCl^- + H_2O \rightleftharpoons HOCl + OH^-$

$$K_b = \frac{[HOCl][OH^-]}{[OCl^-]} = \frac{K_w}{K_a} = \frac{1.0 \times 10^{-14}}{3.0 \times 10^{-8}} = 3.3 \times 10^{-7}$$

$$[OCl^-] = \left(\frac{5.0 \text{ g NaOCl}}{100 \text{ g solution}}\right)\left(\frac{1 \text{ mol NaOCl}}{74.5 \text{ g NaOCl}}\right)\left(\frac{1.0 \text{ g}}{1 \text{ mL}}\right)\left(\frac{1000 \text{ mL}}{1 \text{ L}}\right)$$
$$= 0.67 \text{ M}$$

	$[OCl^-]$	$[HOCl]$	$[OH^-]$
I	0.67	–	–
C	–x	+x	+x
E	0.67 –x	+x	+x

Assume that x << 0.67

$$K_b = \frac{(x)(x)}{0.67} = 3.3\times10^{-7} \qquad x = 4.7\times10^{-4} = [OH^-]$$

$$pOH = 3.3$$

$$pH = 10.7$$

18.76 $HF \rightleftharpoons H^+ + F^-$

$$K_a = \frac{[H^+][F^-]}{[HF]} = 6.8\times10^{-4}$$

	$[HF]$	$[H^+]$	$[F^-]$
I	0.15	–	–
C	–x	+x	+x
E	0.15–x	+x	+x

Substituting the above values for equilibrium concentrations into the mass action expression and assuming that x << 0.15 gives:

$$K_a = \frac{(x)(x)}{0.15} = 6.8\times10^{-4}$$

$x^2 = 1.0 \times 10^{-4}$
$x = 1.0 \times 10^{-2}$ M = $[H^+]$
pH = $-\log[H^+] = -\log(0.010) = 2.00$
% ionization = 0.010/0.15 × 100 % = 6.7 %

Since the % ionization is > 5%, we can not make the simplifying assumption. Consequently, we need to solve for x using a quadratic equation. The mass action expression is:

$$K_a = \frac{(x)(x)}{0.15 - x} = 3.0\times10^{-9}$$

We may rearrange this expression to obtain the following quadratic equation:

$$x^2 + (6.8 \times 10^{-4})x - (1.0 \times 10^{-4}) = 0$$

where a = 1, b = 6.8×10^{-4}, and c = 1.0×10^{-4}. If we substitute these values into the quadratic equation and take the positive root we get x = 0.0097 M.
$[H^+]$ = x = 0.0097 M.
% ionization = 0.0097/0.15 × 100 % = 6.5 %
pH = $-\log[H^+] = -\log(0.0097) = 2.01$

(In this problem, the change in pH caused by simplifying assumptions is very slight.)

18.78 $CN^- + H_2O \rightleftharpoons HCN + OH^-$

$$K_b = \frac{[HCN][OH^-]}{[CN^-]} = \frac{K_w}{K_a} = \frac{1.0\times10^{-14}}{3.0\times10^{-8}} = 1.6\times10^{-5}$$

	$[CN^-]$	[HCN]	$[OH^-]$
I	0.0050	–	–
C	–x	+x	+x
E	0.0050 –x	+x	+x

Assume that x << 0.0050

$$K_b = \frac{(x)(x)}{0.0050} = 1.6\times10^{-5} \qquad x = 2.8\times10^{-4} = [OH^-]$$

However...2.8×10^{-4} is not << 0.0050...

So, we use the method of successive approximations:

$$K_b = \frac{(x)(x)}{0.0050 - 0.00028} = 1.6\times10^{-5} \qquad x = 2.7\times10^{-4}$$

$$Kb = \frac{(x)(x)}{0.0050 - 0.00027} = 1.6\times10^{-5} \qquad x = 2.7\times10^{-4}$$

$x = 2.7\times10^{-4} = [OH^-]$

pOH = 3.56

pH = 10.44

18.80 $K_a = 10^{-pKa} = 10^{-4.92} = 1.2 \times10^{-5}$

$$\text{H–Paba} \rightleftharpoons H^+ + Paba^-$$

$$K_a = \frac{[H^+][Paba^-]}{[H-Paba]} = 1.2\times10^{-5}$$

	[H–Paba]	$[H^+]$	$[Paba^-]$
I	0.030	–	–
C	–x	+x	+x
E	0.030–x	+x	+x

Substituting the above values for equilibrium concentrations into the mass action expression and assuming that x << 0.030 gives:

$$K_a = \frac{[x][x]}{[0.030]} = 1.2 \times 10^{-5}$$

$x^2 = 3.6 \times 10^{-7}$
$x = 6.0 \times 10^{-4}$ M = $[H^+]$
pH = –log$[H^+]$ = –log(6.0×10^{-4}) = 3.22

18.82 $HC_2H_3O_2 \rightleftharpoons H^+ + C_2H_3O_2^-$

$$K_a = \frac{[H^+][C_2H_3O_2^-]}{[HC_2H_3O_2]} = 1.8 \times 10^{-5}$$

	$[HC_2H_3O_2]$	$[H^+]$	$[C_2H_3O_2^-]$
I	0.15	–	0.25
C	–x	+x	+x
E	0.15–x	+x	0.25+x

Substituting these values into the mass action expression gives:

$$K_a = \frac{(x)(0.25 + x)}{0.15 - x} = 1.8 \times 10^{-5}$$

Assume that x << 0.15 M and x << 0.25 M, then;

$$x \times \left(\frac{0.25}{0.15}\right) \approx 1.8 \times 10^{-5}$$

$$x \approx \left(\frac{0.15}{0.25}\right) \times 1.8 \times 10^{-5}$$

$$x \approx 1.1 \text{X} 10^{-5} \text{ M} = [H^+]$$

pH = –log $[H^+]$ = 4.97

18.84 The equilibrium we will consider in this problem is: $NH_3 + H_2O \rightleftharpoons NH_4^+ + OH^-$

$$K_b = \frac{[NH_4^+][OH^-]}{[NH_3]} = 1.8 \times 10^{-5}$$

$$= \frac{(0.45)[OH^-]}{0.25} = 1.8 \times 10^{-5}$$

$[OH^-] = 1.0 \times 10^{-5}$ M
pOH = –log$[OH^-]$ = –log(1.0×10^{-5}) = 5.00
pH = 14.00 – pOH = 14.00 – 5.00 = 9.00

18.86 For a conjugate acid-base pair, $K_a \cdot K_b = 1.00 \times 10^{-14}$. The K_b of ammonia is given in Table 18.2, so we can find the K_a for the ammonium ion:

$$K_a \cdot K_b = 1.00 \times 10^{-14}$$
$$K_a \cdot 1.8 \times 10^{-5} = 1.00 \times 10^{-14}$$
$$K_a = 1.4 \times 10^{-5}$$

Using equation 18.10:

$$[H^+] = K_a \frac{[HA]}{[A^-]}$$

$$[H^+] = K_a \frac{[NH_4^+]}{[NH_3]}$$

$$[H^+] = 1.4 \times 10^{-5} \frac{[0.20]}{[0.25]} = 1.1 \times 10^{-5}$$

Therefore, the initial pH of the buffer = $-\log(1.1 \times 10^{-5}) = 5.0$

Initial amounts in the solution are:
mol NH_3 = (.25 mol/L)(.25 L) = 0.063 mol NH_3
mol NH_4^+ = (.20 mol/L)(.25 L) = 0.050 mol NH_4^+

The added acid (0.0250 L(0.10 mol/L) = 0.00250 mol HCl) will react with the ammonia present in the buffer solution. Assume the added acid reacts completely. For each mole of acid added, one mole of NH_3 is converted to NH_4^+. Since 0.00250 mol of acid is added;

mol $NH_4^+{}_{final}$ = (0.050 + 0.00250) mol = 0.053 mol
mol NH_{3final} = (0.063 – 0.00250) mol = 0.061 mol

The final volume of solution is 250 mL + 25.0 mL = 275 mL.

The final concentrations are:

$[NH_4^+]_{final}$ = 0.053 mol/0.275 L = 0.19 M NH_4^+
$[NH_3]_{final}$ = 0.061 mol/0.275 L = 0.22 M NH_3

So the *changes* in concentrations are:

$\Delta[NH_4^+] = [NH_4^+]_{final} - [NH_4^+]_{initial} = 0.19\ M - 0.20\ M = -\ 0.01\ M$
$\Delta[NH_3] = [NH_3]_{final} - [NH_3]_{initial} = 0.22\ M - 0.25\ M = -\ 0.03\ M$

18.88 The initial pH of the buffer is 4.97 as determined in Exercise 18.82. The added acid, 0.050 mol, will react with the acetate ion present in the buffer solution. Assume the added acid reacts completely. For each mole of acid added, one mole of $C_2H_3O_2^-$ is converted to $HC_2H_3O_2$. Since 0.050 mol of acid is added;

$$[HC_2H_3O_2]_{final} = (0.15 + 0.050)\ M = 0.20\ M$$
$$[C_2H_3O_2^-]_{final} = (0.25 - 0.050)\ M = 0.20\ M$$

Now, substitute these values into the mass action expression to calculate the final $[H^+]$ in solution;

$$\frac{[H^+](0.20)}{(0.20)} = 1.8 \times 10^{-5}$$

$[H^+] = 1.8 \times 10^{-5}$ mol L^{-1} and the pH = 4.74

The pH of the solution changes by 4.74 – 4.93 = – 0.23 pH units upon addition of the acid.

18.90 The initial pH is 9.00 as calculated in Exercise 18.84. For every mole of H^+ added, one mol of NH_3 will be changed to one mol of NH_4^+. Since we added 0.020 mol H^+,

$$[NH_4^+]_{final} = 0.45\ M + 0.020\ M = 0.47\ M$$
$$[NH_3]_{final} = 0.25\ M - 0.020\ M = 0.23\ M$$

Using these new concentrations, we can calculate a new pH:

$$K_b = \frac{[NH_4^+][OH^-]}{[NH_3]} = \frac{(0.47)[OH^-]}{0.23} = 1.8 \times 10^{-5}$$

$[OH^-] = 8.8 \times 10^{-6}$ M, the pOH = 5.06 and the pH = 8.94.

As expected, when an acid is added, the pH decreases. In this problem, the pH decreases by 0.06 pH units from 9.00 to 8.94.

18.92 pOH = 14.00 – pH = 14.00 – 9.25 = 4.75

$$pOH = pK_b + \log\frac{[cation]}{[base]}$$

$$4.75 = 4.74 + \log([NH_4^+]/[NH_3])$$

Therefore, $\log([NH_4^+]/[NH_3]) = 0.01$. Taking the antilog of both sides of this equation gives:

$$[NH_4^+]/[NH_3] = 10^{0.01} = 1$$

The ratio is 1 to 1.

18.94 $$pH = pK_a + \log\frac{[anion]}{[acid]} = pK_a + \log\frac{[A^-]}{[HA]}$$

$$5.00 = 4.74 + \log([NaC_2H_3O_2]/[HC_2H_3O_2])$$
$$[NaC_2H_3O_2]/[HC_2H_3O_2] = 1.8$$
$$[NaC_2H_3O_2] = 1.8 \times [HC_2H_3O_2]$$
$$[NaC_2H_3O_2] = 1.8 \times 0.15 = 0.27 \text{ M}$$

Thus to the 1 L of acetic acid solution we add: 0.27 mol $NaC_2H_3O_2 \times 82.0$ g/mol = 22 g $NaCHO_2$.

18.96 pOH = 14.00 – pH = 14.00 – 10.00 = 4.00

$$pOH = pK_b + \log\frac{[cation]}{[base]}$$

$$4.00 = 4.74 + \log([NH_4^+]/[NH_3])$$

Therefore, $\log([NH_4^+]/[NH_3]) = -0.74$. Taking the antilog of both sides of this equation gives:

$$[NH_4^+]/[NH_3] = 10^{-0.74} = 0.18$$
$$[NH_4^+] = 0.18 \times [NH_3]$$
$$[NH_4^+] = 0.18 \times 0.20 \text{ M}$$
$$[NH_4^+] = 3.6 \times 10^{-2} \text{ M} = [NH_4Cl]$$

Finally, 0.500 L of buffer solution would require:
$0.500 \text{ L} \times 3.6 \times 10^{-2} \text{ mol/L} \times 53.5 \text{ g/mol} = 0.96 \text{ g } NH_4Cl$

18.98 The equilibrium is; $HC_2H_3O_2 \rightleftharpoons H^+ + C_2H_3O_2^-$

$$K_a = \frac{[H^+][C_2H_3O_2^-]}{[HC_2H_3O_2]} = 1.8\times10^{-5}$$

The initial pH is; $\frac{[H^+](0.110)}{0.100} = 1.8\times10^{-5}$, $[H^+] = 1.64 \times 10^{-5}$ M and pH = 4.786.
In this calculation we are able to use either the molar concentration or the number of moles since the volume is constant in this portion of the problem.

In order to calculate the change in pH, we need to determine the concentrations of $HC_2H_3O_2$ and $C_2H_3O_2^-$ after the complete reaction of the added acid. One mole of $C_2H_3O_2^-$ will be consumed for every mole of acid added and one mole of $HC_2H_3O_2$ will be produced. The number of moles of acid added is;

$$\# \text{mol H}^+ = (25.00 \text{ mL HCl})\left(\frac{0.100 \text{ mol HCl}}{1000 \text{ mL HCl}}\right)\left(\frac{1 \text{ mol H}^+}{1 \text{ mol HCl}}\right)$$

$$= 2.50 \times 10^{-3} \text{ mol H}^+$$

The new concentration of $HC_2H_3O_2$ and $C_2H_3O_2^-$ are;

$$[HC_2H_3O_2]_{final} = \frac{(0.100 \text{ mol} - 0.00250 \text{ mol})}{0.525 \text{ L}} = 0.195 \text{ M}$$

$$[C_2H_3O_2^-]_{final} = \frac{(0.110 \text{ mol} + 0.00250 \text{ mol})}{0.525 \text{ L}} = 0.205 \text{ M}$$

Note: The new volume has been used in these calculations.

$$K_a = \frac{[H^+][C_2H_3O_2^-]}{[HC_2H_3O_2]} = \frac{[H^+](0.205)}{(0.195)} = 1.8 \times 10^{-5}$$

$[H^+] = 1.71 \times 10^{-5}$ and pH = 4.766.

Notice that the change in pH is very small in spite of adding a strong acid. If the same amount of HCl were added to water, a completely different effect would be observed.

Since HCl is a strong acid, the $[H^+]$ in a water solution will be the result of the strong acid dissociation. We do, of course, need to account for the dilution. Using the dilution equation, $M_1V_1 = M_2V_2$, we determine the $[H^+] = 0.0167$ mol L^{-1} and the pH = 1.78. The change in pH in this case is 7.00 – 1.78 = 5.22 pH units. A significantly larger change!!

18.100 $H_2C_6H_6O_6 + H_2O \rightleftharpoons H_3O^+ + HC_6H_6O_6^-$ $\quad K_{a1} = 6.7 \times 10^{-5}$
$HC_6H_6O_6^- + H_2O \rightleftharpoons H_3O^+ + C_6H_6O_6^{2-}$ $\quad K_{a2} = 2.7 \times 10^{-12}$

	$[H_2C_6H_6O_6]$	$[H_3O^+]$	$[HC_6H_6O_6^-]$
I	0.15	–	–
C	–x	+x	+x
E	0.15–x	+x	+x

Assume x << 0.15

$$K_a = \frac{[x][x]}{[0.15]} = 6.7\times 10^{-5} \qquad x = 3.2\times 10^{-3}$$

$[H_2C_6H_6O_6] \cong 0.15\ M$

$[H_3O^+] = [HC_6H_6O_6^-] = 3.2\times 10^{-3}\ M$

$[C_6H_6O_6^{2-}] = 2.7\times 10^{-12}\ M$

$pH = 2.50$

18.102 $H_3PO_4 \rightleftharpoons H_2PO_4^- + H^+ \qquad K_{a1} = \frac{[H_2PO_4^-][H^+]}{[H_3PO_4]} = 7.1\times 10^{-3}$

$H_2PO_4^- \rightleftharpoons HPO_4^{2-} + H^+ \qquad K_{a2} = \frac{[HPO_4^{2-}][H^+]}{[H_2PO_4^-]} = 6.3\times 10^{-8}$

$HPO_4^{2-} \rightleftharpoons PO_4^{3-} + H^+ \qquad K_{a3} = \frac{[PO_4^{3-}][H^+]}{[HPO_4^{2-}]} = 4.5\times 10^{-13}$

The following assumptions are made:

$[H^+]_{total} \approx [H^+]_{first\ step}$

$[H_2PO_4^-]_{total} \approx [H_2PO_4^-]_{first\ step}$

$[HPO_4^{2-}]_{total} \approx [HPO_4^{2-}]_{second\ step}$

The first dissociation:

	$[H_3PO_4]$	$[H_2PO_4^-]$	$[H^+]$
I	3.0	–	–
C	–x	+x	+x
E	3.0–x	+x	+x

$$K_{a1} = \frac{[H_2PO_4^-][H^+]}{[H_3PO_4]} = \frac{(x)(x)}{(3.0 - x)} = 7.1\times 10^{-3}$$

Solve for x by successive approximations or by the quadratic equation:

$x = 0.14\ M = [H^+] = [H_2PO_4^-]$

$[H_3PO_4] = 3.0 - 0.14 = 2.9\ M$

$pH = -\log(0.14) = 0.85$

The second dissociation:

	$[H_2PO_4^-]$	$[HPO_4^{2-}]$	$[H^+]$
I	0.14	–	0.14
C	–x	+x	+x
E	0.14–x	+x	0.14+ x

Assume that x << 0.14, therefore $0.14 - x \approx 0.14$ and $0.14 + x \approx 0.14$

$$K_{a2} = \frac{(x)(0.14)}{(0.14)} = 6.3\times10^{-8}$$

$x = 6.3 \times 10^{-8} = [HPO_4^{2-}]$

The third dissociation:

	$[HPO_4^{2-}]$	$[PO_4^{3-}]$	$[H^+]$
I	6.3×10^{-8}	–	0.14
C	–x	+x	+x
E	6.3×10^{-8}–x	+x	0.14+x

Assume that $x << 6.3 \times 10^{-8}$, therefore $6.3 \times10^{-8} - x \approx 6.3 \times10^{-8}$ and $0.14 + x \approx 0.14$

$$K_{a3} = \frac{(x)(0.14)}{(6.3\times10^{-8})} = 4.5\times10^{-13}$$

Solving for x we get, $x = 2.0 \times10^{-19} = [PO_4^{3-}]$

18.104 $H_2C_4H_4O_6 \rightleftharpoons HC_4H_4O_6^- + H^+$ $\quad K_{a1} = \frac{[HC_4H_4O_6^-][H^+]}{[H_2C_4H_4O_6]} = 9.2\times10^{-4}$

$HC_4H_4O_6^- \rightleftharpoons C_4H_4O_6^{2-} + H^+$ $\quad K_{a2} = \frac{[C_4H_4O_6^{2-}][H^+]}{[HC_4H_4O_6^-]} = 4.3\times10^{-5}$

The following assumptions are made:

$[H^+]_{total} \approx [H^+]_{first\ step}$

$[HC_4H_4O_6^-]_{total} \approx [HC_4H_4O_6^-]_{first\ step}$

$[C_4H_4O_6^{2-}]_{total} \approx [C_4H_4O_6^{2-}]_{second\ step}$

The first dissociation:

	$[H_2C_4H_4O_6]$	$[HC_4H_4O_6^-]$	$[H^+]$
I	0.10	–	–
C	–x	+x	+x
E	0.10–x	+x	+x

$$K_{a1} = \frac{[HC_4H_4O_6^-][H^+]}{[H_2C_4H_4O_6]} = \frac{(x)(x)}{(0.10 - x)} = 9.2\times10^{-4}$$

Solve for x by successive approximation:

$x = 0.0091\ M = [H^+] = [HC_4H_4O_6^-]$

The second dissociation:

	$[HC_4H_4O_6^-]$	$[C_4H_4O_6^{2-}]$	$[H^+]$
I	0.0091	–	0.0091
C	–x	+x	+x
E	0.0091–x	+x	0.0091+x

Assume that x << 0.0091, therefore $0.0091 - x \approx 0.0091$ and $0.0091 + x \approx 0.0091$

$$K_{a2} = \frac{(x)(0.0091)}{(0.0091)} = 4.3 \times 10^{-5}$$

$x = 4.3 \times 10^{-5} = [C_4H_4O_6^{2-}]$

*18.106 $\quad H_3PO_3 \rightleftharpoons H_2PO_3^- + H^+ \quad K_{a_1} = \frac{[H_2PO_3^-][H^+]}{[H_3PO_3]} = 1.0 \times 10^{-2}$

$$H_2PO_3^- \rightleftharpoons HPO_3^{2-} + H^+ \quad K_{a_2} = \frac{[HPO_3^{2-}][H^+]}{[H_2PO_3^-]} = 2.6 \times 10^{-7}$$

To simplify the calculation, assume that the second dissociation does not contribute a significant amount of H^+ to the final solution. Solving the equilibrium problem for the first dissociation gives:

	$[H_3PO_3]$	$[H_2PO_3^-]$	$[H^+]$
I	1.0	–	–
C	–x	+x	+x
E	1.0–x	+x	+x

$$K_{a_1} = \frac{[H_2PO_3^-][H^+]}{[H_3PO_3]} = \frac{(x)(x)}{(1.0 - x)} = 1.0 \times 10^{-2}$$

Because K_{a_1} is so large, a quadratic equation must be solved. On doing so we learn that

$x = 0.095$ M $= [H^+] = [H_2PO_3^-]$.

$pH = -\log[H^+] = -\log(0.095) = 1.02$

The $[HPO_3^{2-}]$ may be determined from the second ionization constant.

	$[H_2PO_3^-]$	$[HPO_3^{2-}]$	$[H^+]$
I	0.095	–	0.095
C	–x	+x	+x
E	0.095–x	+x	0.095+x

$$K_{a_2} = \frac{[HPO_3^{2-}][H^+]}{[H_2PO_3^-]} = \frac{(x)(0.95+x)}{(0.095-x)} = 2.6\times10^{-7}$$

If we assume that x is small then $0.095 \pm x \approx 0.095$. Then, $x = [HPO_3^{2-}] = 2.6 \times 10^{-7}$ M.

18.108 The hydrolysis equation is:

$$SO_3^{2-} + H_2O \rightleftharpoons HSO_3^- + OH^- \qquad K_b = \frac{[HSO_3^-][OH^-]}{[SO_3^{2-}]}$$

In order to obtain K_b we will use the relationship $K_w = K_a \times K_b$

$$K_b = \frac{K_w}{K_a} = \frac{1.0\times10^{-14}}{6.6\times10^{-8}} = 1.5\times10^{-7}$$

	$[SO_3^{2-}]$	$[HSO_3^-]$	$[OH^-]$
I	0.12	–	–
C	–x	+x	+x
E	0.12–x	+x	+x

Since K_b is so small, assume that $x << 0.12$. Using the expression for K_b above, and filling in the values from the table, we determine $x = 1.3 \times 10^{-4}$ M $= [HSO_3^-]$.

x is also equal to $[OH^-]$, therefore
$pOH = -\log(1.3 \times 10^{-4}) = 3.87$, $pH = 14.00 - pOH = 10.13$

To find the concentration of H_2SO_3, consider the second ionization equation:

$$HSO_3^- + H_2O \rightleftharpoons H_2SO_3 + OH^- \qquad K_{b_2} = \frac{[H_2SO_3][OH^-]}{[HSO_3^-]}$$

In order to obtain K_{b2} we will use the relationship $K_w = K_a \times K_b$

$$K_{b_2} = \frac{K_w}{K_a} = \frac{1.0\times10^{-14}}{1.2\times10^{-2}} = 8.3\times10^{-13}$$

	$[HSO_3^-]$	$[H_2SO_3]$	$[OH^-]$
I	1.3×10^{-4}	–	1.3×10^{-4}
C	–x	+x	+x
E	1.3×10^{-4}–x	+x	1.3×10^{-4}+x

Since K_{b2} is so small, assume that x << 1.3×10^{-4}, therefore at equilibrium $[HSO_3^-] \approx [OH^-] \approx 1.3 \times 10^{-4}$.

$$K_{b_2} = \frac{[H_2SO_3][OH^-]}{[HSO_3^-]}$$

$$8.3\times10^{-13} = \frac{[H_2SO_3][8.3\times10^{-13}]}{[8.3\times10^{-13}]}$$

x = 8.3×10^{-13} M = $[H_2SO_3]$.

18.110 Only the first hydrolysis needs to be examined. The difficulty is that we are solving for the initial concentration in this problem.

$$CO_3^{2-} + H_2O \rightleftharpoons HCO_3^- + OH^- \qquad K_b = \frac{[HCO_3^-][OH^-]}{[CO_3^{2-}]}$$

In order to obtain K_b we will use the relationship $K_w = K_a \times K_b$

$$K_b = \frac{K_w}{K_a} = \frac{1.0\times10^{-14}}{4.7\times10^{-11}} = 2.1\times10^{-4}$$

	$[CO_3^{2-}]$	$[HCO_3^-]$	$[OH^-]$
I	Z	–	–
C	–x	+x	+x
E	Z–x	+x	+x

In this problem, we know that the pH at equilibrium is 11.62. Using this fact we calculate a pOH = 2.38 and a $[OH^-] = 10^{-pOH} = 4.2 \times 10^{-3}$ M = x. Substitute this into the equilibrium expression and solve for Z, our initial concentration of carbonate ion.

$$K_b = 2.1\times10^{-4} = \frac{[HCO_3^-][OH^-]}{[CO_3^{2-}]} = \frac{(x)(x)}{(\mathbf{Z} - x)} = \frac{(4.2\times10^{-3})^2}{\mathbf{Z} - 4.2\times10^{-3}}$$

Solving this equation we determine that Z = 8.8×10^{-2} M = $[CO_3^{2-}]$. (Note: We could have assumed that the change upon dissociation is small in this example, i.e., x << Z. It is unnecessary to do this for this problem. If you had made this assumption, a value of Z = 8.4×10^{-2} M would be obtained.)

The problem asks for the number of grams of $Na_2CO_3 \cdot 10H_2O$ so convert the concentration to a number of grams: 25.2 grams.

18.112 $C_6H_5O_7^{3-} + H_2O \rightleftharpoons HC_6H_5O_7^{2-} + OH^-$

$$K_b = \frac{[HC_6H_5O_7^{2-}][OH^-]}{[C_6H_5O_7^{3-}]} = \frac{K_w}{K_a} = \frac{1.0\times10^{-14}}{4.0\times10^{-7}} = 2.5\times10^{-8}$$

	$[C_6H_5O_7^{3-}]$	$[HC_6H_5O_7^{2-}]$	$[OH^-]$
I	0.10	–	–
C	–x	+x	+x
E	0.10–x	+x	+x

Assume x << 0.10

$$K_b = \frac{(x)(x)}{0.10} = 2.5\times 10^{-8} \qquad x = 5.0\times10^{-5} = [OH^-]$$

pOH = 4.30
pH = 9.70

*18.114

$$PO_4^{3-} + H_2O \rightleftharpoons HPO_4^{2-} + OH^- \qquad K_{b1} = \frac{K_w}{K_{a3}} = 2.2\times10^{-2}$$

$$HPO_4^{2-} + H_2O \rightleftharpoons H_2PO_4^- + OH^- \qquad K_{b2} = \frac{K_w}{K_{a2}} = 1.6\times10^{-7}$$

$$H_2PO_4^- + H_2O \rightleftharpoons H_3PO_4 + OH^- \qquad K_{b3} = \frac{K_w}{K_{a1}} = 1.4\times10^{-12}$$

By analogy with polyprotic acids, we know that $[H_2PO_4^-] = 1.6\times10^{-7}$.
We need to solve the first equilibrium expression to determine $[HPO_4^{2-}]$.

$$K_{b1} = \frac{[HPO_4^{2-}][OH^-]}{[PO_4^{3-}]} = 2.2\times10^{-2}$$

	$[PO_4^{3-}]$	$[HPO_4^{2-}]$	$[OH^-]$
I	0.50	–	–
C	–x	+x	+x
E	0.50–x	+x	+x

Assume x << 0.50

Use the quadratic equation to solve for x since K_b is so large.

$$K_{b1} = \frac{(x)(x)}{0.50 - x} = 2.2\times 10^{-2}$$

$x = 9.4\times10^{-2}$ M = $[OH^-] = [HPO_4^{2-}]$
pOH = 1.027
pH = 12.97

$[PO_4^{3-}] = 0.50 - 9.4 \times 10^{-2} = 0.41$ M
Solve the third equilibrium expression to determine $[H_3PO_4]$:

$$K_{b3} = \frac{[H_3PO_4][OH^-]}{[H_2PO_4^-]} = 1.4 \times 10^{-12}$$

Substitute the calculated values of $[H_2PO_4^-]$ and $[OH^-]$ and solve for x :

$$K_{b3} = \frac{(9.4 \times 10^{-2})x}{(1.6 \times 10^{-7})} = 1.4 \times 10^{-12}$$

$x = [H_3PO_4] = 2.4 \times 10^{-18}$

18.116 Since HCO_2H and NaOH react in a 1:1 ratio:

$$HCO_2H + NaOH \rightarrow NaHCO_2 + H_2O$$

we can use the equation $V_a \times M_a = V_b \times M_b$ to determine the volume of NaOH that is required to reach the equivalence point, i.e. the point at which the number of moles of NaOH is equal to the number of moles of HCO_2H:

$$V_{NaOH} = 50 \text{ mL} \times 0.10/0.10 = 50 \text{ mL}$$

Thus the final volume at the equivalence point will be 50 + 50 = 100 mL.
The concentration of $NaHCO_2$ would then be:
$0.10 \text{ mol/L} \times 0.050 \text{ L} = 5.0 \times 10^{-3}$ mol $HCO_2H = 5.0 \times 10^{-3}$ mol $NaHCO_2$
$5.0 \times 10^{-3} \text{ mol}/0.100 \text{ L} = 5.0 \times 10^{-2}$ M $NaHCO_2$

The hydrolysis of this salt at the equivalence point proceeds according to the following equilibrium: $HCO_2^- + H_2O \rightleftharpoons HCO_2H + OH^-$

$$K_b = \frac{[HCO_2H][OH^-]}{[HCO_2^-]} = 5.6 \times 10^{-11}$$

	$[HCO_2^-]$	$[HCO_2H]$	$[OH^-]$
I	0.050	–	–
C	–x	+x	+x
E	0.050–x	+x	+x

Substituting the above values for equilibrium concentrations into the mass action expression and assuming x << 0.050 gives:

$$K_b = \frac{(x)(x)}{0.050} = 5.6 \times 10^{-11}$$

$x^2 = 2.8 \times 10^{-12}$
$x = 1.7 \times 10^{-6}$ M $= [OH^-] = [HCO_2H]$
$pOH = -\log[OH^-] = -\log(1.7 \times 10^{-6}) = 5.77$
$pH = 14.00 - pOH = 14.00 - 5.77 = 8.23$

Cresol red would be a good indicator, since it has a color change near the pH at the equivalence point.

18.118

$$\#\,\text{moles } HC_2H_3O_2 = (25.0\text{ mL } HC_2H_3O_2)\left(\frac{0.180\text{ mol } HC_2H_3O_2}{1000\text{ mL } HC_2H_3O_2}\right)$$

$$= 4.50 \times 10^{-3}\text{ moles } HC_2H_3O_2$$

$$\#\,\text{moles } OH^- = (35.0\text{ mL } OH^-)\left(\frac{0.250\text{ mol } OH^-}{1000\text{ mL } OH^-}\right) = 8.75\times10^{-3}\text{ moles } OH^-$$

$$\text{excess } OH^- = 8.75\times10^{-3} - 4.50\times10^{-3} = 4.25\times10^{-3}\text{ moles}$$

$$[OH^-] = \frac{4.25\times10^{-3}\text{ moles}}{(25.0+35.0\text{ mL})\left(\frac{1\text{ L}}{1000\text{ mL}}\right)} = 7.08\times10^{-2}\text{ M}$$

$$pOH = 1.150$$
$$pH = 12.850$$

*18.120

a) $HC_2H_3O_2 \rightleftharpoons H^+ + C_2H_3O_2^-$ $\quad K_a = \frac{[H^+][C_2H_3O_2^-]}{[HC_2H_3O_2]} = 1.8\times10^{-5}$

	$[HC_2H_3O_2]$	$[H^+]$	$[C_2H_3O_2^-]$
I	0.1000	–	–
C	–x	+x	+x
E	0.1000–x	+x	+x

Substituting the above values for equilibrium concentrations into the mass action expression and assuming that x << 0.1000 gives:

$x = [H^+] = 1.342 \times 10^{-3}$ M.

$pH = -\log [H^+] = -\log (1.342 \times 10^{-3}) = 2.8724$.

(b) When NaOH is added, it will react with the acetic acid present decreasing the amount in solution and producing additional acetate ion. Since this a one–to–one reaction, the number of moles of acetic acid will decrease by the same amount as the number of moles of NaOH added and the number of moles of acetate ion will increase by an identical amount. We must determine the number of moles of all ions present and calculate new concentrations accounting for dilution.

$$\# \text{moles } HC_2H_3O_2 = (0.02500 \text{ L solution})\left(\frac{0.1000 \text{ moles } HC_2H_3O_2}{1 \text{ L solution}}\right)$$

$$= 2.500 \times 10^{-3} \text{ moles } HC_2H_3O_2$$

$$\# \text{moles } OH^- = (0.01000 \text{ L solution})\left(\frac{0.1000 \text{ moles } OH^-}{1 \text{ L solution}}\right)$$

$$= 1.000 \times 10^{-3} \text{ moles } OH^-$$

$$[HC_2H_3O_2] = \frac{2.500 \times 10^{-3} \text{ moles} - 1.000 \times 10^{-3} \text{ moles}}{0.02500 \text{ L} + 0.01000 \text{ L}}$$

$$= 4.286 \times 10^{-2} \text{ M } HC_2H_3O_2$$

$$[C_2H_3O_2^-] = \frac{0 \text{ moles} + 1.000 \times 10^{-3} \text{ moles}}{0.02500 \text{ L} + 0.01000 \text{ L}}$$

$$= 2.857 \times 10^{-2} \text{ M } C_2H_3O_2^-$$

$$pH = pK_a + \log \frac{[C_2H_3O_2^-]}{[HC_2H_3O_2]}$$

$$= 4.7447 + \log \frac{(2.857 \times 10^{-2})}{(4.286 \times 10^{-2})}$$

$$= 4.5686$$

(c) When half the acetic acid has been neutralized, there will be equal amounts of acetic acid and acetate ion present in the solution. At this point, $pH = pK_a = 4.7447$.

(d) At the equivalence point, all of the acetic acid will have been converted to acetate ion. The concentration of the acetate ion will be half the original concentration of acetic acid since we have doubled the volume of the solution. We then need to solve the equilibrium problem that results when we have a solution that possesses a $[C_2H_3O_2^-] = 0.05000$ M.

$$C_2H_3O_2^- + H_2O \rightleftharpoons HC_2H_3O_2 + OH^-$$

$$K_b = \frac{[HC_2H_3O_2][OH^-]}{[C_2H_3O_2^-]} = 5.6 \times 10^{-10}$$

	$[C_2H_3O_2^-]$	$[HC_2H_3O_2]$	$[OH^-]$
I	0.05000	–	–
C	–x	+x	+x
E	0.05000–x	+x	+x

Substituting the above values for equilibrium concentrations into the mass action expression and assuming that x << 0.05000 gives: $x = [OH^-] = 5.292 \times 10^{-6}$ M.
$pOH = -\log [OH^-] = -\log (5.292 \times 10^{-6}) = 5.2764$.
$pH = 14.0000 - pOH = 14.0000 - 5.2764 = 8.7236$.

Additional Exercises

18.122 $HC_2H_3O_2 + H_2O \rightleftharpoons H_3O^+ + C_2H_3O_2^-$

$$K_a = \frac{[H_3O^+][C_2H_3O_2^-]}{[HC_2H_3O_2]} = 1.8\times10^{-5}$$

	$[HC_2H_3O_2]$	$[H_3O^+]$	$[C_2H_3O_2^-]$
I	1.0	–	–
C	1.0–x	+x	+x
E	1.0 –x	+x	+x

Assume x << 1.0

$$K_a = \frac{[x][x]}{[1.0]} = 1.8\times10^{-5} \quad x = 4.2\times10^{-3} = [H_3O^+]$$

% ionization = $([H^+]/[\text{total acid}]) \times 100\ \% = (4.2 \text{ x } 10^{-3}/1.0) \times 100 = 0.42\ \%$

At lower concentrations, the only change is the denominator in the final expression for K_a above. Calculating the answers, we have:

1.0 M = 0.42% ionization
0.10 M = 1.3 % ionization
0.010 M = 4.2 % ionization

As a weak acid becomes more dilute, its % ionization goes up.

18.124 When these two solutions are mixed, the HCl will react with the ammonia producing NH_4^+. Since HCl is the limiting reactant, we will have an excess of ammonia. The resulting solution will be a buffer as it consists of ammonia, a weak base, and ammonium ion, its conjugate acid.

We need to determine the amount of NH_4^+ that is produced and the amount of NH_3 that remains:

$$\#\,\text{mol H}^+ = (100\ \text{mL HCl})\left(\frac{0.500\ \text{mol HCl}}{1000\ \text{mL HCl}}\right)\left(\frac{1\ \text{mol H}^+}{1\ \text{mol HCl}}\right) = 0.0500\ \text{mol H}^+$$

$$\#\,\text{mol NH}_3 = \#\,\text{mol initially} - \text{mol H}^+\ \text{added}$$

$$= (300\ \text{mL HCl})\left(\frac{0.500\ \text{mol HCl}}{1000\ \text{mL HCl}}\right)\left(\frac{1\ \text{mol H}^+}{1\ \text{mol HCl}}\right) - 0.0500\ \text{mol H}^+$$

$$= 0.100\ \text{mol NH}_3$$

$$\#\,\text{mol NH}_4{}^+ = \#\,\text{mol H}^+ = 0.0500\ \text{mol NH}_4{}^+$$

$$\text{pOH} = \text{pK}_b + \log\frac{[\text{NH}_4{}^+]}{[\text{NH}_3]} = 4.74 - 0.30 = 4.44$$

$$\text{pH} = 14.00 - \text{pOH} = 9.56$$

18.126 a) We first need to determine the concentration of the components of the solution.

$$[\text{HC}_2\text{H}_3\text{O}_2] = \frac{\#\,\text{mol HC}_2\text{H}_3\text{O}_2}{\#\,\text{L soln}}$$

$$= \left(\frac{15.0\ \text{g HC}_2\text{H}_3\text{O}_2}{0.750\ \text{L}}\right)\left(\frac{1\ \text{mol HC}_2\text{H}_3\text{O}_2}{60.05\ \text{g HC}_2\text{H}_3\text{O}_2}\right)$$

$$= 0.333\ \text{M}$$

$$[\text{NaC}_2\text{H}_3\text{O}_2] = \frac{\#\,\text{mol NaC}_2\text{H}_3\text{O}_2}{\#\,\text{L soln}}$$

$$= \left(\frac{25.0\ \text{g NaC}_2\text{H}_3\text{O}_2}{0.750\ \text{L}}\right)\left(\frac{1\ \text{mol NaC}_2\text{H}_3\text{O}_2}{82.03\ \text{g NaC}_2\text{H}_3\text{O}_2}\right)$$

$$= 0.406\ \text{M}$$

$$\text{pH} = \text{pK}_a + \log\frac{[\text{anion}]}{[\text{acid}]} = 4.745 + \log\frac{0.406}{0.333}$$

$$= 4.831$$

b) The added base will react with the acid present using it up and producing additional acetate ion. To solve this problem, determine the number of moles of base added and subtract this amount from the amount of acid present. Add the same amount to the amount of acetate ion present. Be sure to account for dilution when you determine the new concentrations.

$$\# \text{mol OH}^- = (25.00 \text{ mL NaOH})\left(\frac{0.25 \text{ mol NaOH}}{1000 \text{ mL NaOH}}\right)\left(\frac{1 \text{ mol OH}^-}{1 \text{ mol NaOH}}\right)$$

$$= 0.0063 \text{ mol OH}^-$$

Consequently, the # mol of acetic acid will decrease by 0.0063 and the # of moles of acetate ion will increase by 0.0063. The new concentrations are:

$$[HC_2H_3O_2] = \frac{\# \text{mol } HC_2H_3O_2}{\# \text{L soln}}$$

$$= \left(\frac{(0.250 - 0.0063) \text{mol } HC_2H_3O_2}{0.775 \text{ L}}\right)$$

$$= 0.314 \text{ M}$$

$$[C_2H_3O_2^-] = \frac{\# \text{mol } C_2H_3O_2^-}{\# \text{L soln}}$$

$$= \left(\frac{(0.305 + 0.0063) \text{mol } C_2H_3O_2^-}{0.775 \text{ L}}\right)$$

$$= 0.402 \text{ M}$$

$$\text{pH} = \text{pK}_a + \log\frac{[\text{anion}]}{[\text{acid}]} = 4.745 + \log\frac{0.402}{0.314}$$

$$= 4.852$$

c) This problem is identical to the previous except that we are adding acid to the original solution. Consequently, the acid concentration will increase and the acetate ion concentration will decrease.

$$\# \text{mol H}^+ = (25.0 \text{ mL HCl})\left(\frac{0.40 \text{ mol HCl}}{1000 \text{ mL HCl}}\right)\left(\frac{1 \text{ mol H}^+}{1 \text{ mol HCl}}\right)$$

$$= 0.010 \text{ mol H}^+$$

$$[HC_2H_3O_2] = \frac{\# \text{mol } HC_2H_3O_2}{\# \text{L soln}}$$

$$= \left(\frac{(0.250 + 0.010) \text{mol } HC_2H_3O_2}{0.775 \text{ L}}\right)$$

$$= 0.335 \text{ M}$$

$$\left[C_2H_3O_2^-\right] = \frac{\#\,\text{mol}\; C_2H_3O_2^-}{\#\,\text{L soln}}$$

$$= \left(\frac{(0.305 - 0.010)\,\text{mol}\; C_2H_3O_2^-}{0.775\,\text{L}}\right)$$

$$= 0.381\,\text{M}$$

$$\text{pH} = \text{pK}_a + \log\frac{[\text{anion}]}{[\text{acid}]} = 4.745 + \log\frac{0.381}{0.335}$$

$$= 4.801$$

*18.128

a) Formic acid $pK_a = 3.74$

b) Start by determining the ratio of acid and base in the buffer solution.

$$\text{pH} = \text{pK}_a + \log\frac{[\text{NaCHO}_2]}{[\text{HCHO}_2]}$$

$$\frac{[\text{NaCHO}_2]}{[\text{HCHO}_2]} = 10^{\text{pH - pKa}} = 10^{(3.80 - 3.74)}$$

$$= 1.15$$

Since the volume is constant, the ratio of moles $NaCHO_2$ to moles $HCHO_2$ = 1.15. We need a minimum of 5.0×10^{-3} mol of $HCHO_2$. So,

$$\#\,\text{g HCHO}_2 = (5.0\times10^{-3}\ \text{mol HCHO}_2)\left(\frac{46.0\ \text{g HCHO}_2}{1\ \text{mol HCHO}_2}\right)$$

$$= 0.230\ \text{g HCHO}_2$$

$$\#\text{g NaCHO}_2 = (5.7\times10^{-3}\ \text{mol NaCHO}_2)\left(\frac{68.0\ \text{g NaCHO}_2}{1\ \text{mol NaCHO}_2}\right)$$

$$= 0.390\ \text{g NaCHO}_2$$

$$\text{c)}\,[\text{HCHO}_2] = \frac{5.0\times10^{-3}\ \text{mol}}{0.750\ \text{L}} = 6.7 \times 10^{-3}\ \text{M}$$

$$[\text{NaCHO}_2] = \frac{5.7\times10^{-3}\ \text{mol}}{0.750\ \text{L}} = 7.6 \times 10^{-3}\ \text{M}$$

d)

$$[H_3O^+] = 10^{-pH} = 1.6\times10^{-4}$$

$$\#\text{ moles } H_3O^+ = (0.750\text{ L})\left(\frac{1.6\times10^{-4}\text{ mol}}{1\text{ L}}\right) = 1.2\times10^{-4}\text{ moles}$$

add 5.0×10^{-3} moles $\#\text{ moles } H_3O^+ = 5.1\times10^{-3}$ moles

$$[H_3O^+] = \frac{5.1\times10^{-3}\text{ mol}}{0.750\text{ L}} = 6.8\times10^{-3}\text{ M}$$

pH = 2.17

e) From part (d) we know there are 1.2×10^{-4} moles H_3O^+ before adding the base. After adding 5.0×10^{-3} moles NaOH the acid will be neutralized and there will be 4.9×10^{-3} moles excess base.

$$[OH^-] = \frac{4.9\times10^{-3}\text{ mol}}{0.750\text{ L}} = 6.5\times10^{-3}\text{ M}$$

pOH = 2.18

pH = 11.82

*18.130

Recall that $\Delta t_f = K_f m$, where K_f is the molal freezing point depression constant for the solvent and *m* is the molality of the solution. We need to determine the molality of this solution. In order to do this, we need to determine the concentrations of the ions which result from the dissociation of the weak acid. Recall that each ion in the solution helps to lower the freezing point.

$$HC_2HO_2Cl_2 \rightleftharpoons H^+ + C_2HO_2Cl_2^-$$

$$K_a = \frac{[H^+][C_2HO_2Cl_2^-]}{[HC_2HO_2Cl_2]} = 5.0\times10^{-2}$$

	$[HC_2HO_2Cl_2]$	$[H^+]$	$[C_2HO_2Cl_2^-]$
I	0.50	–	–
C	–x	+x	+x
E	0.50–x	+x	+x

$$K_a = \frac{(x)(x)}{(0.50 - x)} = 5.0\times10^{-2}$$

Because the value of K_a is relatively large, we must solve a quadratic equation in order to determine the H^+ concentration. Doing so gives:

$x = 0.14\text{ M} = [H^+] = [C_2HO_2Cl_2^-]$

$[HC_2HO_2Cl_2] = 0.50 - x = 0.36\text{ M}$

We now calculate the molality of the components of the solution. Because this is an aqueous solution with a density of 1.0 g/mL, the molality and the molarity have the same value: $m_{H^+} = 0.14$, $m_{C_2HO_2Cl_2^-} = 0.14$ and $m_{HC_2HO_2Cl_2} = 0.36$.

The total molality is 0.14 + 0.14 + 0.36 = 0.64 *m*.

The freezing point depression constant for water equals –1.86 °C/*m*.

$\Delta t_f = (-1.86\ ^\circ C/m)(0.64\ m) = -1.19\ ^\circ C$.

The freezing point of the solution is –1.19 °C.

*18.132

a) $HSO_4^- + H_2O \rightleftharpoons H_3O^+ + SO_4^{2-}$ $\quad K_a = \frac{[SO_4^{2-}][H_3O^+]}{[HSO_4^-]}$

b)

	$[HSO_4^-]$	$[H_3O^+]$	$[SO_4^{2-}]$
I	0.010	–	–
C	–x	+x	+x
E	0.010–x	+x	+x

$$K_a = \frac{[x][x]}{0.010 - x} = 1.0 \times 10^{-2}$$

Solve the quadratic equation to get $x = 6.2 \times 10^{-3} = [H_3O^+]$

c) $K_a = \frac{[x][x]}{0.010} = 1.0 \times 10^{-2}$ $\quad x = 1.0 \times 10^{-2}$ M

d) The exact answer (obtained in part b) is 38% smaller than the approximate solution.

18.134

$$pH = pK_a - \log\frac{[CO_2]}{[HCO_3^-]}$$

$$[HCO_3^-] = [CO_2]10^{pH - pK_a}$$

$$[HCO_3^-] = (0.022)10^{(7.35-6.35)}$$

$$= 0.22\ M$$

*18.136

This exercise is an example of a titration using a strong acid and a strong base. Prior to reaching the equivalence point, there will be an excess of acid present. The amount of acid present may be determined by subtracting the number of moles of base added from the amount of acid initially present. We may then use the dilution equation, $M_1 \times V_1 = M_2 \times V_2$, to determine the new concentration of acid and then calculate the pH. After the equivalence point, there is an excess of base. The amount in excess is determined by subtracting the inital amount of acid present from the amount of base added. Again, the dilution equation may be used to determine the resulting concentration.

Start by determining the amount of H^+ initially present:

$$\#\,\text{moles } H^+ = (25.00 \text{ mL solution})\left(\frac{0.1000 \text{ moles } H^+}{1000 \text{ mL solution}}\right)$$

$$= 2.500 \times 10^{-3} \text{ moles } H^+$$

a) Initially, there is only 0.1000 M HCl present: $[H^+] = 0.1000$ M, $pH = -\log[H^+] = 1.0000$.

b) 10.00 mL added base

$$\#\,\text{moles } OH^- = (10.00 \text{ mL solution})\left(\frac{0.1000 \text{ moles } OH^-}{1000 \text{ mL solution}}\right)$$

$$= 1.000 \times 10^{-3} \text{ moles } OH^-$$

$$[H^+] = \frac{2.500 \times 10^{-3} \text{ moles} - 1.000 \times 10^{-3} \text{ moles}}{0.02500 \text{ L} + 0.01000 \text{ L}}$$

$$= 4.286 \times 10^{-2} \text{ M } H^+$$

$$pH = -\log[H^+] = 1.3680$$

c) 24.90 mL added base

$$\#\,\text{moles } OH^- = (24.90 \text{ mL solution})\left(\frac{0.1000 \text{ moles } OH^-}{1000 \text{ mL solution}}\right)$$

$$= 2.490 \times 10^{-3} \text{ moles } OH^-$$

$$[H^+] = \frac{2.500 \times 10^{-3} \text{ moles} - 2.490 \times 10^{-3} \text{ moles}}{0.02500 \text{ L} + 0.02490 \text{ L}}$$

$$= 2.004 \times 10^{-4} \text{ M } H^+$$

$$pH = -\log[H^+] = 3.6981$$

d) 24.99 mL added base

$$\# \text{moles } OH^- = (24.99 \text{ mL solution})\left(\frac{0.1000 \text{ moles } OH^-}{1000 \text{ mL solution}}\right)$$

$$= 2.499 \times 10^{-3} \text{ moles } OH^-$$

$$[H^+] = \frac{2.500 \times 10^{-3} \text{ moles} - 2.499 \times 10^{-3} \text{ moles}}{0.02500 \text{ L} + 0.02499 \text{ L}}$$

$$= 2.000 \times 10^{-5} \text{ M } H^+$$

$$pH = -\log[H^+] = 4.6990$$

e) 25.00 mL added base

The solution now has equal moles of acid and base, so it is neutral. (However, at this point there still remain hydrogen ions from the autoionization of water.)

$$pH = 7.000$$

f) 25.01 mL added base

At this point, any excess base is not neutralized by HCl, so we can assume any excess over 2.500 x 10^{-3} mol is the total hydroxide ion concentration in the solution:

$$\# \text{moles } OH^- = (25.01 \text{ mL solution})\left(\frac{0.1000 \text{ moles } OH^-}{1000 \text{ mL solution}}\right)$$

$$= 2.501 \times 10^{-3} \text{ moles } OH^-$$

$$[OH^-] = \frac{2.501 \times 10^{-3} \text{ moles} - 2.500 \times 10^{-3} \text{ moles}}{0.02500 \text{ L} + 0.02501 \text{ L}}$$

$$= 2.000 \times 10^{-5} \text{ M } OH^-$$

$$pOH = -\log[OH^-] = 4.699$$

$$pH = 14.00 - pOH = 9.301$$

g) 25.01 mL added base

At this point, any excess base is not neutralized by HCl, so we can assume any excess over 2.500 x 10^{-3} mol is the total hydroxide ion concentration in the solution:

$$\# \text{moles OH}^- = (25.10 \text{ mL solution})\left(\frac{0.1000 \text{ moles OH}^-}{1000 \text{ mL solution}}\right)$$

$$= 2.510 \times 10^{-3} \text{ moles OH}^-$$

$$[\text{OH}^-] = \frac{2.510 \times 10^{-3} \text{ moles} - 2.500 \times 10^{-3} \text{ moles}}{0.02500 \text{ L} + 0.02510 \text{ L}}$$

$$= 1.996 \times 10^{-4} \text{ M OH}^-$$

pOH = –log[OH$^-$] = 3.700

pH = 14.00 – pOH = 10.300

h) 26.00 mL added base

At this point, any excess base is not neutralized by HCl, so we can assume any excess over 2.500 x 10^{-3} mol is the total hydroxide ion concentration in the solution:

$$\# \text{moles OH}^- = (26.00 \text{ mL solution})\left(\frac{0.1000 \text{ moles OH}^-}{1000 \text{ mL solution}}\right)$$

$$= 2.600 \times 10^{-3} \text{ moles OH}^-$$

$$[\text{OH}^-] = \frac{2.600 \times 10^{-3} \text{ moles} - 2.500 \times 10^{-3} \text{ moles}}{0.02500 \text{ L} + 0.02600 \text{ L}}$$

$$= 1.961 \times 10^{-3} \text{ M OH}^-$$

pOH = –log[OH$^-$] = 2.708

pH = 14.00 – pOH = 11.292

i) 50.00 mL added base

At this point, any excess base is not neutralized by HCl, so we can assume any excess over 2.500 x 10^{-3} mol is the total hydroxide ion concentration in the solution:

$$\# \text{moles OH}^- = (50.00 \text{ mL solution})\left(\frac{0.1000 \text{ moles OH}^-}{1000 \text{ mL solution}}\right)$$

$$= 5.000 \times 10^{-3} \text{ moles OH}^-$$

$$[\text{OH}^-] = \frac{5.000 \times 10^{-3} \text{ moles} - 2.500 \times 10^{-3} \text{ moles}}{0.02500 \text{ L} + 0.05000 \text{ L}}$$

$$= 0.03333 \text{ M OH}^-$$

pOH = –log[OH$^-$] = 1.477

pH = 14.00 – pOH = 12.523

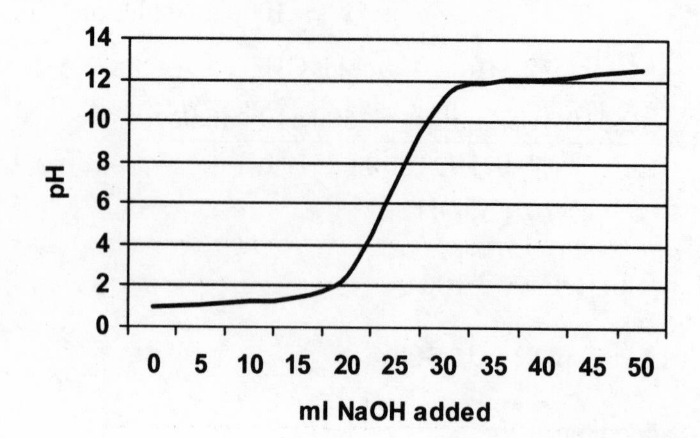

Chapter 19

Practice Exercises

19.1 a) $K_{sp} = [Ba^{2+}][C_2O_4^{2-}]$ b) $K_{sp} = [Ag^+]^3[PO_4^{3-}]$

19.2 $TlI \rightleftharpoons Tl^+ + I^-$

	$[Tl^+]$	$[I^-]$
I	–	–
C	+x	+x
E	+x	+x

$K_{sp} = x^2 = (1.8 \text{ x } 10^{-5})^2 = 3.2 \text{ x } 10^{-10}$

19.3 $PbF_2(s) \rightleftharpoons Pb^{2+}(aq) + 2F^-(aq)$ $K_{sp} = [Pb^{2+}][F^-]^2$

$K_{sp} = (2.15\times10^{-3})(2(2.15\times10^{-3}))^2 = 3.98\times10^{-8}$

19.4 $CoCO_3(s) \rightleftharpoons Co^{2+}(aq) + CO_3^{2-}(aq)$

	$[Co^{2+}]$	$[CO_3^{2-}]$
I	–	0.10
C	$+1.0 \times 10^{-9}$	$+ 1.0 \times 10^{-9}$
E	$+1.0 \times 10^{-9}$	$0.10 + 1.0 \times 10^{-9}$

Substituting the above values for equilibrium concentrations into the expression for K_{sp} gives:

$K_{sp} = [Co^{2+}][CO_3^{2-}] = (1.0 \times 10^{-9})(0.10 + 1.0 \times 10^{-9}) = 1.0 \times 10^{-10}$

19.5 $PbF_2(s) \rightleftharpoons Pb^{2+}(aq) + 2F^-(aq)$

	$[Pb^{2+}]$	$[F^-]$
I	0.10	–
C	$+ 3.1 \times 10^{-4}$	$+ 2(3.1 \times 10^{-4})$
E	$0.10 + 3.1 \times 10^{-4}$	$+ 6.2 \times 10^{-4}$

Substituting the above values for equilibrium concentrations into the expression for K_{sp} gives:

$K_{sp} = [Pb^{2+}][F^-]^2 = [0.10 + 3.1 \times 10^{-4}][6.2 \times 10^{-4}]^2$

Now $(0.10 + 3.1 \times 10^{-4})$ is also ≈ 0.10:
Hence, $K_{sp} = (0.10)(6.2 \times 10^{-4})^2 = 3.9 \times 10^{-8}$

19.6 a) $AgBr(s) \rightleftharpoons Ag^+(aq) + Br^-(aq)$ $K_{sp} = [Ag^+][Br^-] = 5.0 \times 10^{-13}$

	$[Ag^+]$	$[Br^-]$
I	–	–
C	+x	+x
E	+x	+x

Substituting the above values for equilibrium concentrations into the expression for K_{sp} gives:

$K_{sp} = 5.0 \times 10^{-13} = [Ag^+][Br^-] = (x)(x)$
$x = \sqrt{5.0 \times 10^{-13}} = 7.1 \times 10^{-7}$
Thus the solubility is 7.1×10^{-7} M AgBr.

b) $Ag_2CO_3(s) \rightleftharpoons 2Ag^+(aq) + CO_3^{2-}(aq)$ $K_{sp} = [Ag^+]^2[CO_3^{2-}] = 8.1 \times 10^{-12}$

	$[Ag^+]$	$[CO_3^{2-}]$
I	–	–
C	+ 2x	+x
E	+ 2x	+x

Substituting the above values for equilibrium concentrations into the expression for K_{sp} gives:

$K_{sp} = 8.1 \times 10^{-12} = [Ag^+]^2[CO_3^{2-}] = (2x)^2(x)$ and $4x^3 = 8.1 \times 10^{-12}$

$x = \sqrt[3]{(8.1 \times 10^{-12})/4} = 1.3 \times 10^{-4}$
Thus the molar solubility is 1.3×10^{-4} M Ag_2CO_3.

19.7 $AgI(s) \rightleftharpoons Ag^+(aq) + I^-(aq)$ $K_{sp} = [Ag^+][I^-] = 8.3 \times 10^{-17}$

	$[Ag^+]$	$[I^-]$
I	–	0.20
C	+x	+x
E	+x	0.20 + x

Substituting the above values for equilibrium concentrations into the expression for K_{sp} gives:

$K_{sp} = 8.3 \times 10^{-17} = [Ag^+][I^-] = (x)(0.20 + x)$

We know that the value of K_{sp} is very small, and it suggests the simplifying assumption that $(0.20 + x) \approx 0.20$:

Hence, $8.3 \times 10^{-17} = (0.20)x$, and $x = 4.2 \times 10^{-16}$. The assumption that $(0.20 + x) \approx 0.20$ is seen to be valid indeed.

Thus 4.2×10^{-16} M of AgI will dissolve in a 0.20 M NaI solution.

In pure water,
$K_{sp} = 8.3 \times 10^{-17} = [Ag^+][I^-] = (x)(x)$
$x = [AgI(aq)] = 9.1 \times 10^{-9}$ M (much more soluble)

19.8 $Fe(OH)_3(s) \rightleftharpoons Fe^{3+}(aq) + 3OH^-(aq) K_{sp} = [Fe^{3+}][OH^-]^3 = 1.6 \times 10^{-39}$

	$[Fe^{3+}]$	$[OH^-]$
I	–	0.050
C	+ x	+ 3x
E	+ x	0.050 + 3x

Substituting the above values for equilibrium concentrations into the expression for K_{sp} gives:

$$K_{sp} = 1.6 \times 10^{-39} = [Fe^{3+}][OH^-]^3 = (x)[0.050 + 3x]^3$$

We try to simplify by making the approximation that (0.050 + 3x) □ 0.050:

$$1.6 \times 10^{-39} = (x)(0.050)^3 \text{ or } x = 1.3 \times 10^{-35}$$

Clearly the assumption that $(0.050 + 3x) \approx 0.050$ is justified.
Thus 1.3×10^{-35} M of $Fe(OH)_3$ will dissolve in a 0.050 M sodium hydroxide solution.

19.9 The expression for K_{sp} is $K_{sp} = [Ca^{2+}][SO_4^{2-}] = 2.4 \times 10^{-5}$ and the ion product for this solution would be:
$[Ca^{2+}][SO_4^{2-}] = (2.5 \times 10^{-3})(3.0 \times 10^{-2}) = 7.5 \times 10^{-5}$

Since the ion product is larger than the value of K_{sp}, a precipitate will form.

19.10 The solubility product constant is $K_{sp} = [Ag^+]^2[CrO_4^{2-}] = 1.2 \times 10^{-12}$ and the ion product for this solution would be:
$[Ag^+]^2[CrO_4^{2-}] = (4.8 \times 10^{-5})^2(3.4 \times 10^{-4}) = 7.8 \times 10^{-13}$

Since the ion product is smaller than the value of K_{sp}, no precipitate will form.

19.11 We expect $PbSO_4(s)$ since nitrates are soluble.

Because two solutions are to be mixed together, there will be a dilution of the concentrations of the various ions, and the diluted ion concentrations must be used. In general, on dilution, the following relationship is found for the concentrations of the initial solution (M_i) and the concentration of the final solution (M_f): $M_iV_i = M_fV_f$

Thus the final or diluted concentrations are:

$$[Pb^{2+}] = (1.0\times10^{-3}\ M)\left(\frac{100.0\ mL}{200.0\ mL}\right) = 5.0\times10^{-4}\ M$$

$$[SO_4^{\ 2-}] = (2.0\times10^{-3}\ M)\left(\frac{100.0\ mL}{200.0\ mL}\right) = 1.0\times10^{-3}\ M$$

The value of the ion product for the final (diluted) solution is:
$[Pb^{2+}][SO_4^{\ 2-}] = (5.0 \times 10^{-4})(1.0 \times 10^{-3}) = 5.0 \times 10^{-7}$

Since this is smaller than the value of K_{sp} (6.3×10^{-7}), a precipitate of $PbSO_4$ is not expected.

19.12 We expect a precipitate of $PbCl_2$ since nitrates are soluble.
We proceed as in Practice Exercise 11. $M_iV_i = M_fV_f$

$$[Pb^{2+}] = (0.10\ M)\left(\frac{50.0\ mL}{70.0\ mL}\right) = 0.071\ M$$

$$[Cl^-] = (0.040\ M)\left(\frac{20.0\ mL}{70.0\ mL}\right) = 0.011\ M$$

The value of the ion product for such a solution would be:
$[Pb^{2+}][Cl^-]^2 = (7.1 \times 10^{-2})(1.1 \times 10^{-2})^2 = 8.6 \times 10^{-6}$

Since the ion product is smaller than K_{sp}, a precipitate of $PbCl_2$ is not expected.

19.13 Consulting Table 19.2, we find that Fe^{2+} is much more soluble in acid than Hg^{2+}. We want to make the H^+ concentration large enough to prevent FeS from precipitating, but small enough that HgS *does* precipitate. First, we calculate the highest pH at which FeS will remain soluble, by using K_{spa} for FeS. (Recall that a saturated solution of $H_2S = 0.10$ M.)

$$K_{spa} = \frac{[Fe^{2+}][H_2S]}{[H^+]^2} = \frac{[0.010][0.10]}{[H^+]^2} = 6\times10^2$$

$[H^+] = 0.0013$ M
$pH = -\log [H^+] = 2.9$

Since Fe^{2+} is much more soluble in acid than Hg^{2+} we already know that this pH will precipitate HgS, but we can check it by using K_{spa} for HgS:

$$K_{spa} = \frac{[Hg^{2+}][H_2S]}{[H^+]^2} = \frac{[0.010][0.10]}{[H^+]^2} = 2\times10^{-32}$$

$[H^+] = 2.2 \times 10^{14}$ M

(This concentration is impossibly high, but it tells us that this much acid would be required to dissolve HgS at these concentrations.)

19.14 Follow the exact procedure outlined in Example 19.10.

$K_{sp} = [Ca^{2+}][CO_3^{2-}] = 4.5 \times 10^{-9}$

$K_{sp} = [Ni^{2+}][CO_3^{2-}] = 1.3 \times 10^{-7}$

$NiCO_3$ is more soluble and will precipitate when:

$$[CO_3^{2-}] = \frac{K_{sp}}{[Ni^{2+}]} = \frac{1.3\times10^{-7}}{0.10} = 1.3\times10^{-6}$$

$CaCO_3$ will precipitate when:

$$[CO_3^{2-}] = \frac{K_{sp}}{[Ca^{2+}]} = \frac{4.5\times10^{-9}}{0.10} = 4.5\times10^{-8}$$

$CaCO_3$ will precipitate and $NiCO_3$ will not precipitate if $[CO_3^{2-}]$ .4.5×10^{-8} and $[CO_3^{2-}] < 1.3 \times 10^{-6}$. Now, using the equation highlighted in example 19.10 we get:

$$[H^+]^2 = (2.4\times10^{-17})\left(\frac{0.030}{[CO_3^{2-}]}\right) \quad NiCO_3 \text{ will precipitate if}$$

$$[H^+]^2 = (2.4\times10^{-17})\left(\frac{0.030}{1.3 \times 10^{-6}}\right) = 5.5\times10^{-13}$$

$$[H^+] = 7.4\times10^{-7} \qquad pH = 6.13$$

$CaCO_3$ will precipitate if :

$$[H^+]^2 = (2.4\times10^{-17})\left(\frac{0.030}{4.5\times10^{-8}}\right) = 1.6\times10^{-11}$$

$$[H^+] = 4.0\times10^{-6} \qquad pH = 5.40$$

So $CaCO_3$ will precipitate and $NiCO_3$ will not if the pH is maintained between pH = 5.40 and pH = 6.13

19.15 The overall equilibrium is $AgCl(s) + 2NH_3(aq) \rightleftharpoons Ag(NH_3)_2^+(aq) + Cl^-(aq)$

$$K_c = \frac{[Ag(NH_3)_2^+][Cl^-]}{[NH_3]^2}$$

In order to obtain a value for K_c for this reaction, we need to use the expressions for K_{sp} of AgCl(s) and the K_{form} of $Ag(NH_3)_2^+$:

$$K_{sp} = [Ag^+][Cl^-] = 1.8\times10^{-10}$$

$$K_{form} = \frac{[Ag(NH_3)_2^+]}{[Ag^+][NH_3]^2} = 1.6\times10^7$$

$$K_c = K_{sp}\times K_{form} = \frac{[Ag(NH_3)_2^+][Cl^-]}{[NH_3]^2} = 2.9\times10^{-3}$$

Now we may use an equilibrium table for the reaction in question:

	$[NH_3]$	$[Ag(NH_3)_2^+]$	$[Cl^-]$
I	0.10	–	–
C	–2x	+x	+x
E	0.10–2x	x	x

Substituting these values into the mass action expression gives:

$$K_c = 2.9\times10^{-3} = \frac{(x)(x)}{(0.10-2x)^2}$$

Take the square root of both sides to get $0.054 = \frac{(x)}{(0.10-2x)}$

Solving for x we get, $x = 4.9\times10^{-3}$ M. The molar solubility of AgCl in 0.10 M NH_3 is therefore 4.9×10^{-3} M.

In order to determine the solubility in pure water, we simply look at K_{sp}

$$AgCl(s) \rightleftharpoons Ag^+(aq) + Cl^-(aq) \qquad K_{sp} = [Ag^+][Cl^-] = 1.8 \times 10^{-10}$$

At equilibrium; $[Ag^+] = [Cl^-] = 1.3 \times 10^{-5}$. Hence the the molar solubility of AgCl in 0.10 M NH_3 is about 380 times greater than in pure water.

19.16 We will use the information gathered for the last problem. Specifically,

$$AgCl(s) + 2NH_3(aq) \rightleftharpoons Ag(NH_3)_2^+(aq) + Cl^-(aq)$$

$$K_c = \frac{[Ag(NH_3)_2^+][Cl^-]}{[NH_3]^2} = 2.9\times10^{-3}$$

If we completely dissolve 0.20 mol of AgCl, the equilibrium [Cl^-] and [$Ag(NH_3)_2^+$] will be 0.20 M in a one liter container. The question asks, therefore, what amount of NH_3 must be initially present so that the equilibrium concentration of Cl^- is 0.20M?

	[NH_3]	[$Ag(NH_3)_2^+$]	[Cl^-]
I	Z	–	–
C	–2x	+x	+x
E	Z–2x	x	x

$$K_c = 2.9\times10^{-3} = \frac{(x)(x)}{(Z-2x)^2}$$

Take the square root of both sides to get;

$$0.054 = \frac{x}{Z-2x} = \frac{0.20}{Z-0.40}$$

We have substituted the known value of x. Solving for Z we get, Z = 4.1 M

Consequently, we would need to add 4.1 moles of NH_3 to a one liter container of 0.20 M AgCl in order to completely dissolve the AgCl.

Review Problems

19.14 a) $CaF_2(s) \rightleftharpoons Ca^{2+} + 2F^-$ $\qquad K_{sp} = [Ca^{2+}][F^-]^2$

b) $Ag_2CO_3(s) \rightleftharpoons 2Ag^+ + CO_3^{2-}$ $\qquad K_{sp} = [Ag^+]^2[CO_3^{2-}]$

c) $PbSO_4(s) \rightleftharpoons Pb^{2+} + SO_4^{2-}$ $\qquad K_{sp} = [Pb^{2+}][SO_4^{2-}]$

d) $Fe(OH)_3(s) \rightleftharpoons Fe^{3+} + 3OH^-$ $\qquad K_{sp} = [Fe^{3+}][OH^-]^3$

e) $PbI_2(s) \rightleftharpoons Pb^{2+} + 2I^-$ $\qquad K_{sp} = [Pb^{2+}][I^-]^2$

f) $Cu(OH)_2(s) \rightleftharpoons Cu^{2+} + 2OH^-$ $\qquad K_{sp} = [Cu^{2+}][OH^-]^2$

19.16

$$\#\text{moles } BaSO_4 = (0.00245 \text{ g } BaSO_4)\left(\frac{1 \text{ mole } BaSO_4}{233.3906 \text{ g } BaSO_4}\right)$$

$$\#\text{moles } BaSO_4 = 1.05\times10^{-5} \text{ moles}$$

$$[Ba^{2+}] = [SO_4^{2-}] = 1.05\times10^{-5} \text{ M}$$

$$K_{sp} = [Ba^{2+}][SO_4^{2-}] = (1.05\times10^{-5})^2 = 1.10\times10^{-10}$$

19.18 $BaSO_3 (s) \rightleftharpoons Ba^{2+} + SO_3^{2-}$ $K_{sp} = [Ba^{2+}][SO_3^{2-}]$

$K_{sp} = (0.10)(8.0 \times 10^{-6}) = 8.0 \times 10^{-7}$

(In this problem, all of the Ba^{2+} comes from the $BaCl_2$.)

19.20 $Ag_3PO_4(s) \rightleftharpoons 3Ag^+ + PO_4^{3-}$ $K_{sp} = [Ag^+]^3[PO_4^{3-}]$

$K_{sp} = (3(1.8\times10^{-5}))^3(1.8\times10^{-5}) = 2.8 \times 10^{-18}$

19.22 $PbBr_2(s) \rightleftharpoons Pb^{2+} + 2Br^-$ $K_{sp} = [Pb^{2+}][Br^-]^2$

	$[Pb^{2+}]$	$[Br^-]$
I	–	–
C	+ x	+ 2x
E	x	2x

$K_{sp} = (x)(2x)^2 = 4x^3 = 2.1 \times 10^{-6}$, $x = \sqrt[3]{\frac{2.1 \times 10^{-6}}{4}} = 8.1 \times 10^{-3}$ M

19.24 For every mole of CO_3^{2-} produced, 2 moles of Ag^+ will be produced. Let x = $[CO_3^{2-}]$ at equilibrium and $[Ag^+] = 2x$ at equilibrium. $K_{sp} = [Ag^+]^2[CO_3^{2-}] = (2x)^2(x) = 4x^3$. Solving we find $x = 1.3 \times 10^{-4}$. Thus, the molar solubility of Ag_2CO_3 is 1.3×10^{-4} M.

19.26 To solve this problem, determine the molar solubility for each compound.

LiF: let $x = [Li^+] = [F^-]$ $K_{sp} = [Li^+][F^-] = x^2 = 1.7 \times 10^{-3}$

$x = 4.1 \times 10^{-2}$ M = molar solubility of LiF.

BaF_2: let $x = [Ba^{2+}]$, $[F^-] = 2x$ $K_{sp} = [Ba^{2+}][F^-]^2 = (x)(2x)^2 = 1.7 \times 10^{-6}$

$4x^3 = 1.7 \times 10^{-6}$, and $x = 7.5 \times 10^{-3}$ M = molar solubility of BaF_2.

From the data above, LiF has a greater molar solubility than BaF_2.

19.28 First determine the molar solubility of the MX salt.

Let $x = [M^+] = [X^-]$, $K_{sp} = [M^+][X^-] = (x)(x) = 3.2 \times 10^{-10}$

$x = 1.8 \times 10^{-5}$ M. This is the equilibrium concentration of the two ions.

For the MX_3 salt, let x = equilibrium concentration of M^{3+}, $[X^-] = 3x$.

$K_{sp} = [M^{3+}][X^-]^3 = (x)(3x)^3 = 27x^4$. The value of x in this expression is the value determined in the first part of this problem.

So, $K_{sp} = (27)(1.8 \times 10^{-5})^4 = 2.8 \times 10^{-18}$

19.30 $CaSO_4(s) \rightleftharpoons Ca^{2+}(aq) + SO_4^{2-}(aq)$ $\quad K_{sp} = [Ca^{2+}][SO_4^{2-}]$

let $x = [Ca^{2+}] = [SO_4^{2-}]$ $\quad K_{sp} = x^2 = 2.4 \times 10^{-5}$ and $x = 4.9 \times 10^{-3}$ M.

The molar solubility of $CaSO_4$ is 4.9×10^{-3} M.

19.32 a) $CuCl(s) \rightleftharpoons Cu^+(aq) + Cl^-(aq)$ $\quad K_{sp} = [Cu^+][Cl^-]$

	$[Cu^+]$	$[Cl^-]$
I	–	–
C	+x	+x
E	x	x

$K_{sp} = x^2 = 1.9 \times 10^{-7}$ $\quad \therefore$ x = molar solubility = 4.4×10^{-4} M

b) $CuCl(s) \rightleftharpoons Cu^+(aq) + Cl^-(aq)$ $\quad K_{sp} = [Cu^+][Cl^-]$

	$[Cu^+]$	$[Cl^-]$
I	–	0.0200
C	+x	+x
E	x	0.0200+x

$K_{sp} = (x)(0.0200+x) = 1.9 \times 10^{-7}$ $\quad$ Assume that $x << 0.0200$

$\therefore$ x = molar solubility = 9.5×10^{-6} M

c) $CuCl(s) \rightleftharpoons Cu^+(aq) + Cl^-(aq)$ $\quad K_{sp} = [Cu^+][Cl^-]$

	$[Cu^+]$	$[Cl^-]$
I	–	0.200
C	+x	+x
E	x	0.200+x

$K_{sp} = (x)(0.200+x) = 1.9 \times 10^{-7}$ $\quad$ Assume that $x << 0.200$

$\therefore$ x = molar solubility = 9.5×10^{-7} M

d) $CuCl(s) \rightleftharpoons Cu^+(aq) + Cl^-(aq)$ $\quad K_{sp} = [Cu^+][Cl^-]$

Note that the Cl^- concentration equals (2)(0.150 M) since two moles of Cl^- are produced for every mole of $CaCl_2$.

	$[Cu^+]$	$[Cl^-]$
I	–	0.300
C	+x	+x
E	x	0.300+x

$K_{sp} = (x)(0.300+x) = 1.9 \times 10^{-7}$ Assume that $x << 0.300$

$\therefore$ x = molar solubility = 6.3×10^{-7} M

19.34 $Mg(OH)_2(s) \rightleftharpoons Mg^{2+}(aq) + 2\ OH^-(aq)$ $K_{sp} = [Mg^{2+}][OH^-]^2$

pH = 12.50

pOH = 14.00 − pH = 1.50

$[OH^-] = 10^{-1.50} = 0.0316$ M

	$[Mg^{2+}]$	$[OH^-]$
I	–	0.0316
C	+x	+2x
E	x	0.0316 + 2x

We assume that $2x << 0.0316$, so that $0.0316 + 2x \approx 0.0316$, then we enter the equilibrium values of the above table into the K_{sp} expression:

$K_{sp} = [Mg^{2+}][OH^-]^2$

$7.1 \times 10^{-12} = x(0.0316)^2$

x = molar solubility = 8.4×10^{-5} M

19.36 $Ag_2CrO_4\ (s) \rightleftharpoons 2Ag^+\ (aq) + CrO_4^{2-}(aq)$ $K_{sp} = [Ag^+]^2[CrO_4^{2-}]$

a)

	$[Ag^+]$	$[CrO_4^{2-}]$
I	0.200	–
C	0.200+2x	+x
E	0.200+2x	x

$K_{sp} = (0.200+2x)^2(x)$ Assume that $2x << 0.200$

$1.2 \times 10^{-12} = (0.200)^2(x)$ $x = 3.0 \times 10^{-11}$

The molar solubility is 3.0×10^{-11} M.

b)

	$[Ag^+]$	$[CrO_4^{2-}]$
I	–	0.200
C	+2x	+x
E	2x	0.200 + x

$K_{sp} = (2x)^2(0.200 + x)$ Assume that $x << 0.200$

$1.2 \times 10^{-12} = 4x^2(0.200)$ $x = 1.2 \times 10^{-6}$

The molar solubility is 1.2×10^{-6} M.

19.38 $CaSO_4\ (s) \rightleftharpoons Ca^{2+}\ (aq) + SO_4^{2-}(aq)$ $K_{sp} = [Ca^{2+}][SO_4^{2-}]$

let $x = [Ca^{2+}]$, $[SO_4^{2-}] = 0.015\ M + x$ $K_{sp} = (x)(0.015 + x) = 2.4 \times 10^{-5}$

Solving the resulting quadratic equation gives $x = 1.5 \times 10^{-3}$ M.
The molar solubility of $CaSO_4$ in 0.015 M $CaCl_2$ is 1.5×10^{-3} M.

(Note: If we had assumed that $x << 0.015$, as is usually the method we use to solve these problems, we would have determined that $x = 0.0016$ M. This is slightly larger than 10% of the value 0.015 so our usual assumption is not valid in this problem.)

19.40 $HC_2H_3O_2 + H_2O \rightleftharpoons H_3O^+ + C_2H_3O_2^-$

$$K_a = \frac{[H_3O^+][C_2H_3O_2^-]}{[HC_2H_3O_2]} = 1.8 \times 10^{-5}$$

First, we calculate the % ionization in water (no added sodium acetate):

	$[HC_2H_3O_2]$	$[H_3O^+]$	$[C_2H_3O_2^-]$
I	0.10	–	–
C	0.10–x	+x	+x
E	0.10 –x	x	+x

Assume $x << 0.10$

$$K_a = \frac{[x][x]}{[0.10]} = 1.8 \times 10^{-5} \quad x = 1.3 \times 10^{-3} = [H_3O^+]$$

% ionization = $([H^+]/[\text{total acid}]) \times 100 = (1.3 \times 10^{-3}/0.10) \times 100\ \% =$ *1.3 %*

Now we calculate the % ionization using 0.050 mol/0.500L = 0.10 M for the concentration of sodium acetate:

	$[HC_2H_3O_2]$	$[H_3O^+]$	$[C_2H_3O_2^-]$
I	0.10	–	0.10
C	0.10–x	+x	+x
E	0.10 –x	x	0.10 + x

Assume x << 0.10

$$K_a = \frac{[x][0.10]}{[0.10]} = 1.8 \times 10^{-5} \quad x = 1.8 \times 10^{-5} = [H_3O^+]$$

% ionization = ($[H^+]$/[total acid]) × 100 % = (1.8 × 10^{-5}/0.10) ×100 = *0.018 %*

So the % ionization decreased by (1.3 – 0.018) = 1.3 %, using correct significant figures. (This does not mean it has no dissociation, but that the dissociation is very small compared to 1.3%.)

Using the $[H^+]$ values above, the pH initially was:
$-\log [H^+] = -\log [1.3 \times 10^{-3}] = 2.9$

After addition of the sodium acetate it was:
$-\log [H^+] = -\log [1.8 \times 10^{-5}] = 4.7$

The pH changes by (4.7 – 2.9) = +1.8 pH units

*19.42 $Fe(OH)_2(s) \rightleftharpoons Fe^{2+}(aq) + 2\ OH^-(aq)$ $\qquad K_{sp} = [Fe^{2+}][OH^-]^2$

mol OH^- = 2.20 g NaOH(1 mol/40.01 g NaOH) = 0.0550 mol NaOH

$[OH^-]$ = mol OH^-/L solution = 0.0550 mol/0.250 L = 0.22 M

	$[Fe^{2+}]$	$[OH^-]$
I	–	0.22
C	+x	+2x
E	x	0.22 + 2x

We assume that x << 0.22, so that 0.22 + 2x ≈ 0.22, then we enter the equilibrium values of the above table into the K_{sp} expression:

$K_{sp} = [Fe^{2+}][OH^-]^2$
$7.9 \times 10^{-16} = x(0.22)^2$
x = molar solubility = 1.6×10^{-14} M

Next, we must determine how many moles of $Fe(OH)_2$ are formed in the reaction. This is a limiting reactant problem.

The number of moles of OH^- is 0.0550 (see above).
The number of moles of Fe^{2+} is (.250 L)(0.10 mol/L) = 0.025 mol

From the balanced equation at the top, we need two OH^- for every one Fe^{2+}. This would be 2(0.025 mol) = 0.050 mol OH^-. Looking at the molar quantities above, we have more than enough OH^- so, Fe^{2+} is our limiting reactant:

0.025 mol $Fe(OH)_2$ will form in 0.25 L solution. If dissolved, this would be a concentration of 0.025 mol/0.25 L = 0.10 M. But from above, the maximum molar solubility of is 1.6×10^{-14} M.

This means that remainder of $Fe(OH)_2$ in excess of this value precipitates:
$0.10 - 1.6 \times 10^{-14} \approx 0.10$ M.

This works out to 0.25 L(0.10 mol/L) = 0.025 mol $Fe(OH)_2$(89.8 g/mol)
= 2.2 g solid $Fe(OH)_2$ (essentially all of it).

The concentration of Fe^{2+} in the final solution is at its maximum,
1.6×10^{-14} M.

19.44 $Fe(OH)_2(s) \rightleftharpoons Fe^{2+}(aq) + 2\ OH^-(aq)$ $\qquad K_{sp} = [Fe^{2+}][OH^-]^2$

pH = 9.50
pOH = 14.00 − pH = 4.50
$[OH^-] = 10^{-4.50} = 3.16 \times 10^{-5}$ M

	$[Fe^{2+}]$	$[OH^-]$
I	–	3.16×10^{-5}
C	+x	+2x
E	x	$3.16 \times 10^{-5} + 2x$

Since K_{sp} for iron(II) hydroxide is so tiny, we can safely assume that $2x << 3.16 \times 10^{-5}$, so that $(3.16 \times 10^{-5}) + 2x \approx 3.16 \times 10^{-5}$, then we enter the equilibrium values of the above table into the K_{sp} expression:

$K_{sp} = [Fe^{2+}][OH^-]^2$
$7.9 \times 10^{-16} = x(3.16 \times 10^{-5})^2$
x = molar solubility = 7.9×10^{-7} M

19.46 In order for a precipitate to form, the value of the reaction quotient, Q, must be greater than the value of K_{sp}. For $PbCl_2$, $K_{sp} = 1.7 \times 10^{-5}$ (see Table).

$Q = [Pb^{2+}][Cl^-]^2 = (0.0150)(0.0120)^2 = 2.16 \times 10^{-6}$. Since $Q < K_{sp}$, no precipitate will form.

19.48 To solve this problem, determine the value for Q and apply LeChâtelier's Principle.

(a) $[Pb^{2+}] = (50.0\text{ mL})(0.0100\text{ moles/L})/(100.0\text{ mL}) = 5.00 \times 10^{-3}$
$[Br^-] = (50.0\text{ mL})(0.0100\text{ moles/L})/(100.0\text{ mL}) = 5.00 \times 10^{-3}$
$Q = [Pb^{2+}][Br^-]^2 = (5.00 \times 10^{-3})(5.00 \times 10^{-3})^2 = 1.25 \times 10^{-7}$
For $PbBr_2$, $K_{sp} = 2.1 \times 10^{-6}$
Since $Q < K_{sp}$, no precipitate will form.

(b) $[Pb^{2+}] = (50.0\text{ mL})(0.0100\text{ moles/L})/(100.0\text{ mL}) = 5.00 \times 10^{-3}$
$[Br^-] = (50.0\text{ mL})(0.100\text{ moles/L})/(100.0\text{ mL}) = 5.00 \times 10^{-2}$
$Q = [Pb^{2+}][Br^-]^2 = (5.00 \times 10^{-3})(5.00 \times 10^{-2})^2 = 1.25 \times 10^{-5}$
For $PbBr_2$, $K_{sp} = 2.1 \times 10^{-6}$
Since $Q > K_{sp}$, yes, a precipitate will form.

19.50 $AgCl(s) \rightleftharpoons Ag^+ + Cl^-$ $\quad K_{sp} = [Ag^+][Cl^-] = 1.8 \times 10^{-10}$
$AgI(s) \rightleftharpoons Ag^+ + I^-$ $\quad K_{sp} = [Ag^+][I^-] = 8.3 \times 10^{-17}$

When $AgNO_3$ is added to the solution, AgI will precipitate before any AgCl does due to the lower solubility of AgI. In order to answer the question, i.e., what is the $[I^-]$ when AgCl first precipitates, we need to find the minimum concentration of Ag^+ that must be added to precipitate AgCl.

Let $x = [Ag^+]$; $K_{sp} = (x)(0.050) = 1.8 \times 10^{-10}$; $x = 3.6 \times 10^{-9}$ M

When the AgCl starts to precipitate, the solution will have a $[Ag^+]$ of 3.6×10^{-9} M. Now we ask, what is the $[I^-]$ if $[Ag^+] = 3.6 \times 10^{-9}$ M?
So, $K_{sp} = [Ag^+][I^-] = (3.6 \times 10^{-9})(x) = 8.3 \times 10^{-17}$; $x = 2.3 \times 10^{-8}\text{ M} = [I^-]$

19.52 The less soluble substance is PbS. We need to determine the minimum $[H^+]$ at which CoS will precipitate.

$$K_{spa} = \frac{[Co^{2+}][H_2S]}{[H^+]^2} = \frac{(0.010)(0.1)}{[H^+]^2} = 0.5 \text{ (from Table 19.3)}$$

$$[H^+] = \sqrt{\frac{(0.010)(0.1)}{0.5}}$$
$$= 0.045\text{ M}$$

pH = –log[H^+] = 1.35. At a pH lower than 1.35, PbS will precipitate and CoS will not. At larger values of pH, both PbS and CoS will precipitate.

19.54 $Cu(OH)_2(s) \rightleftharpoons Cu^{2+}(aq) + 2\ OH^-(aq)$

$$K_{sp} = [Cu^{2+}][OH^-]^2$$
$$4.8\times10^{-20} = [0.10][OH^-]^2$$
$$[OH^-] = 6.9 \times 10^{-10}\ M$$
$$pOH = -\log[OH^-] = -\log[6.9 \times 10^{-10}] = 9.2$$
$$pH = 14.00 - pOH = 4.8$$

4.8 is the pH *below* which all the $Cu(OH)_2$ will be soluble.

$Mn(OH)_2(s) \rightleftharpoons Mn^{2+}(aq) + 2\ OH^-(aq)$

$$K_{sp} = [Mn^{2+}][OH^-]^2$$
$$1.6\times10^{-13} = [0.10][OH^-]^2$$
$$[OH^-] = 1.3 \times 10^{-6}\ M$$
$$pOH = -\log[OH^-] = -\log[1.3 \times 10^{-6}] = 5.9$$
$$pH = 14.00 - pOH = 8.1$$

8.1 is the pH *below* which all the $Mn(OH)_2$ will be soluble.

Therefore, from 4.8-8.1 $Mn(OH)_2$ will be soluble, but some $Cu(OH)_2$ will precipitate out of solution.

19.56 a) $Cu^{2+}(aq) + 4Cl^-(aq) \rightleftharpoons CuCl_4^{2-}(aq)$ $\qquad K_{form} = \dfrac{[CuCl_4^{2-}]}{[Cu^{2+}][Cl^-]^4}$

b) $Ag^+(aq) + 2I^-(aq) \rightleftharpoons AgI_2^-(aq)$ $\qquad K_{form} = \dfrac{[AgI_2^-]}{[Ag^+][I^-]^2}$

c) $Cr^{3+}(aq) + 6NH_3(aq) \rightleftharpoons Cr(NH_3)_6^{3+}(aq)$ $K_{form} = \dfrac{[Cr(NH_3)_6^{3+}]}{[Cr^{3+}][NH_3]^6}$

19.58 a) $Co(NH_3)_6^{3+}(aq) \rightleftharpoons Co^{3+}(aq) + 6NH_3(aq)$ $K_{inst} = \dfrac{[Co^{3+}][NH_3]^6}{[Co(NH_3)_6^{3+}]}$

b) $HgI_4^{2-}(aq) \rightleftharpoons Hg^{2+}(aq) + 4I^-(aq)$ $\qquad K_{inst} = \dfrac{[Hg^{2+}][I^-]^4}{[HgI_4^{2-}]}$

c) $Fe(CN)_6^{4-}(aq) \rightleftharpoons Fe^{2+}(aq) + 6CN^-(aq)$ $K_{inst} = \frac{[Fe^{2+}][CN^-]^6}{[Fe(CN)_6^{4-}]}$

19.60 $K_c = K_{sp} \cdot K_{form} = (1.7 \times 10^{-5})(2.5 \times 10^{1}) = 4.3 \times 10^{-4}$

19.62 There are two events in this net process: one is the formation of a complex ion (an equilibrium which has an appropriate value for K_{form}), and the other is the dissolving of $Fe(OH)_3$, which is governed by K_{sp} for the solid.

$$Fe(OH)_3(s) \rightleftharpoons Fe^{3+}(aq) + 3OH^-(aq) \qquad K_{sp} = [Fe^{3+}][OH^-]^3 = 1.6 \times 10^{-39}$$

$$Fe^{3+}(aq) + 6CN^-(aq) \rightleftharpoons Fe(CN)_6^{3-}(aq) \qquad K_{form} = \frac{[Fe(CN)_6^{3-}]}{[Fe^{3+}][CN^-]^6} = 1.0 \times 10^{31}$$

The net process is:

$$Fe(OH)_3(s) + 6CN^-(aq) \rightleftharpoons Fe(CN)_6^{3-}(aq) + 3OH^-(aq)$$

The equilibrium constant for this process should be:

$$K_c = \frac{[Fe(CN)_6^{3-}][OH^-]^3}{[CN^-]^6}$$

The numerical value for the above K_c is equal to the product of K_{sp} for $Fe(OH)_3(s)$ and K_{form} for $Fe(CN)_6^{3-}$, as can be seen by multiplying the mass action expressions for these two equilibria: $K_c = K_{form} \times K_{sp} = 1.6 \times 10^{-8}$

Because K_{form} is so very large, we can assume that all of the dissolved iron ion is present in solution as the complex, thus: $[Fe(CN)_6^{3-}] = 0.11$ mol/1.2 L = 0.092 M. Also the reaction stoichiometry shows that each iron ion that dissolves gives 3 OH^- ions in solution, and we have: $[OH^-] = 0.092 \times 3 = 0.28$ M. We substitute these values into the K_c expression and rearrange to get:

$$[CN^-] = \sqrt[6]{\frac{[Fe(CN)_6^{3-}][OH^-]^3}{K_c}}$$

$$= \sqrt[6]{\frac{(0.092)(0.28)^3}{1.6 \times 10^{-8}}}$$

Thus we arrive at the concentration of cyanide ion that is required in order to satisfy the mass action requirements of the equilibrium: $[CN^-] = 7.1$ mol L^{-1}. Since this concentration of CN^- must be present in 1.2 L, the number of moles of cyanide that are required is: 7.1 mol $L^{-1} \times 1.2$ L = 8.5 mol CN^-.

Additionally, a certain amount of cyanide is needed to form the complex ion. The stoichiometry requires six times as much cyanide ion as iron ion. This is 0.11 moles × 6 = 0.66 mol. This brings the total required cyanide to (8.5 + 0.66) = 9.2 mol.

9.2 mol x 49.0 g/mol = 450 g NaCN are required.

19.64 The applicable equilibria are as follows:

$AgI(s) \rightleftharpoons Ag^+(aq) + I^-(aq)$ $\qquad K_{sp} = [Ag^+][I^-] = 8.3 \times 10^{-17}$

$Ag^+(aq) + 2I^-(aq) \rightleftharpoons AgI_2^-(aq)$ $\qquad K_{form} = \dfrac{[AgI_2^-]}{[Ag^+][I^-]^2} = 1 \times 10^{11}$

When a solution of AgI_2^- is diluted, all of the concentrations of the species in K_{form} above decrease. However, the decrease of $[I^-]$ has more effect on equilibrium because its expression is *squared.* Hence, the denominator is decreased more than the numerator in the reaction quotient, Q. The system reacts according to Le Châtelier's Principle, by moving to the left (toward reactants) to increase the value of $[I^-]$.

As the system moves to the left, more Ag^+ is created, which has an effect on the first equilibrium above. Again, Le Châtelier's Principle causes the reaction to move to the left to re-establish equilibrium, which produces AgI(s) precipitate.

The two equations above may be combined and Kc found as follows:

$AgI(s) + I^-(aq) \rightleftharpoons AgI_2^-(aq)$ $\qquad K_c = \dfrac{[AgI_2^-]}{[I^-]} = K_{sp} \cdot K_{form} = 8.3 \times 10^{-6}$

To answer the second question, we make a table and fill in what we know. We begin with 1.0 M I^-. This is reduced by some amount (x) as it reacts with the silver ions, and $[AgI_2^-]$ is increased by the same amount:

	$[I^-]$	$[AgI_2^-]$
I	1.0	–
C	–x	+x
E	1.0 – x	x

Now we insert the equilibrium values into the above equation:

$$K_c = \frac{[AgI_2^-]}{[I^-]} = 8.3 \times 10^{-6}$$

$$K_c = \frac{[x]}{[1.0 - x]} = 8.3 \times 10^{-6}$$
$$x = 8.3 \times 10^{-6}$$

This value represents the change in concentration of I^- which, from the balanced equation, equals the change in concentration of AgI(s). The given volume is 100 mL, which allows us to find the amount of AgI reacting:

$0.100\ L(8.3 \times 10^{-6}\ mol/L) = 8.3 \times 10^{-7}$ mol AgI

8.3×10^{-7} mol AgI(234.8 g/mol) = 1.9×10^{-4} g AgI

19.66 Recall that $K_{inst} = 1/K_{form}$.

$Zn(OH)_2\ (s) \rightleftharpoons Zn^{2+}(aq) + 2OH^-(aq) \qquad K_{sp} = [Zn^{2+}][OH^-]^2 = 3.0 \times 10^{-16}$

$Zn^{2+}(aq) + 4NH_3\ (aq) \rightleftharpoons Zn(NH_3)_4^{2+}(aq) \qquad K_{form} = \frac{[Zn(NH_3)_4^{2+}]}{[Zn^{2+}][NH_3]^4} = ?$

Combined, this is:

$Zn(OH)_2\ (s) + 4NH_3\ (aq) \rightleftharpoons Zn(NH_3)_4^{2+}(aq) + 2OH^-(aq)$

$$K_c = \frac{[Zn(NH_3)_4^{2+}][OH^-]^2}{[NH_3]^4}$$

	$[NH_3]$	$[Zn(NH_3)_4^{2+}]$	$[OH^-]$
I	1.0	–	–
C	–4x	+x	+2x
E	1.0 – 4x	x	2x

$$K_c = \frac{[x][2x]^2}{[1.0 - 4x]^4}$$

The problem gives the molar solubility of $Zn(OH)_2$ as 5.7×10^{-3} M. This means in one liter of 1.0 M NH_3, $x = 5.7 \times 10^{-3}$ moles. Substituting this value in for x, we get $K_c = 8.1 \times 10^{-7}$.

$$K_c = K_{sp} \cdot K_{form}$$
$$8.1 \times 10^{-7} = 3.0 \times 10^{-16} \cdot K_{form}$$
$$K_{form} = 2.7 \times 10^{9}$$

$K_{inst} = 1/K_{form}$
$K_{inst} = 1/(2.7 \times 10^9) = 3.7 \times 10^{-10}$

Additional Exercises

19.68 We must first calculate the solubility in terms of # mols/L, i.e.,

$$\# \text{mol}/\text{L} = \left(7.05\times10^{-3}\ \text{g}/\text{L}\right)\left(\frac{1\ \text{mol Mg(OH)}_2}{58.32\ \text{g Mg(OH)}_2}\right) = 1.21\times10^{-4}\ \text{M}$$

Next, use this to establish the individual ion concentrations based on the equilibrium:

$Mg(OH)_2 \rightleftharpoons Mg^{2+} + 2OH^-$
$[Mg^{2+}] = 1.21\times10^{-4}$ M
$[OH^-] = 2.42\times10^{-4}$ M

Finally, calculate K_{sp} using the standard expression:

$$K_{sp} = [Mg^{2+}][OH^-]^2 = (1.21\times10^{-4})(2.42\times10^{-4})^2 = 7.09 \times 10^{-12}$$

19.70 $Mg(OH)_2 \rightleftharpoons Mg^{2+} + 2OH^-$

$K_{sp} = [Mg^{2+}][OH^-]^2 = 7.1 \times 10^{-12}$
$K_{sp} = [x][2x]^2 = 7.1 \times 10^{-12}$
$K_{sp} = 4x^3 = 7.1 \times 10^{-12}$
$x = 1.2 \times 10^{-4}$

$[OH^-] = 2x = 2(1.2 \text{ x } 10^{-4}) = 2.4 \times 10^{-4}$
$pOH = -\log [OH^-] = -\log [2.4 \times 10^{-4}] = 3.6$
$pH = 14.00 - pOH = 14.00 - 3.9 = 10.4$

*19.72 In this problem, we have two simultaneous equilibria occurring:

$Mn(OH)_2(s) \rightleftharpoons Mn^{2+}(aq) + 2OH^-(aq)$

$$K_{sp} = [Mn^{2+}][OH^-]^2 = 1.6 \times 10^{-13}$$

$Fe^{2+}(aq) + 2OH^-(aq) \rightleftharpoons Fe(OH)_2(s)$

$$K_c = \frac{1}{[Fe^{2+}][OH^-]^2} = 1/(7.9 \times 10^{-16})$$
$$= 1.3 \times 10^{15}$$

The second equilibrium represents the opposite equation from that of K_{sp}. Therefore, its value is $1/K_{sp}$ for $Fe(OH)_2$.

Combined, and omitting spectator ions, this is:

$$Mn(OH)_2(s) + Fe^{2+}(aq) \rightleftharpoons Mn^{2+}(aq) + Fe(OH)_2(s)$$

$$K_c = \frac{[Mn^{2+}]}{[Fe^{2+}]} = K_{sp\,(Mn)} \cdot K_{c\,(Fe)} = (1.6\times10^{-13})(1.3\times10^{15}) = 208$$

	$[Fe^{2+}]$	$[Mn^{2+}]$
I	0.100	–
C	–x	+x
E	0.100–x	x

$$K_c = \frac{[Mn^{2+}]}{[Fe^{2+}]}$$
$$208 = \frac{[x]}{[0.100 - x]}$$
$$20.8 - 208x = x$$
$$20.8 = 209x$$
$$x = 0.0995$$

Therefore, $[Fe^{2+}] = 0.100 - x = 0.001$ M and $[Mn^{2+}] = 0.0995$ M

Since K_{sp} for $Fe(OH)_2$ and $Mn(OH)_2$ are so small, we assume there is almost no free hydroxide ion present and therefore the pH would remain neutral, around 7.

*19.74 To solve this problem, recognize that for a solution having a density of 1.00 g mL^{-1}, 1 ppm = 1 mg L^{-1}. Therefore, the initial hard water solution has a concentration of 278 mg Ca^{2+} / 1 L solution. Converting to molar concentration:

$$\frac{\#\,mol\,Ca^{2+}}{L\,solution} = \left(\frac{278\,mg\,Ca^{2+}}{1\,L\,solution}\right)\left(\frac{1\,g\,Ca^{2+}}{1000\,mg\,Ca^{2+}}\right)\left(\frac{1\,mol\,Ca^{2+}}{40.078\,g\,Ca^{2+}}\right)$$
$$= 6.94\times10^{-3}\,M\,Ca^{2+}$$

The concentration of CO_3^{2-} is:

$$\frac{\#\,mol\,CO_3^{2-}}{L\,solution} = \left(\frac{1.00\,g\,Na_2CO_3}{1\,L\,solution}\right)\left(\frac{1\,mol\,Na_2CO_3}{105.99\,g\,Na_2CO_3}\right)\left(\frac{1\,mol\,CO_3^{2-}}{1\,mol\,Na_2CO_3}\right)$$
$$= 9.43\times10^{-3}\,M\,CO_3^{2-}$$

Comparing the concentrations of Ca^{2+} and CO_3^{2-}, we observe that Ca^{2+} is the limiting reactant. Because of the small value of Ksp, we can assume that $CaCO_3$ will precipitate using all of the available Ca^{2+} and leaving $(9.43 \times 10^{-3} - 6.94 \times 10^{-3}) = 2.49 \times 10^{-3}$ M CO_3^{2-}. The question now becomes, how much Ca^{2+} will be present in a solution having a $[CO_3^{2-}] = 2.49 \times 10^{-3}$ M? Use the solubility product constant for K_{sp} to answer this question.

$$K_{sp} = 4.5 \times 10^{-9} = [Ca^{2+}][CO_3^{2-}]$$

$$[Ca^{2+}] = \frac{K_{sp}}{[CO_3^{2-}]} = \frac{4.5 \times 10^{-9}}{2.49 \times 10^{-3}} = 1.8 \times 10^{-6}\ \text{M}$$

Converting back to units of ppm (mg L^{-1}) we get:

$$\#\,\text{ppm}\ Ca^{2+} = \left(\frac{1.8 \times 10^{-6}\ \text{mol}\ Ca^{2+}}{1\ \text{L solution}}\right)\left(\frac{40.08\ \text{g}\ Ca^{2+}}{1\ \text{mol}\ Ca^{2+}}\right)\left(\frac{1000\ \text{mg}}{1\ \text{g}}\right)$$

$$= 7.2 \times 10^{-2}\ \text{ppm}\ Ca^{2+}$$

*19.76 a) $Mg(OH)_2(s) \rightleftharpoons Mg^{2+} + 2OH^-$

$NH_4^+ + OH^- \rightleftharpoons NH_3 + H_2O$

b) The NH_4^+ reacts with any OH^- produced in the dissociation of $Mg(OH)_2$ thereby shifting the equilibrium to the right.

c) We want all of the 0.10 mol $Mg(OH)_2$ to go into solution. Using the K_{sp} value for $Mg(OH)_2$, we may find the hydroxide ion concentration under these conditions:

$$Mg(OH)_2(s) \rightleftharpoons Mg^{2+} + 2OH^-$$

$$K_{sp} = [Mg^{2+}][OH^-]^2$$

$$7.1 \times 10^{-12} = [0.10][OH^-]^2$$

$$[OH^-] = 8.4 \times 10^{-6}$$

Now we can use this value in the following, simultaneous equilibrium:

$$NH_4^+(aq) + OH^-(aq) \rightleftharpoons H_2O + NH_3(aq)$$

$$K_c = 1/K_{b\,NH_3} = 1/1.8 \times 10^{-5} = 5.6 \times 10^4$$

$$K_c = \frac{[NH_3]}{[OH^-][NH_4^+]} = \frac{[0.20]}{[8.4 \times 10^{-6}][NH_4^+]} = 5.6 \times 10^4$$

(We know that $[NH_3] = 0.20$ M because in the equation below 2 moles of ammonia are formed for every one mole of magnesium ion:

$$Mg(OH)_2(s) + 2NH_4^+(aq) \rightleftharpoons Mg^{2+}(aq) + 2H_2O + 2NH_3(aq))$$

Solving for $[NH_4^+]$, we get 0.43 M.

So the total $[NH_3] + [NH_4^+] = 0.20 + 0.43 = 0.63$ M

One must therefore add 0.63 mol NH_4Cl to a liter of solution.

d) The resulting solution will contain 0.20 mol of NH_3. Solve the weak base equilibrium problem for NH_3. The pH = 11.28.

19.78 First, we must calculate the mass of $CaSO_4$ dissolved. The volume is a cylinder, with $V = h\pi r^2$:

$h = 0.50$ in (2.54 cm/1 inch) = 1.27 cm

$r = ½$ diameter = 0.50 cm

Therefore:

$V = 1.0\ cm^3$, and

mass = $1.0\ cm^3(0.97\ g/cm^3) = 0.97$ g

However, because it is a hydrate ($CaSO_4{\cdot}2H_2O$), plaster is only (136.2/172.2 = 0.79) 79% calcium sulfate, therefore the true mass of $CaSO_4$ is:

Mass = 0.97(.79) = 0.77 g $CaSO_4$

In moles, this is 0.77 g $CaSO_4$ = (1 mol/136.2g) = 0.0056 mol $CaSO_4$

Now we must calculate the volume of water necessary to dissolve 0.77 g of $CaSO_4$. We start by finding its molar solubility:

$$K_{sp} = [Ca^{2+}][SO_4^{2-}] = 2.4 \times 10^{-5}$$

$$K_{sp} = [x][x] = 2.4 \times 10^{-5}$$

$x = 4.9 \times 10^{-3}$ mol/L

So the volume of water needed is:

0.0056 mol(1 L/4.9×10^{-3} mol) = 1.2 L

Finally, we find the amount of time needed to produce this much water:

1.2 L (1 day/2.00 L) = 0.60 days, or about 14 hours.

19.80 a) $K_{form} = \dfrac{1}{K_{inst}} = \dfrac{1}{5.6\times10^{-2}} = 1.8 \times 10^2$

b) Because this value is small, this ion is much less stable than those listed in the table.

*19.82 Let x = mols of PbI_2 that dissolve per liter;

Let y = mols of $PbBr_2$ that dissolve per liter.

Then, at equilibrium, we have

$[Pb^{2+}] = x+y$, $[I^-] = 2x$ and $[Br^-] = 2y$

We know:

$PbBr_2(s) \rightleftharpoons Pb^{2+} + 2Br^-$ $\quad K_{sp} = 2.1 \times 10^{-6} = [Pb^{2+}][Br^-]^2$

$PbI_2(s) \rightleftharpoons Pb^{2+} + 2I^-$ $\quad K_{sp} = 7.9 \times 10^{-9} = [Pb^{2+}][I^-]^2$

Substituting we get: $2.1 \times 10^{-6} = (x+y)(2y)^2$ and $7.9 \times 10^{-9} = (x+y)(2x)^2$

Solving for x and y we find:

$x = 4.85 \times 10^{-4}$

$y = 7.91 \times 10^{-3}$

Thus, $[Pb^{2+}] = 8.40 \times 10^{-3}$ M, $[I^-] = 9.70 \times 10^{-4}$ M and $[Br^-] = 1.58 \times 10^{-2}$ M

(Note: $[Br^-] > [I^-]$ because $PbBr_2$ is more soluble than PbI_2.)

*19.84 We want all of the 0.10 mol $Mg(OH)_2$ to go into solution. Using the K_{sp} value for $Mg(OH)_2$, we may find the hydroxide ion concentration under these conditions:

$$Mg(OH)_2(s) \rightleftharpoons Mg^{2+} + 2OH^-$$

$$K_{sp} = [Mg^{2+}][OH^-]^2$$

$$7.1\times10^{-12} = [0.10][OH^-]^2$$

$$[OH^-] = 8.4\times10^{-6}$$

Now we can use this value in the following, simultaneous equilibrium:

$$NH_4^+(aq) + OH^-(aq) \rightleftharpoons H_2O + NH_3(aq)$$

$$K_c = 1/K_{b_{NH_3}} = 1/1.8 \times 10^{-5} = 5.6 \times 10^4$$

$$K_c = \frac{[NH_3]}{[OH^-][NH_4^+]} = \frac{[0.20]}{[8.4\times10^{-6}][NH_4^+]} = 5.6\times10^4$$

(We know that $[NH_3] = 0.20$ M because in the equation below 2 moles of ammonia are formed for every one mole of magnesium ion:

$$Mg(OH)_2(s) + 2NH_4^+(aq) \rightleftharpoons Mg^{2+}(aq) + 2H_2O + 2NH_3(aq))$$

Solving for $[NH_4^+]$, we get 0.43 M.

So the total $[NH_3] + [NH_4^+] = 0.20 + 0.43 = 0.63$ M

One must therefore add 0.63 mol NH_4Cl to a liter of solution.

*19.86 First, let's examine the question to make clear what is happening. A solution contains 0.20 M Ag^+ ions and 0.10 M acetate ions. The ion product of these two (0.020) is less than K_{sp} for silver acetate (2.3 x 10^{-3}), so the silver acetate remains in solution. There are also H^+ ions (H_3O^+) in the solution as a result of the following equilibrium (OAc^- will symbolize acetate):

$$H_2O + HOAc \rightleftharpoons H_3O^+(aq) + OAc^-(aq)$$

The amount of H_3O^+ may be found by using the K_a for acetic acid:

	[HOAc]	$[H_3O^+]$	$[OAc^-]$
I	0.10	–	–
C	–x	+x	+x
E	0.10 – x	x	x

$$K_a = \frac{[H_3O^+][OAc^-]}{[HOAc]}$$

$$1.8\times10^{-5} = \frac{[x][x]}{[0.10 - x]}$$

$$x \approx 1.3 \text{ x } 10^{-3}$$

So $[H_3O^+] = 1.3 \times 10^{-3}$ M.

However, when F^- is added to the solution (in the form of KF), the following equilibrium takes place:

$$F^-(aq) + H_3O^+(aq) \rightleftharpoons H_2O + HF$$

This depletes H_3O^+ ions from the solution, which causes the first equilibrium above to move to the right, producing more acetate ions. When the acetate ion concentration hits some minimum value (determined by K_{sp}) silver acetate will precipitate. That value may be found as follows:

$$K_{sp} = [Ag^+][OAc^-]$$
$$2.3\times10^{-3} = [0.20][x]$$
$$x = 0.012 \text{ mol/L}$$

So the problem becomes...How many grams of KF must be added such that the acetate concentration increases to 0.012 M? This is now a simultaneous equilibrium problem:

$$H_2O + HOAc \rightleftharpoons H_3O^+(aq) + OAc^-(aq) \qquad K_a = 1.8\times10^{-5}$$

$$F^-(aq) + H_3O^+(aq) \rightleftharpoons H_2O + HF \qquad K_c' = \frac{1}{K_a} = \frac{1}{6.5\times10^{-4}} = 1.5\times10^3$$

(Note the the above equation is simply the reverse of that for the K_a of HF, so K_c' = $1/K_a$.)

Combined, this becomes:

$F^-(aq) + HOAc \rightleftharpoons HF + OAc^-(aq)$

$$K_c = K_a \cdot K_b = (1.8\times10^{-5})(1.5\times10^{3}) = 2.7\times10^{-2}$$

Recall that we have already found the initial concentrations of OAc^- and HOAc above. Using this information, and the fact that we want the final $[OAc^-]$ to be 1.2×10^{-2}, we can begin to fill out the table below.

	$[F^-]$	[HOAc]	[HF]	$[OAc^-]$
I	x	0.099	–	1.3×10^{-3}
C	– 0.011	– 0.011	+ 0.011	+ 0.011
E	?	0.088	0.011	0.012

$$K_c = \frac{[HF][OAc^-]}{[F^-][HOAc]}$$
$$2.7\times10^{-2} = \frac{[0.011][0.012]}{[F^-][0.088]}$$
$$[F^-] = 0.056 \text{ M}$$

Placing this value into the table as the equilibrium concentration of $[F^-]$, we find the initial $[F^-]$ must be 0.056 + 0.011 = 0.067 M.

Therefore the amount of KF needed in the 200 mL solution is:

0.200 L(0.067 mol KF/L)(58.01 g KF/1 mol KF) = 0.78 g KF

*19.88 At its simplest, this is only a K_{sp} problem.

$$K_{sp} = [Mn^+][OH^-]^2$$
$$1.6\times10^{-13} = [0.10][OH^-]^2$$
$$[OH^-] = 1.3\times10^{-6}$$

When $[OH^-] = 1.3 \times 10^{-6}$ M, $Mn(OH)_2$ precipitates.

0.100 L(2.0 mol/L) = 0.20 mol NH_3 are added to 400 mL solution which would make an initial concentration of 0.20 mol/0.500 L = 0.40 M NH_3. The following equilibrium is set up:

$H_2O + NH_3 \rightleftharpoons NH_4^+(aq) + OH^-(aq) \qquad K_b = 1.8\times10^{-5}$

The problem tells us that *all* of the Sn is precipitated as $Sn(OH)_2$:

$Sn^{2+}(aq) + 2OH^-(aq) \rightleftharpoons Sn(OH)_2(s)$

This immediately uses the first 2(0.10M) = 0.20 M OH^- which is produced from the reaction of ammonia with water above, using 0.20 M NH_3. This effectively brings our initial concentration of NH_3 to 0.20 M.

Now we determine how much NH_3 will produce $[OH^-] = 1.3 \times 10^{-6}$ M.

$H_2O + NH_3 \rightleftharpoons NH_4^+(aq) + OH^-(aq) \qquad K_b = 1.8\times10^{-5}$

	$[NH_3]$	$[NH_4^+]$	$[OH^-]$
I	x	–	1.0×10^{-7}
C	-1.2×10^{-6}	$+1.2 \times 10^{-6}$	$+1.2 \times 10^{-6}$
E	?	1.2×10^{-6}	1.3×10^{-6}

$$K_b = \frac{[NH_4^+][OH^-]}{[NH_3]}$$

$$1.8\times10^{-5} = \frac{[1.2\times10^{-6}][1.3\times10^{-6}]}{[NH_3]}$$

$$[NH_3] = 8.7 \times 10^{-8}$$

Therefore the initial $[NH_3]$ should be $8.7 \times 10^{-8} + 1.2 \times 10^{-6} = 1.3 \times 10^{-6}$. So we want to reduce $[NH_3]$ by $0.20 - 1.3 \times 10^{-6} = 0.1999$ M, essentially by 0.20 M. This would require adding equimolar amounts of HCl, or:

0.500 L(0.20 mol NH_3/L)(1 mol HCl/1 mol NH_3)(36.5 g HCl/1 mol HCl)
= 3.7 g HCl.

(The difficulty here arises from the fact that such a small amount of OH^- is required to precipitate the Mg^{2+} from solution that even a minimal amount of NH_3 produces enough hydroxide ion to do so.)

*19.90 $Fe(OH)_3$ has such an exceedingly small K_{sp}, there is virtually no dissociation in pure water. Therefore, the pH would be expected to be about 7.

(If the calculations are done, this is borne out:

$$K_{sp} = [Fe^{3+}][OH^-]^3$$
$$1.6\times10^{-39} = [x][3x]^3$$
$$x = 8.8\times10^{-11}$$
$$[OH^-] = 2.6\times10^{-10}$$

This number is 1,000 times smaller than the hydroxide ion already present in pure water.)

Chapter 20

Practice Exercises

20.1 a) $q>0, w>0$ b) $q<0, w<0$ c) $q<0, w>0$ d) $q>0, w<0$
In case (b), $q<0$ and $w<0$ so ΔE is the most negative.

20.2 $q = -p\Delta V = -(14.0 \text{ atm})(12.0 \text{ L} - 1.0 \text{ L}) = -154 \text{ L atm}$

$\Delta E = w + q = 0$

Therefore, $q = +154$ L atm.

(The energy is converted into heat; since the heat does not leave the system the temperature increases.)

20.3 $\Delta E = q - p\Delta V$ since $q = 0$
$\Delta E = -p\Delta V$
but ΔV is negative for a compression so ΔE increases and T increases.
Energy is added to the system in the form of work.

20.4 $\Delta E° = \Delta H° - \Delta nRT = -217.1 \text{ kJ} - (-1 \text{ mol})(8.314 \text{ J mol}^{-1} \text{ K}^{-1})(298 \text{ K})$
$= -217.1 \text{ kJ} + 2.48 \text{ kJ}$
$= -214.6 \text{ kJ}$

% Difference = $(2.48/217) \times 100\ \% = 1.14\ \%$

20.5 a) ΔS is negative since the products have a lower entropy, i.e. a lower freedom of movement.
b) ΔS is positive since the products have a higher entropy, i.e. a higher freedom of movement.

20.6 a) ΔS is negative since there are less gas molecules. (The product is also more complex, indicating an increase in order.)
b) ΔS is negative since there are less gas molecules. (The product is also more complex, indicating an increase in order.)

20.7 a) ΔS is negative since there is a change from a gas phase to a liquid phase. (The product is also more complex, indicating an increase in order.)
b) ΔS is negative since there are less gas molecules. (The product is also more complex, indicating an increase in order.)
c) ΔS is positive since the particles go from an ordered, crystalline state to a more disordered, aqueous state.

20.8 $\Delta S° = (\text{sum } S°[\text{products}]) - (\text{sum } S°[\text{reactants}])$

a) $\Delta S° = \{S°[H_2O(\ell)] + S°[CaCl_2(s)]\} - \{S°[CaO(s)] + 2S°[HCl(g)]\}$

$\Delta S° = \{1 \text{ mol} \times (69.96 \text{ J mol}^{-1} \text{ K}^{-1}) + 1 \text{ mol} \times (114 \text{ J mol}^{-1} \text{ K}^{-1})\}$

$- \{1 \text{ mol} \times (40 \text{ J mol}^{-1} \text{ K}^{-1}) + 2 \text{ mol} \times (186.7 \text{ J mol}^{-1} \text{ K}^{-1})\}$

$\Delta S° = -229 \text{ J/K}$

b) $\Delta S° = \{S°[C_2H_6(g)]\} - \{S°[H_2(g)] + S°[C_2H_4(g)]\}$

$\Delta S° = \{1 \text{ mol} \times (229.5 \text{ J mol}^{-1} \text{ K}^{-1})\}$

$- \{1 \text{ mol} \times (130.6 \text{ J mol}^{-1} \text{ K}^{-1}) + 1 \text{ mol} \times (219.8 \text{ J mol}^{-1} \text{ K}^{-1})\}$

$\Delta S° = -120.9 \text{ J/K}$

20.9 First, we calculate $\Delta S°$, using the data:

$\Delta S° = \{2S°[Fe_2O_3(s)]\} - \{3S°[O_2(g)] + 4S°[Fe(s)]\}$

$\Delta S° = \{2 \text{ mol} \times (90.0 \text{ J mol}^{-1} \text{ K}^{-1})\}$

$- \{3 \text{ mol} \times (205.0 \text{ J mol}^{-1} \text{ K}^{-1}) + 4 \text{ mol} \times (27 \text{ J mol}^{-1} \text{ K}^{-1})\}$

$\Delta S° = -543 \text{ J/K} = -0.543 \text{ kJ/mol}$

Next, we calculate $\Delta H°$ using the data :

$\Delta H° = (\text{sum } \Delta H_f°[\text{products}]) - (\text{sum } \Delta H_f°[\text{reactants}])$

$\Delta H° = \{2\Delta H_f°[Fe_2O_3(s)]\} - \{3\Delta H_f°[O_2(g)] + 4\Delta H_f°[Fe(s)]\}$

$\Delta H° = \{2 \text{ mol} \times (-822.2 \text{ kJ/mol})\} - \{3 \text{ mol} \times (0.0 \text{ kJ/mol}) + 4 \times (0.0 \text{ kJ/mol})\}$

$\Delta H° = -1644 \text{ kJ}$

The temperature is 25.0 + 273.15 = 298.15 K, and the calculation of $\Delta G°$ is as follows: $\Delta G° = \Delta H° - T\Delta S° = -1644 \text{ kJ} - (298.15 \text{ K})(-0.543 \text{ kJ/K}) = -1482 \text{ kJ}$

20.10 We calculate $\Delta G°_{rxn}$, using the data from Table 20.2:

a) $\Delta G°_{rxn} = \{2\Delta G_f°[NO_2(g)]\} - \{2\Delta G_f°[NO(g)] + \Delta G_f°[\ O_2(g)]\}$

$\Delta G°_{rxn} = \{2 \text{ mol} \times (+51.84 \text{ kJ mol}^{-1})\}$

$- \{2 \text{ mol} \times (+86.69 \text{ kJ mol}^{-1}) + 1 \text{ mol} \times (0 \text{ kJ mol}^{-1})\}$

$\Delta G°_{rxn} = -69.7 \text{ kJ/mol}$

b) $\Delta G°_{rxn} = \{\Delta G_f°[CaCl_2(s)] + 2\Delta G_f°[H_2O(g)]\ \}$

$- \{\Delta G_f°[Ca(OH)_2(s)] + 2\Delta G_f°[\ HCl(g)]\}$

$\Delta G°_{rxn} = \{1 \text{ mol} \times (-750.2 \text{ kJ mol}^{-1}) + 2 \text{ mol} \times (-228.6 \text{ kJ mol}^{-1})\}$

$- \{1 \text{ mol}\times(-896.76 \text{ kJ mol}^{-1}) + 2 \text{ mol}\times(-95.27 \text{ kJ mol}^{-1})\}$

$\Delta G°_{rxn} = -120.1 \text{ kJ/mol}$

20.11 The maximum amount of work that is available is the free energy change for the process, in this case, the standard free energy change, $\Delta G°$, since the process occurs at 25 °C.

$4Al(s) + 3O_2(g) \rightarrow 2Al_2O_3(s)$

$\Delta G° = (\text{sum } \Delta G_f°[\text{products}]) - (\text{sum } \Delta G_f°[\text{reactants}])$
$\Delta G° = 2\Delta G_f°[Al_2O_3(s)] - \{3\Delta G_f°[O_2(g)] + 4\Delta G_f°[Al(s)]\}$
$\Delta G° = 2 \text{ mol} \times (-1576.4 \text{ kJ/mol}) - \{3 \text{ mol} \times (0.0 \text{ kJ/mol}) + 4 \text{ mol} \times (0.0 \text{ kJ/mol})\}$
$\Delta G° = -3152.8$ kJ, for the reaction as written.

This calculation conforms to the reaction *as written*. This means that the above value of $\Delta G°$ applies to the equation involving *4 mol* of Al. The conversion to give energy *per mole* of aluminum is then: –3152.8 kJ/4 mol Al = –788 kJ/mol

The maximum amount of energy that may be obtained is thus 788 kJ.

20.12 For the vaporization process in particular, and for any process in general, we have:

$$\Delta G = \Delta H - T\Delta S$$

If the temperature is taken to be that at which equilibrium is obtained, that is the temperature of the boiling point (where liquid and vapor are in equilibrium with one another), then we also have the result that ΔG is equal to zero:

$$\Delta G = 0 = \Delta H - T\Delta S, \text{ or } T_{eq} = \Delta H/\Delta S$$

We know ΔH to be 60.7 kJ/mol; we need the value for ΔS in units kJ mol^{-1} K^{-1}:

$\Delta S° = (\text{sum } S°[\text{products}]) - (\text{sum } S°[\text{reactants}])$
$\Delta S° = S°[Hg(g)] - S°[Hg(\ell)]$
$\Delta S° = (175 \times 10^{-3} \text{ kJ mol}^{-1} \text{ K}^{-1}) - (76.1 \times 10^{-3} \text{ kJ mol}^{-1} \text{ K}^{-1})$
$\Delta S° = 98.9 \times 10^{-3} \text{ kJ mol}^{-1} \text{ K}^{-1}$

$T_{eq} = 60.7 \text{ kJ/mol} \div 98.9 \times 10^{-3} \text{ kJ/mol K} = 614 \text{ K } (341 \text{ °C})$

20.13 $\Delta G° = (\text{sum } \Delta G_f°[\text{products}]) - (\text{sum } \Delta G_f°[\text{reactants}])$
$\Delta G° = 2\Delta G_f°[SO_3(g)] - \{2\Delta G_f°[SO_2(g)] + \Delta G_f°[O_2(g)]\}$
$\Delta G° = 2 \text{ mol} \times (-370.4 \text{ kJ/mol}) - \{2 \text{ mol} \times (-300.4 \text{ kJ/mol}) + (0.0 \text{ kJ/mol})\}$
$\Delta G° = -140.0$ kJ/mol

Since the sign of $\Delta G°$ is negative, the reaction should be spontaneous.

20.14 $\Delta G° = \Delta H° - T\Delta S°$

$\Delta G° = \{2\Delta G_f°[HCl(g)] + \Delta G_f°[CaCO_3(s)]\}$
$\quad - \{\Delta G_f°[CaCl_2(s)] + \Delta G_f°[H_2O(g)] + \Delta G_f°[CO_2(g)]\}$

$\Delta G° = \{2 \text{ mol} \times (-95.27 \text{ kJ/mol}) + 1 \text{ mol} \times (-1128.8 \text{ kJ/mol})\}$
$\quad - \{1 \text{ mol} \times (-750.2 \text{ kJ/mol}) + 1 \text{ mol} \times (-228.6 \text{ kJ/mol}) +$
$\quad 1 \text{ mol}\times(-394.4 \text{ kJ/mol})\}$

$\Delta G° = +53.9 \text{ kJ}$

Since $\Delta G°$ is positive, the reaction is not spontaneous, and we do not expect to see products formed from reactants.

20.15 First, we compute the standard free energy change for the reaction, based on the:

$\Delta G° = (\text{sum } \Delta G_f°[\text{products}]) - (\text{sum } \Delta G_f°[\text{reactants}])$

$\Delta G° = \{\Delta G_f°[H_2O(g)] + \Delta G_f°[CO_2(g)] + \Delta G_f°[Na_2CO_3(s)]\} -$
$\quad \{2\Delta G_f°[NaHCO_3(s)]\}$

$\Delta G° = \{1 \text{ mol} \times (-228.6 \text{ kJ/mol}) + 1 \text{ mol} \times (-394.4 \text{ kJ/mol})$
$\quad + 1 \text{ mol} \times (-1048 \text{ kJ/mol})\} - \{2 \text{ mol} \times (-851.9 \text{ kJ/mol})\}$

$\Delta G° = +33 \text{ kJ}$

Next, we determine values for $\Delta H°$ and $\Delta S°$:

$\Delta H° = (\text{sum } \Delta H_f°[\text{products}]) - (\text{sum } \Delta H_f°[\text{reactants}])$

$\Delta H° = \{\Delta H_f°[H_2O(g)] + \Delta H_f°[CO_2(g)] + \Delta H_f°[Na_2CO_3(s)]\} -$
$\{2\Delta H_f°[NaHCO_3(s)]\}$

$\Delta H° = \{1 \text{ mol} \times (-241.8 \text{ kJ/mol}) + 1 \text{ mol} \times (-393.5 \text{ kJ/mol})$
$\quad + 1 \text{ mol} \times (-1131 \text{ kJ/mol})\} - \{2 \text{ mol} \times (-947.7 \text{ kJ/mol})\}$

$\Delta H° = +129 \text{ kJ}$

$\Delta S° = (\text{sum } S°[\text{products}]) - (\text{sum } S°[\text{reactants}])$

$\Delta S° = \{S°[H_2O(g)] + S°[CO_2(g)] + S°[Na_2CO_3(s)]\} - \{2S°[NaHCO_3(s)]\}$

$\Delta S° = \{1 \text{ mol} \times (188.7 \text{ J mol}^{-1} \text{ K}^{-1}) + 1 \text{ mol} \times (213.6 \text{ J mol}^{-1} \text{ K}^{-1})$
$\quad + 1 \text{ mol} \times (136 \text{ J mol}^{-1} \text{ K}^{-1})\} - \{2 \text{ mol} \times (102 \text{ J mol}^{-1} \text{ K}^{-1})\}$

$\Delta S° = 334 \text{ J/K} = 0.334 \text{ kJ/K}$

Next, we assume that both $\Delta H°$ and $\Delta S°$ are independent of temperature, and use these values to determine ΔG at a temperature of 200 + 273 = 473 K:

$\Delta G°_{473} = \Delta H° - T\Delta S° = 129 \text{ kJ} - (473 \text{ K})(0.334 \text{ kJ/K}) = -29 \text{ kJ}$

At the lower of these two temperatures (25 °C), the reaction has a positive value of ΔG. At the higher of these two temperatures (200 °C), the reaction has a negative value of ΔG. Thus ΔG becomes more negative as the temperature is raised, so the reaction becomes increasingly more favorable as the temperature is increased. In other words, the position of the equilibrium will be shifted more towards products at the higher temperature.

20.16 Using the data provided we may write:

$$\Delta G = \Delta G^\circ + RT \ln\left(\frac{P_{N_2O_4}}{P_{NO_2}{}^2}\right)$$

$$= -5.40\times10^3 \text{ J mol}^{-1} + (8.314 \text{ J mol}^{-1}\text{ K}^{-1})(298 \text{ K}) \ln\left(\frac{0.25 \text{ atm}}{(0.60 \text{ atm})^2}\right)$$

$$= -5.40\times10^3 \text{ J mol}^{-1} + (-9.03\times10^2 \text{ J mol}^{-1})$$

$$= -6.30\times10^3 \text{ J mol}^{-1}$$

Since ΔG is negative, the forward reaction is spontaneous and the reaction will proceed to the right.

20.17 $\Delta G^\circ = -RT \ln K_p$
$\Delta G^\circ = -(8.314 \text{ J K}^{-1}\text{ mol}^{-1})(25 + 273 \text{ K}) \times \ln(6.9 \times 10^5) = -33 \times 10^3 \text{ J}$
$\Delta G^\circ = -33 \text{ kJ}$

20.18 $\Delta G^\circ = -RT \ln K_p$
$3.3 \times 10^3 \text{ J} = -(8.314 \text{ J K}^{-1}\text{ mol}^{-1})(298 \text{ K}) \times \ln(K_p)$
$\ln(K_p) = -3.3 \times 10^3 \text{ J}/[(8.314 \text{ J K}^{-1}\text{ mol}^{-1})(298 \text{ K})] = -1.3$
Taking the antilog of both sides of the above equation gives: $K_p = 0.26$

20.19 $\Delta G^\circ = \Delta H^\circ - T\Delta S^\circ = -92.4 \times 10^3 \text{ J} - (323 \text{ K})(-198.3 \text{ J/K}) = -2.83 \times 10^4 \text{ J}$
$\Delta G = -RT \ln K$
Thus, $-2.83 \times 10^4 \text{ J} = -(8.314 \text{ J K}^{-1}\text{ mol}^{-1})(323 \text{ K}) \times \ln(K_p)$
$\ln(K_p) = -2.38 \times 10^4 \text{ J}/[(-8.314 \text{ J K}^{-1}\text{ mol}^{-1})(323 \text{ K})] = 10.5$
Taking the antilog of both sides of the above equation gives: $K_p = 3.84 \times 10^4$

Review Problems

20.54 $\Delta E = q + w = 300 \text{ J} + 700 \text{ J} = +1000 \text{ J}$

The overall process is endothermic, meaning that the internal energy of the system increases. Notice that both terms, q and w, contribute to the increase in internal energy of the system; the system gains heat (+q) and has work done on it (+w).

20.56 work = $P \times \Delta V$

The total pressure is atmospheric pressure plus that caused by the hand pump:

$P = (30.0 + 14.7)\ \text{lb/in}^2 = 44.7\ \text{lb/in}^2$

Converting to atmospheres we get:

$P = 44.7\ \text{lb/in}^2 \times 1\ \text{atm}/14.7\ \text{lb/in}^2 = 3.04\ \text{atm}$

Next we convert the volume change in units in^3 to units L:

$24.0\ \text{in}^3 \times (2.54\ \text{cm/in})^3 \times 1\ \text{L}/1000\ \text{cm}^3 = 0.393\ \text{L}$

Hence $P \times \Delta V = (3.04\ \text{atm})(0.393\ \text{L}) = 1.19\ \text{L·atm}$

$1.19\ \text{L·atm} \times 101.3\ \text{J/L·atm} = 121\ \text{J}$

20.58 We use the data supplied in Appendix C.

a) $3PbO(s) + 2NH_3(g) \rightarrow 3Pb(s) + N_2(g) + 3H_2O(g)$

$\Delta H° = \{3\Delta H_f°[Pb(s)] + \Delta H_f°[N_2(g)] + 3\Delta H_f°[H_2O(g)]\}$
$- \{3\Delta H_f°[PbO(s)] + 2\Delta H_f°NH_3(g)]\}$

$\Delta H° = \{3\ \text{mol} \times (0\ \text{kJ/mol}) + 1\ \text{mol} \times (0\ \text{kJ/mol}) + 3\ \text{mol} \times (-241.8\ \text{kJ/mol})\}$
$- \{3\ \text{mol} \times (-217.3\ \text{kJ/mol}) + 2\ \text{mol} \times (-46.19\ \text{kJ/mol})\}$

$\Delta H° = +\ 24.58\ \text{kJ}$

$\Delta E = \Delta H° - \Delta nRT$

$\Delta E = 24.58\ \text{kJ} - (+2\ \text{mol})(8.314\ \text{J/mol K})(10^{-3}\ \text{kJ/J})(298\ \text{K}) = 19.6\ \text{kJ}$

b) $NaOH(s) + HCl(g) \rightarrow NaCl(s) + H_2O(\ell)$

$\Delta H° = \{\Delta H_f°[NaCl(s)] + \Delta H_f°[H_2O(\ell)]\} - \{\Delta H_f°[NaOH(s)] + \Delta H_f°[HCl(g)]\}$

$\Delta H° = \{1\ \text{mol} \times (-411.0\ \text{kJ/mol}) + 1\ \text{mol} \times (-285.9\ \text{kJ/mol})\}$
$- \{1\ \text{mol} \times (-426.8\ \text{kJ/mol}) + 1\ \text{mol} \times (-92.30)\}$

$\Delta H° = -178\ \text{kJ}$

$\Delta E = \Delta H° - \Delta nRT$

$\Delta E = -178\ \text{kJ} - (-1)(8.314\ \text{J/mol K})(10^{-3}\ \text{kJ/J})(298\ \text{K}) = -175\ \text{kJ}$

c) $Al_2O_3(s) + 2Fe(s) \rightarrow Fe_2O_3(s) + 2Al(s)$

$\Delta H° = \{\Delta H_f°[Fe_2O_3(s)] + 2\Delta H_f°[Al(s)]\} - \{\Delta H_f°[\ Al_2O_3(s)] + 2\Delta H_f°[Fe(s)]\}$

$\Delta H° = \{1\ \text{mol} \times (-822.3\ \text{kJ/mol}) + 2\ \text{mol} \times (\text{o kJ/mol})\}$
$- \{1\ \text{mol} \times (-1669.8\ \text{kJ/mol}) + 2\ \text{mol} \times (0\ \text{kJ/mol})\}$

$\Delta H° = 847.6\ \text{kJ}$

$\Delta E = \Delta H°$ (847.6 kJ), since the value of Δn for this reaction is zero.

d) $2CH_4(g) \rightarrow C_2H_6(g) + H_2(g)$
$\Delta H° = \{\Delta H_f°[C_2H_6(g)] + \Delta H_f°[H_2(g)]\} - \{2\Delta H_f°[CH_4(g)]\}$
$\Delta H° = \{1 \text{ mol} \times (-84.667 \text{ kJ/mol}) + 1 \text{ mol} \times (0.0 \text{ kJ/mol})\}$
$- \{2 \text{ mol} \times (-74.848 \text{ kJ/mol})\}$
$H° = 65.029 \text{ kJ}$

$\Delta E = \Delta H°$ (65.029 kJ), since the value of Δn for this reaction is zero.

20.60 In general, we have the equation:
$\Delta H° = (\text{sum } \Delta H_f°[\text{products}]) - (\text{sum } \Delta H_f°[\text{reactants}])$

a) $\Delta H° = \{\Delta H_f°[CaCO_3(s)]\} - \{\Delta H_f°[CO_2(g)] + \Delta H_f°[CaO(s)]\}$
$\Delta H° = \{1 \text{ mol} \times (-1207 \text{ kJ/mol})\}$
$- \{1 \text{ mol} \times (-394 \text{ kJ/mol}) + 1 \text{ mol} \times (-635.5 \text{ kJ/mol})\}$
$\Delta H° = -178 \text{ kJ}$, $\therefore$ favored.

b) $\Delta H° = \{\Delta H_f°[C_2H_6(g)]\} - \{\Delta H_f°[C_2H_2(g)] + 2\Delta H_f°[H_2(g)]\}$
$\Delta H° = \{1 \text{ mol} \times (-84.5 \text{ kJ/mol})\}$
$- \{1 \text{ mol} \times (227 \text{ kJ/mol}) + 2 \text{ mol} \times (0.0 \text{ kJ/mol})\}$
$\Delta H° = -311 \text{ kJ}$, $\therefore$ favored.

c) $\Delta H° = \{\Delta H_f°[Fe_2O_3(s)] + 3\Delta H_f°[Ca(s)]\}$
$- \{2\Delta H_f°[Fe(s)] + 3\Delta H_f°[CaO(s)]\}$
$\Delta H° = \{1 \text{ mol} \times (-822.2 \text{ kJ/mol}) + 3 \text{ mol} \times (0.0 \text{ kJ/mol})\}$
$- \{2 \text{ mol} \times (0.0 \text{ kJ/mol}) + 3 \text{ mol} \times (-635.5 \text{ kJ/mol})\}$
$\Delta H° = +1084.3 \text{ kJ}$, $\therefore$ not favored

d) $\Delta H° = \{\Delta H_f°[H_2O(\ell)] + \Delta H_f°[CaO(s)]\} - \{\Delta H_f°[Ca(OH)_2(s)]\}$
$\Delta H° = \{1 \text{ mol} \times (-285.9 \text{ kJ/mol}) + 1 \text{ mol} \times (-635.5 \text{ kJ/mol})\}$
$- \{1 \text{ mol} \times (-986.59 \text{ kJ/mol})\}$
$\Delta H° = +65.2 \text{ kJ}$, $\therefore$ not favored

e) $\Delta H° = \{2\Delta H_f°[HCl(g)] + \Delta H_f°[Na_2SO_4(s)]\}$
$- \{2\Delta H_f°[NaCl(s)] + \Delta H_f°[H_2SO_4(\ell)]\}$
$\Delta H° = \{2 \text{ mol} \times (-92.30 \text{ kJ/mol}) + 1 \text{ mol} \times (-1384.5 \text{ kJ/mol})\}$
$- \{2 \text{ mol} \times (-411.0 \text{ kJ/mol}) + 1 \text{ mol} \times (-811.32 \text{ kJ/mol})\}$
$\Delta H° = +64.2 \text{ kJ}$, $\therefore$ not favored

20.62 The probability is given by the number of possibilities that lead to the desired arrangement, divided by the total number of possible arrangements.

We list each of the possible results, heads (H) or tails (T) for each of the coins, and systematically write down all of the distinct arrangements:

There is only one arrangement that gives four heads: HHHH.
There is only one arrangement that gives four tails: TTTT
Four distinct arrangements can lead to three heads and one tail:

HHHT, THHH, HTHH, HHTH

There are similarly four distinct arrangements that lead to three tails and one head:

TTTH, HTTT, THTT, TTHT

There are six distinct arrangements that can lead to two heads and two tails:

HHTT, TTHH, THHT, THTH, HTTH, HTHT

Hence, the probability of all heads (HHHH) is 1 in 16 or $1/16 = 0.0625$.
The probability of two heads and two tails is 6 in 16 or $6/16 = 0.375$.

20.64 a) negative – since the number of moles of gaseous material decreases.
b) negative – since the number of moles of gaseous material decreases.
c) negative – since the number of moles of gas decreases.
d) positive – since a gas appears where there formerly was none.

20.66 $\Delta S° = (\text{sum } S°[\text{products}]) - (\text{sum } S°[\text{reactants}])$

a) $\Delta S° = \{2S°[NH_3(g)]\} - \{3S°[H_2(g)] + S°[N_2(g)]\}$
$\Delta S° = \{2 \text{ mol} \times (192.5 \text{ J mol}^{-1} \text{ K}^{-1})\} - \{3 \text{ mol} \times (130.6 \text{ J mol}^{-1} \text{ K}^{-1}) + 1 \text{ mol} \times (191.5 \text{ J mol}^{-1} \text{ K}^{-1})\}$
$\Delta S° = -198.3 \text{ J/K}$, $\therefore$ not favored

b) $\Delta S° = \{S°[CH_3OH(\ell)]\} - \{2S°[H_2(g)] + S°[CO(g)]\}$
$\Delta S° = \{1 \text{ mol} \times (126.8 \text{ J mol}^{-1} \text{ K}^{-1})\} - \{2 \text{ mol} \times (130.6 \text{ J mol}^{-1} \text{ K}^{-1}) + 1 \text{ mol} \times (197.9 \text{ J mol}^{-1} \text{ K}^{-1})\}$
$\Delta S° = -332.3 \text{ J/K}$, $\therefore$ not favored

c) $\Delta S° = \{6S°[H_2O(g)] + 4S°[CO_2(g)]\} - \{7S°[O_2(g)] + 2S°[C_2H_6(g)]\}$
$\Delta S° = \{6 \text{ mol} \times (188.7 \text{ J mol}^{-1} \text{ K}^{-1}) + 4 \text{ mol} \times (213.6 \text{ J mol}^{-1} \text{ K}^{-1})\} - \{7 \text{ mol} \times (205.0 \text{ J mol}^{-1} \text{ K}^{-1}) + 2 \text{ mol} \times (229.5 \text{ J mol}^{-1} \text{ K}^{-1})\}$
$\Delta S° = +92.6 \text{ J/K}$, $\therefore$ favorable

d) $\Delta S° = \{2S°[H_2O(\ell)] + S°[CaSO_4(s)]\} - \{S°[H_2SO_4(\ell)] + S°[Ca(OH)_2(s)]\}$
$\Delta S° = \{2 \text{ mol} \times (69.96 \text{ J mol}^{-1} \text{ K}^{-1}) + 1 \text{ mol} \times (107 \text{ J mol}^{-1} \text{ K}^{-1})\} - \{1 \text{ mol} \times (157 \text{ J mol}^{-1} \text{ K}^{-1}) + 1 \text{ mol} \times (76.1 \text{ J mol}^{-1} \text{ K}^{-1})\}$
$\Delta S° = +14 \text{ J/K}$, $\therefore$ favorable

e) $\Delta S° = \{2S°[N_2(g)] + S°[SO_2(g)]\} - \{2S°[N_2O(g)] + S°[S(s)]\}$
$\Delta S° = \{2 \text{ mol} \times (191.5 \text{ J mol}^{-1} \text{ K}^{-1}) + 1 \text{ mol} \times (248 \text{ J mol}^{-1} \text{ K}^{-1})\}$
$- \{2 \text{ mol} \times (220.0 \text{ J mol}^{-1} \text{ K}^{-1}) + 1 \text{ mol} \times (31.8 \text{ J mol}^{-1} \text{ K}^{-1})\}$
$\Delta S° = +159.6$ J/K, ∴ favorable

20.68 The entropy change that is designated $\Delta S_f°$ is that which corresponds to the reaction in which one mole of a substance is formed from elements in their standard states. Since the value is understood to correspond to the reaction forming one mole of a single pure substance, the units may be written either J K^{-1} or $\text{J mol}^{-1} \text{ K}^{-1}$.

a) $2C(s) + 2H_2(g) \rightarrow C_2H_4(g)$
$\Delta S° = \{S°[C_2H_4(g)]\} - \{2S°[C(s)] + 2S°[H_2(g)]\}$
$\Delta S° = \{1 \text{ mol} \times (219.8 \text{ J mol}^{-1} \text{ K}^{-1})\}$
$- \{2 \text{ mol} \times (5.69 \text{ J mol}^{-1} \text{ K}^{-1}) + 2 \text{ mol} \times (130.6 \text{ J mol}^{-1} \text{ K}^{-1})\}$
$\Delta S° = -52.8$ J/K or $-52.8 \text{ J mol}^{-1} \text{ K}^{-1}$

b) $N_2(g) + 1/2O_2(g) \rightarrow N_2O(g)$
$\Delta S° = \{S°[N_2O(g)]\} - \{S°[N_2(g)] + 1/2S°[O_2(g)]\}$
$\Delta S° = \{1 \text{ mol} \times (220.0 \text{ J mol}^{-1} \text{ K}^{-1})\} - \{1 \text{ mol} \times (191.5 \text{ J mol}^{-1} \text{ K}^{-1})$
$+ 1/2 \text{ mol} \times (205.0 \text{ J mol}^{-1} \text{ K}^{-1})\}$
$\Delta S° = -74.0$ J/K or $-74.0 \text{ J mol}^{-1} \text{ K}^{-1}$

c) $Na(s) + 1/2Cl_2(g) \rightarrow NaCl(s)$
$\Delta S° = \{S°[NaCl(s)]\} - \{1/2S°[Cl_2(g)] + S°[Na(s)]\}$
$\Delta S° = \{1 \text{ mol} \times (72.38 \text{ J mol}^{-1} \text{ K}^{-1})\} - \{1/2 \text{ mol} \times 223.0 \text{ J mol}^{-1} \text{ K}^{-1})$
$+ 1 \text{ mol} \times (51.0 \text{ J mol}^{-1} \text{ K}^{-1})\}$
$\Delta S° = -90.1$ J/K or $-90.1 \text{ J mol}^{-1} \text{ K}^{-1}$

d) $Ca(s) + S(s) + 3O_2(g) + 2H_2(g) \rightarrow CaSO_4{\cdot}2H_2O(s)$
$\Delta S° = \{S°[CaSO_4{\cdot}2H_2O(s)]\} - \{2S°[H_2(g)] + 3S°[O_2(g)] + S°[S(s)]$
$+ S°[Ca(s)]\}$
$\Delta S° = \{1 \text{ mol} \times (194.0 \text{ J mol}^{-1} \text{ K}^{-1})\} - \{2 \text{ mol} \times (130.6 \text{ J mol}^{-1} \text{ K}^{-1})$
$+ 3 \text{ mol} \times (205.0 \text{ J mol}^{-1} \text{ K}^{-1}) + 1 \text{ mol} \times (31.8 \text{ J mol}^{-1} \text{ K}^{-1})$
$+ 1 \text{ mol} \times (154.8 \text{ J mol}^{-1} \text{ K}^{-1})\}$
$\Delta S° = -868.9$ J/K or $-868.9 \text{ J mol}^{-1} \text{ K}^{-1}$

e) $2H_2(g) + 2C(s) + O_2(g) \rightarrow HC_2H_3O_2(\ell)$
$\Delta S° = \{S°[HC_2H_3O_2(\ell)]\} - \{2S°[H_2(g)] + 2S°[C(s)] + S°[O_2(g)]\}$
$\Delta S° = \{1 \text{ mol} \times (160 \text{ J mol}^{-1} \text{ K}^{-1})\} - \{2 \text{ mol} \times (130.6 \text{ J mol}^{-1} \text{ K}^{-1})$
$+ 2 \text{ mol} \times (5.69 \text{ J mol}^{-1} \text{ K}^{-1}) + 1 \text{ mol} \times (205.0 \text{ J mol}^{-1} \text{ K}^{-1})\}$
$\Delta S° = -318$ J/K or $-318 \text{ J mol}^{-1} \text{ K}^{-1}$

20.70 $\Delta S° = (\text{sum } S°[\text{products}]) - (\text{sum } S°[\text{reactants}])$

$\Delta S° = \{2S°[HNO_3(\ell)] + S°[NO(g)]\} - \{3S°[NO_2(g)] + S°[H_2O(\ell)]\}$

$\Delta S° = \{2 \text{ mol} \times (155.6 \text{ J mol}^{-1} \text{ K}^{-1}) + 1 \text{ mol} \times (210.6 \text{ J mol}^{-1} \text{ K}^{-1})\}$
$- \{3 \text{ mol} \times (240.5 \text{ J mol}^{-1} \text{ K}^{-1}) + 1 \text{ mol} \times (69.96 \text{ J mol}^{-1} \text{ K}^{-1})\}$

$\Delta S° = -269.7 \text{ J/K}$

20.72 The quantity $\Delta G_f°$ applies to the equation in which one mole of pure phosgene is produced from the naturally occurring forms of the elements:

$$C(s) + 1/2O_2(g) + Cl_2(g) \rightarrow COCl_2(g),\ \Delta G_f° = ?$$

We can determine $\Delta G_f°$ if we can find values for $\Delta H_f°$ and $\Delta S_f°$, because $\Delta G° = \Delta H° - T\Delta S°$.

The value of $\Delta S_f°$ is determined using S° for phosgene in the following way:

$\Delta S_f° = \{S°[COCl_2(g)]\} - \{S°[C(s)] + 1/2S°[O_2(g)] + S°[Cl_2(g)]\}$

$\Delta S_f° = \{1 \text{ mol} \times (284 \text{ J mol}^{-1} \text{ K}^{-1})\} - \{1 \text{ mol} \times (5.69 \text{ J mol}^{-1} \text{ K}^{-1})$
$+ 1/2 \text{ mol} \times (205.0 \text{ J mol}^{-1} \text{ K}^{-1}) + 1 \text{ mol} \times (223.0 \text{ J mol}^{-1} \text{ K}^{-1})\}$

$\Delta S_f° = -47 \text{ J mol}^{-1} \text{ K}^{-1}$ or -47 J/K

$\Delta G_f° = \Delta H_f° - T\Delta S_f° = -223 \text{ kJ/mol} - (298 \text{ K})(-0.047 \text{ kJ/mol K})$
$= -209 \text{ kJ/mol}$

20.74 $\Delta G° = (\text{sum } \Delta G_f°[\text{products}]) - (\text{sum } \Delta G_f°[\text{reactants}])$

a) $\Delta G° = \{\Delta G_f°[H_2SO_4(\ell)]\} - \{\Delta G_f°[H_2O(\ell)] + \Delta G_f°[SO_3(g)]\}$

$\Delta G° = \{1 \text{ mol} \times (-689.9 \text{ kJ/mol})\} - \{1 \text{ mol} \times (-237.2 \text{ kJ/mol})$
$+ 1 \text{ mol} \times (-370.4 \text{ kJ/mol})\}$

$\Delta G° = -82.3 \text{ kJ}$

b) $\Delta G° = \{2\Delta G_f°[NH_3(g)] + \Delta G_f°[H_2O(\ell)] + \Delta G_f°[CaCl_2(s)]\}$
$- \{\Delta G_f°[CaO(s)] + 2\Delta G_f°[NH_4Cl(s)]\}$

$\Delta G° = \{2 \text{ mol} \times (-16.7 \text{ kJ/mol}) + 1 \text{ mol} \times (-237.2 \text{ kJ/mol})$
$+ 1 \text{ mol} \times (-750.2 \text{ kJ/mol})\} - \{1 \text{ mol} \times (-604.2 \text{ kJ/mol})$
$+ 2 \text{ mol} \times (-203.9 \text{ kJ/mol})\}$

$\Delta G° = -8.8 \text{ kJ}$

c) $\Delta G° = \{\Delta G_f°[H_2SO_4(\ell)] + \Delta G_f°[CaCl_2(s)]\} - \{\Delta G_f°[CaSO_4(s)]$
$+ 2\Delta G_f°[HCl(g)]\}$

$\Delta G° = \{1 \text{ mol} \times (-689.9 \text{ kJ/mol}) + 1 \text{ mol} \times (-750.2 \text{ kJ/mol})\}$
$- \{1 \text{ mol} \times (-1320.3 \text{ kJ/mol}) + 2 \text{ mol} \times (-95.27 \text{ kJ/mol})\}$

$\Delta G° = +70.7 \text{ kJ}$

d) $\Delta G° = \{\Delta G_f°[C_2H_5OH(\ell)]\} - \{\Delta G_f°[H_2O(g)] + \Delta G_f°[C_2H_4(g)]\}$

$\Delta G° = \{1 \text{ mol} \times (-174.8 \text{ kJ/mol})\} - \{1 \text{ mol} \times (-228.6 \text{ kJ/mol}) + 1 \text{ mol} \times (68.12 \text{ kJ/mol})\}$

$\Delta G° = -14.3 \text{ kJ}$

e) $\Delta G° = \{2\Delta G_f°[H_2O(\ell)] + \Delta G_f°[SO_2(g)] + \Delta G_f°[CaSO_4(s)]\} - \{2\Delta G_f°[H_2SO_4(\ell)] + \Delta G_f°[Ca(s)]\}$

$\Delta G° = \{2 \text{ mol} \times (-237.2 \text{ kJ/mol}) + 1 \text{ mol} \times (-300 \text{ kJ/mol}) + 1 \text{ mol} \times (-1320.3 \text{ kJ/mol})\} - \{2 \text{ mol} \times (-689.9 \text{ kJ/mol}) + 1 \text{ mol} \times (0.0 \text{ kJ/mol})\}$

$\Delta G° = -715.3 \text{ kJ}$

20.76 $CaSO_4{\cdot}\frac{1}{2}H_2O(s) + 3/2H_2O \rightarrow CaSO_4{\cdot}2H_2O(s)$

$\Delta G° = (\text{sum } \Delta G_f°[\text{products}]) - (\text{sum } \Delta G_f°[\text{reactants}])$

$\Delta G° = \{\Delta G_f°[CaSO_4{\cdot}2H_2O(s)]\} - \{\Delta G_f°[CaSO_4{\cdot}\frac{1}{2}H_2O(s)] + 3/2\Delta G_f°[H_2O(\ell)]\}$

$\Delta G° = \{1 \text{ mol} \times (-1795.7 \text{ kJ/mol})\} - \{1 \text{ mol} \times (-1435.2 \text{ kJ/mol}) + 1.5 \text{ mol} \times (-237.2 \text{ kJ/mol})\}$

$\Delta G° = -4.7 \text{ kJ}$

20.78 Multiply the reverse of the second equation by 2 (remembering to multiply the associated free energy change by −2), and add the result to the first equation:

$4NO(g) \rightarrow 2N_2O(g) + O_2(g),$ $\Delta G° = -139.56 \text{ kJ}$

$4NO_2(g) \rightarrow 4NO(g) + 2O_2(g),$ $\Delta G° = +139.40 \text{ kJ}$

$4NO_2 \rightarrow 3O_2(g) + 2N_2O(g),$ $\Delta G° = -0.16 \text{ kJ}$

This result is the reverse of the desired reaction, which must then have $\Delta G° = +0.16$ kJ.

20.80 The maximum work obtainable from a reaction is equal in magnitude to the value of ΔG for the reaction. Thus, we need only determine $\Delta G°$ for the process:

$\Delta G° = (\text{sum } \Delta G_f°[\text{products}]) - (\text{sum } \Delta G_f°[\text{reactants}])$

$\Delta G° = \{3\Delta G_f°[H_2O(g)] + 2\Delta G_f°[CO_2(g)]\} - \{3\Delta G_f°[O_2(g)] + \Delta G_f°[C_2H_5OH(\ell)]\}$

$\Delta G° = \{3 \text{ mol} \times (-228.6 \text{ kJ/mol}) + 2 \text{ mol} \times (-394.4 \text{ kJ/mol})\} - \{3 \text{ mol} \times (0.0 \text{ kJ/mol})\ 1 \text{ mol} \times (-174.8 \text{ kJ/mol})\}$

$\Delta G° = -1299.8 \text{ kJ}$

Therefore the maximum work available is 1,299.8 kJ.

20.82 At equilibrium, $\Delta G = 0 = \Delta H - T\Delta S$

$T_{eq} = \Delta H/\Delta S$, and assuming that ΔS is independent of temperature, we have:

$T_{eq} = (31.4 \times 10^3 \text{ J mol}^{-1}) \div (94.2 \text{ J mol}^{-1} \text{ K}^{-1}) = 333 \text{ K } (60\ ^\circ\text{C})$

20.84 At equilibrium, $\Delta G = 0 = \Delta H - T\Delta S$

Thus $\Delta H = T\Delta S$, and if we assume that both ΔH and ΔS are independent of temperature, we have:

$\Delta S = \Delta H/T_{eq} = (37.7 \times 10^3 \text{ J/mol}) \div (99.3 + 273.15 \text{ K})$

$\Delta S = 101 \text{ J mol}^{-1} \text{ K}^{-1}$

20.86 The reaction is spontaneous if its associated value for ΔG° is negative.

$\Delta G^\circ = (\text{sum } \Delta G_f^\circ[\text{products}]) - (\text{sum } \Delta G_f^\circ[\text{reactants}])$

$\Delta G^\circ = \{\Delta G_f^\circ[HC_2H_3O_2(\ell)] + \Delta G_f^\circ[H_2O(\ell)] + \Delta G_f^\circ[NO(g)] + \Delta G_f^\circ[NO_2(g)]\}$

$\quad - \{\Delta G_f^\circ[C_2H_4(g)] + 2\Delta G_f^\circ[HNO_3(\ell)]\}$

$\Delta G^\circ = \{1 \text{ mol} \times (-392.5 \text{ kJ/mol}) + 1 \text{ mol} \times (-237.2 \text{ kJ/mol})$

$\quad + 1 \text{ mol} \times (86.69 \text{ kJ/mol}) + 1 \text{ mol} \times (51.84 \text{ kJ/mol})\}$

$\quad - \{1 \text{ mol} \times (68.12 \text{ kJ/mol}) + 2 \text{ mol} \times (-79.9 \text{ kJ/mol})\}$

$\Delta G^\circ = -399.5 \text{ kJ}$

Yes, the reaction is spontaneous.

20.88 $\Delta G^\circ_T = \Delta H^\circ - T\Delta S^\circ$, where $T = 373$ K in all cases, and where the values of ΔH° and ΔS° are obtained in the usual manner, i.e.:

$\Delta H^\circ = (\text{sum } \Delta H_f^\circ[\text{products}]) - (\text{sum } \Delta H_f^\circ[\text{reactants}])$

$\Delta S^\circ = (\text{sum } S^\circ[\text{products}]) - (\text{sum } S^\circ[\text{reactants}])$

$\Delta H^\circ = \{\Delta H_f^\circ[C_2H_6(g)]\} - \{\Delta H_f^\circ[H_2(g)] + \Delta H_f^\circ[C_2H_4(g)]\}$

$\Delta H^\circ = \{1 \text{ mol} \times (-84.5 \text{ kJ/mol})\}$

$\quad - \{1 \text{ mol} \times (0.0 \text{ kJ/mol}) + 1 \text{ mol} \times (51.9 \text{ kJ/mol})\}$

$\Delta H^\circ = -136.4 \text{ kJ}$

$\Delta S^\circ = \{S^\circ[C_2H_6(g)]\} - \{S^\circ[H_2(g)] + S^\circ[C_2H_4(g)]\}$

$\Delta S^\circ = \{1 \text{ mol} \times (229.5 \text{ J mol}^{-1} \text{ K}^{-1})\} - \{1 \text{ mol} \times (130.6 \text{ J mol}^{-1} \text{ K}^{-1})$

$\quad + 1 \text{ mol} \times (219.8 \text{ J mol}^{-1} \text{ K}^{-1})\}$

$\Delta S^\circ = -120.9 \text{ J/K} = -0.1209 \text{ kJ/K}$

$\Delta G^\circ_{373} = \Delta H^\circ - T\Delta S^\circ = -136.4 \text{ kJ} - (373 \text{ K})(-0.1209 \text{ kJ/K}) = -91.3 \text{ kJ}$

20.90 a) $\Delta G° = \{2 \times \Delta G_f°[POCl_3(g)]\} - \{2 \times \Delta G_f°[PCl_3(g)] + 2 \times \Delta G_f°[O_2(g)]\}$

$\Delta G° = \{2 \text{ mol} \times (-1019 \text{ kJ/mol})\}$
$- \{2 \text{ mol} \times (-267.8 \text{ kJ/mol}) + 1 \text{ mol} \times (0 \text{ kJ/mol})\}$

$\Delta G° = -1502 \text{ kJ} = -1.502 \times 10^6 \text{ J}$

$-1.502 \times 10^6 \text{ J} = -RT\ln K_p = -(8.314 \text{ J/K mol})(298 \text{ K}) \times \ln K_p$

$\ln K_p = 606 \;\therefore\; \log K_p = 263$, and $K_p = 10^{263}$.

b) $\Delta G° = \{2 \times \Delta G_f°[SO_2(g)] + 1 \times \Delta G_f°[O_2(g)]\} - \{2 \times \Delta G_f°[SO_3(g)]\}$

$\Delta G° = \{2 \text{ mol} \times (-300 \text{ kJ/mol}) + 1 \text{ mol} \times (0 \text{ kJ/mol})\} - \{2 \text{ mol} \times (-370 \text{ kJ/mol})\}$

$\Delta G° = 140 \text{ kJ} = 1.40 \times 10^5 \text{ J}$

$1.40 \times 10^5 \text{ J} = -RT\ln K_p = -(8.314 \text{ J/K mol})(298 \text{ K}) \times \ln K_p$

$\ln K_p = -56.5$ and $K_p = 2.90 \times 10^{-25}$

20.92 $\Delta G° = -RT \ln K_p$

$-9.67 \times 10^3 \text{ J} = -(8.314 \text{ J/K mol})(1273 \text{ K}) \times \ln K_p$

$\ln K_p = 0.914 \;\therefore\; K_p = 2.49$

$$Q = \frac{[N_2O][O_2]}{[NO_2][NO]} = \frac{(0.015)(0.0350)}{(0.0200)(0.040)} = 0.66$$

Since the value of Q is less than the value of K, the system is not at equilibrium and must shift to the right to reach equilibrium.

20.94 $\Delta G° = -RT \ln K_p$

$-50.79 \times 10^3 \text{ J} = -(8.314 \text{ J K}^{-1} \text{ mol}^{-1})(298 \text{ K}) \times \ln K_p$

$\ln K_p = 20.5$

Taking the anti ln of both sides (e^x) of this equation gives: $K_p = 8.00 \times 10^8$. This is a favorable reaction, since the equilibrium lies far to the side favoring products and is worth studying as a method for methane production.

20.96 If $\Delta G° = 0$, $K_c = 1$. If we start with pure products, the value of Q will be infinite (there are zero reactants) and, since $Q > K_c$, the equilibrium will shift towards the reactants, i.e., the pure products will decompose to their elements.

20.98 $\Delta G° = (\text{sum } \Delta G_f°[\text{products}]) - (\text{sum } \Delta G_f°[\text{reactants}])$

$\Delta G° = \{\Delta G_f°[N_2(g)] + 2\Delta G_f°[CO_2(g)]\} - \{2\Delta G_f°[NO(g)] + 2\Delta G_f°[CO(g)]\}$

$\Delta G° = \{1 \text{ mol} \times (0.0 \text{ kJ/mol}) + 2 \text{ mol} \times (-394.4 \text{ kJ/mol})\}$
$- \{2 \text{ mol} \times (86.69 \text{ kJ/mol}) + 2 \text{ mol} \times (-137.3 \text{ kJ/mol})\}$

$\Delta G° = -687.6 \text{ kJ}$

$\Delta G° = -RT \ln K_p$

$-687.6 \times 10^3 \text{ J} = -(8.314 \text{ J K}^{-1} \text{ mol}^{-1})(298 \text{ K}) \ln K_p$

$\ln K_p = 278$ and $\log K_p = 121 \;\therefore K_p = 10^{121}$

20.100 This requires the breaking of three N–H single bonds:

$$NH_3 \rightarrow N + 3H$$

The enthalpy of atomization of NH_3 is thus three times the average N–H single bond energy: 3×391 kJ/mol = 1.17×10^3 kJ/mol.

20.102 The heat of formation for ethanol vapor describes the following change:

$2C(s) + 3H_2(g) + \frac{1}{2}O_2(g) \rightarrow C_2H_5OH(g)$

This can be arrived at by adding the following thermochemical equations, using data from Table 20.3:

$$
\begin{array}{ll}
3H_2(g) \rightarrow 6H(g) & \Delta H_1° = (6)217.89 \text{ kJ} = 1307.34 \text{ kJ} \\
2C(s) \rightarrow 2C(g) & \Delta H_2° = (2)716.67 \text{ kJ} = 1{,}433.34 \text{ kJ} \\
\frac{1}{2}O_2(g) \rightarrow O(g) & \Delta H_3° = (1)249.17 \text{ kJ} = 249.17 \text{ kJ} \\
\underline{6H(g) + 2C(g) + O(g) \rightarrow C_2H_5OH(g)} & \underline{\Delta H°_4 = x}
\end{array}
$$

$3H_2(g) + 2C(s) + \frac{1}{2}O_2(g) \rightarrow C_2H_5OH(g) \qquad \Delta H_f° = (2989.85 + x)$ kJ

Since $\Delta H_f°$ is given as −235.3 kJ…

$-235.3 = 2989.85 + x$
$x = -3225.2$ kJ

$\Delta H°_{atom}$ is the reverse reaction, so the sign will change:

$\Delta H°_{atom} = 3225.2$ kJ

The sum of all the bond energies in the molecule should be equal to the atomization energy:

$\Delta H°_{atom}$ = 1(C-C bond) + 5(C-H bonds) + 1(O-H bond) + 1(C-O bond)

We use values from Table 20.4:

3225.2 kJ = 1(348 kJ) + 5(412 kJ) + 1(463 kJ) + 1(C-O bond)

C-O bond energy = 354 kJ/mol

20.104 There are two C=S double bonds to be considered:

$\Delta H_f°$ = sum($\Delta H_f°$[gaseous atoms]) – sum(average bond energies in the molecule)
115.3 kJ/mol = $[716.67 + 2 \times 276.98] - [2 \times \text{C=S}]$
The C=S double bond energy is therefore given by the equation:
C=S = $-(115.3 - 716.67 - 2 \times 276.98) \div 2 = 577.7$ kJ/mol

20.106 There are six S—F bonds in the molecule:

$\Delta H_f° = \text{sum}(\Delta H_f°[\text{gaseous atoms}]) - \text{sum}(\text{average bond energies in the molecule})$
$-1096 \text{ kJ/mol} = [277.0 + 6 \times 79.14] - [6 \times \text{S—F}]$
$\text{S—F} = (1096 + 277.0 + 6 \times 79.14) \div 6 = 308.0 \text{ kJ/mol}$

20.108 See the method of review problems 20.102 through 20.106.
$\Delta H_f° = \text{sum}(\Delta H_f°[\text{gaseous atoms}]) - \text{sum}(\text{average bond energies in the molecule})$

$\Delta H_f°[C_2H_2(g)] = [2 \times 716.7 + 2 \times 218.0] - [2 \times 412 + 960]$
$= 85 \text{ kJ/mol}$

20.110 The heat of formation of CF_4 should be more exothermic than that of CCl_4 because more energy is released on formation of a C—F bond than on formation of a C—Cl bond. Also, less energy is needed to form gaseous F atoms than to form gaseous Cl atoms.

Additional Exercises

20.112 First, we calculate a value for ΔH and ΔS, using the data of Appendix C. Next, we calculate a value for $\Delta G_T°$, using the equation $\Delta G = \Delta H - T\Delta S$. Last, we calculate a value for K_p, using the equation $\Delta G = -RT \ln K_p$. (Recall that $\ln(x) = 2.303 \times \log(x)$.)

a) $\Delta H° = \{2 \times \Delta H_f°[POCl_3(g)]\} - \{2 \times \Delta H_f°[PCl_3(g)] + 1 \times \Delta H_f°[O_2(g)]\}$
$\Delta H° = \{2 \text{ mol} \times (-1109.7 \text{ kJ/mol})\}$
$- \{2 \text{ mol} \times (-287.0 \text{ kJ/mol}) + 1 \text{ mol} \times (0 \text{ kJ/mol})\}$
$\Delta H° = -1645.4 \text{ kJ} = -1.6454 \times 10^6 \text{ J}$

$\Delta S° = \{2 \times S°[POCl_3(g)]\} - \{2 \times S°[PCl_3(g)] + 1 \times S°[O_2(g)]\}$
$\Delta S° = \{2 \text{ mol} \times (646.5 \text{ J mol}^{-1} \text{ K}^{-1})\}$
$- \{2 \text{ mol} \times (311.8 \text{ J mol}^{-1} \text{ K}^{-1}) + 1 \text{ mol} \times (205.0 \text{ J mol}^{-1} \text{ K}^{-1})\}$
$\Delta S° = 464.4 \text{ J K}^{-1}$

$\Delta G_{573}° = -1.6454 \times 10^6 \text{ J} - (573 \text{ K})(464.4 \text{ J K}^{-1}) = -1.912 \times 10^6 \text{ J}$
$-1.912 \times 10^6 \text{ J} = -RT\ln K_p$
$-1.912 \times 10^6 \text{ J} = -(8.314 \text{ J/K mol})(573 \text{ K}) \ln K_p$
$\ln K_p = 401.3$ and $\log K_p = 174.2$ $\therefore K_p = 10^{174}$

b) $\Delta H° = \{2 \times \Delta H_f°[SO_2(g)] + 1 \times \Delta H_f°[O_2(g)]\} - \{2 \times \Delta H_f°[SO_3(g)]\}$
$\Delta H° = \{2\ mol \times (-297\ kJ/mol) + 1\ mol \times (0\ kJ/mol)\}$
$- \{2\ mol \times (-396\ kJ/mol)\}$
$\Delta H° = 198\ kJ = 1.98 \times 10^5\ J$

$\Delta S° = \{2 \times S°[SO_2(g)] + 1 \times S°[O_2(g)]\} - \{2 \times S°[SO_3(g)]\}$
$\Delta S° = \{2\ mol \times (248.5\ J\ mol^{-1}\ K^{-1}) + 1\ mol \times (205.0\ J\ mol^{-1}\ K^{-1})\}$
$- \{2\ mol \times (256.2\ J\ mol^{-1}\ K^{-1})\}$
$\Delta S° = 190\ J\ K^{-1}$

$\Delta G_{573}° = 1.98 \times 10^5\ J - (573\ K)(190\ J\ K^{-1}) = 8.94 \times 10^4\ J$
$8.94 \times 10^4\ J = -RT\ln K_p$
$8.94 \times 10^4\ J = -(8.314\ J/K\ mol)(573\ K) \ln K_p$
$\ln K_p = -18.8$ and $K_p = 6.8 \times 10^{-9}$

*20.114 Under reversible conditions, the expansion of an ideal gas produces the maximum amount of work that can be obtained from the change. Such a situation cannot be accomplished by any real process. In an ideal gas, the gas particles are non-interacting. In a real gas, the interaction between the individual particles must be overcome so that the gas may expand. It requires an input of energy to overcome these attractive forces.

20.116 No work ($P \times \Delta V$) is accomplished in a bomb calorimeter because there is no change in volume. At constant pressure, $w = -\Delta n_{gas}RT = -(18\ mol - 15\ mol)(8.314\ J\ mol^{-1}\ K^{-1})(298\ K) = -7.43 \times 10^3\ J = -7.43\ kJ$

*20.118 As in exercise 20.117, we solve for the volume after the first expansion to a pressure of 2 atm:

$$V_2 = \frac{nRT}{P_2} = \frac{(0.817\ mol)(0.0821\ \frac{L\ atm}{mol\ K})(298\ K)}{2\ atm} = 10.0\ L$$

The work after the first stage of expansion is therefore:
$w_1 = -(2\ atm)(10.0\ L - 5.00\ L) = -10.0\ L\ atm.$
The work performed during the second stage of expansion is:
$w_2 = -(1\ atm)(20.0\ L - 10.0\ L) = -10.0\ L\ atm.$
The sum for the whole stepwise process is $w_1 + w_2 = -20.0\ L\ atm$. Thus, we see that more work is performed in this multi–step expansion than in a single expansion.

20.120 First, review the information provided. Since the salt dissolved, the process is spontaneous, and the sign for ΔG is negative. The mixture feels cool, indicating that this is an endothermic process and ΔH must be positive. Dissolving the solid salt increases the disorder of the system which indicates ΔS is positive. The general equation from which we must work is $\Delta G = \Delta H - T\Delta S$. Using this equation, the magnitude of $T\Delta S$ must be larger than the magnitude of ΔH in order to obtain a negative value for ΔG. However, the magnitude of ΔH is almost certainly larger than that of ΔS as can be appreciated from the fact that ΔS° values are given in J and those of ΔH°_f are given in kJ.

20.122 There can be only one temperature for which ΔG has precisely the value zero, i.e., an equilibrium condition exists.

20.124 glucose + phosphate $\rightarrow$ glucose–phosphate + H_2O $\quad \Delta G = 13.13$ kJ
ATP + H_2O $\rightarrow$ ADP + phosphate $\quad \Delta G = -32.22$ kJ

Add these two reactions together to get:

glucose + ATP $\rightarrow$ glucose–phosphate + ADP $\quad \Delta G = -19.09$ kJ

20.126 We need to calculate the amount of energy produced when one gallon of each of these fuels is burned;

$$\#\,\text{mol ethanol} = (3.78\times10^3\ \text{mL ethanol})\left(\frac{0.7893\ \text{g ethanol}}{1\ \text{mL ethanol}}\right)\left(\frac{1\ \text{mole ethanol}}{46.07\ \text{g ethanol}}\right)$$

$$= 64.8\ \text{moles ethanol}$$

$$\#\,\text{kJ} = (64.8\ \text{moles ethanol})\left(\frac{-1299.8\ \text{kJ}}{1\ \text{mole ethanol}}\right)$$

$$= 8.42 \times 10^4\ \text{kJ}$$

$$\#\,\text{mol octane} = (3.78\times10^3\ \text{mL octane})\left(\frac{0.7025\ \text{g octane}}{1\ \text{mL octane}}\right)\left(\frac{1\ \text{mole octane}}{114.23\ \text{g octane}}\right)$$

$$= 23.2\ \text{moles octane}$$

$$\#\,\text{kJ} = (23.2\ \text{moles octane})\left(\frac{-5307\ \text{kJ}}{1\ \text{mole octane}}\right) = 1.23\times10^5\ \text{kJ}$$

$$= 1.23 \times 10^5\ \text{kJ}$$

In spite of the large number of moles of ethanol in one gallon of liquid, the energy produced from the combustion of a gallon of octane is greater than the amount produced when one gallon of ethanol is burned.

20.128 $C(s) + 2Cl_2(g) \rightarrow CCl_4(\ell)$

$\downarrow(1) \quad \downarrow(2)\ (3) \quad \uparrow(4)$

$C(g) + 4Cl(g) \rightarrow CCl_4(g)$

Step 1: $\Delta H = 716.67$ kJ

Step 2: $\Delta H = 4(121.5 \text{ kJ}) = 485.88$ kJ

Step 3: $\Delta H_f°(CCl_4(g)) = 4(BE(C–Cl)) = 4(338 \text{ kJ}) = –1350$ kJ

Step 4: $\Delta H_{cond} = –29.9$ kJ

ΔH = the sum of the four steps = –179 kJ

Chapter 21

Practice Exercises

21.1 anode: $Mg(s) \rightarrow Mg^{2+}(aq) + 2e^-$
cathode: $Fe^{2+}(aq) + 2e^- \rightarrow Fe(s)$
cell notation: $Mg(s) \mid Mg^{2+}(aq) \parallel Fe^{2+}(aq) \mid Fe(s)$

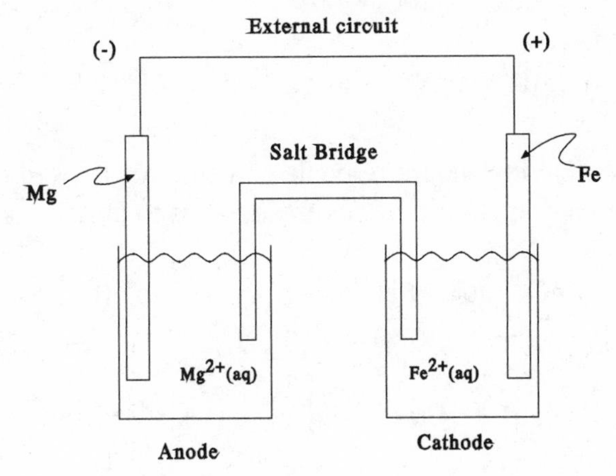

21.2 anode: $Al(s) \rightarrow Al^{3+}(aq) + 3e^-$
cathode: $Pb^{2+}(aq) + 2e^- \rightarrow Pb(s)$
overall: $3Pb^{2+}(aq) + 2Al(s) \rightarrow 2Al^{3+}(aq) + 3Pb(s)$

21.3 $E°_{cell} = E°_{substance\ reduced} - E°_{substance\ oxidized}$

$1.93\ V = (-0.44\ V) - E°_{Mg2+}$
$E°_{Mg2+} = -0.44 - 1.93 = -2.37\ V$

This agrees exactly with Table 21.1.

21.4 The half–reaction with the more positive value of E° (listed higher in Table 21.1) will occur as a reduction. The half–reaction having the less positive (more negative) value of E° (listed lower in Table 21.1) will be reversed and occur as an oxidation.

$Br_2(aq) + 2e^- \rightarrow 2Br^-(aq)$ reduction
$H_2SO_3(aq) + H_2O \rightarrow SO_4^{2-}(aq) + 4H^+(aq) + 2e^-$ oxidation
$Br_2(aq) + H_2SO_3(aq) + H_2O \rightarrow 2Br^-(aq) + SO_4^{2-}(aq) + 4H^+(aq)$

21.5 Either tin(II) or iron(III) will be reduced, depending on which way the reaction proceeds. Iron(III) is listed higher than tin(II) in Table 21.1 (it has a greater reduction potential), so we would expect that the reaction would not be spontaneous in the direction shown.

21.6 The half–reaction having the more positive value for E° will occur as a reduction. The other half–reaction should be reversed, so as to appear as an oxidation.

$NiO_2(s) + 2H_2O + 2e^- \rightarrow Ni(OH)_2(s) + 2OH^-(aq)$	reduction
$Fe(s) + 2OH^-(aq) \rightarrow 2e^- + Fe(OH)_2(s)$	oxidation
$NiO_2(s) + Fe(s) + 2H_2O \rightarrow Ni(OH)_2(s) + Fe(OH)_2(s)$	net reaction

$E°_{cell} = E°_{substance\ reduced} - E°_{substance\ oxidized}$
$E°_{cell} = E°_{NiO_2} - E°_{Fe}$
$E°_{cell} = 0.49 - (-0.88) = 1.37\ V$

21.7 The half–reaction having the more positive value for E° will occur as a reduction. The other half–reaction should be reversed, so as to appear as an oxidation.

Reduction: $3 \times [MnO_4^-(aq) + 8H^+(aq) + 5e^- \rightarrow Mn^{2+}(aq) + 4H_2O]$
Oxidation: $5 \times [Cr(s) \rightarrow Cr^{3+}(aq) + 3e^-]$
Net reaction:
$3MnO_4^-(aq) + 24H^+(aq) + 5Cr(s) \rightarrow 5Cr^{3+}(aq) + 3Mn^{2+}(aq) + 12H_2O$

$E°_{cell} = E°_{substance\ reduced} - E°_{substance\ oxidized}$
$E°_{cell} = E°_{MnO_4^-} - E°_{Cr}$
$E°_{cell} = 1.51\ V - (-0.74\ V) = 2.25\ V$

21.8 A reaction will occur spontaneously in the forward direction if the value of E° is positive. We therefore evaluate E° for each reaction using:

$$E°_{cell} = E°_{substance\ reduced} - E°_{substance\ oxidized}$$

a) $Br_2(aq) + 2e^- \rightarrow 2Br^-(aq)$ reduction
$Cl_2(aq) + 2H_2O \rightarrow 2HOCl(aq) + 2H^+(aq) + 2e^-$ oxidation

$E°_{cell} = E°_{Br_2} - E°_{Cl_2}$
$E°_{cell} = 1.07\ V - (1.36\ V) = -0.29\ V$, ∴ nonspontaneous

b) $2Cr^{3+}(aq) + 6e^- \rightarrow 2Cr(s)$ reduction
$3Zn(s) \rightarrow 3Zn^{2+}(aq) + 6e^-$ oxidation
$E°_{cell} = E°_{Cr^{3+}} - E°_{Zn}$
$E°_{cell} = -0.74\ V - (-0.76\ V) = +0.02\ V$, ∴ spontaneous

21.9 It was stated in Practice Exercise 3 that the reaction in Practice Exercise 1 had a standard cell potential of 1.93 V, or 1.93 J/C. Since 2 mole of e^- are involved, i.e., n = 2, we have:

$\Delta G° = -nFE°_{cell} = -(2)(96{,}500\ C)(1.93\ V) = -3.72 \times 10^5\ J = -3.72 \times 10^2\ kJ$

The cell potential in Practice Exercise 2 may be found by the difference in the standard reduction potentials of the substances involved:

$E^\circ_{cell} = -0.13\text{ V} - (-1.66\text{ V}) = +1.53\text{ V}$

6 moles of e^- are involved, so n = 2. Now we have:

$\Delta G^\circ = -nFE^\circ_{cell} = -(6)(96{,}500\text{ C})(1.53\text{ V}) = -8.86 \times 10^5\text{ J} = -8.86 \times 10^2\text{ kJ}$

21.10 Using Equation 21.7,

$$E^\circ_{cell} = \frac{RT}{nF}\ln K_c$$

$$-0.46\text{ V} = \frac{(8.314\text{ J mol}^{-1}\text{K}^{-1})(298\text{ K})}{2(96{,}500\text{ C mol}^{-1})}\ln K_c$$

$$\ln K_c = -35.83$$

Taking the antilog (e^x) of both sides of the above equation gives $K_c = 2.7 \times 10^{-16}$.

This very small value for the equilibrium constant means that the products of the reaction are not formed spontaneously. The equilibrium lies far to the left, favoring reactants, and we do not expect much product to form.

21.11

$Zn(s)\ Zn^{2+}(aq) + 2e^-$ oxidation

$Cu^{2+}(aq) + 2e^- \rightarrow 2Cu(s)$ reduction

$E^\circ_{cell} = E^\circ_{Cu^{2+}} - E^\circ_{Zn}$

$E^\circ_{cell} = +0.34\text{ V} - (-0.76\text{ V}) = +1.10\text{ V}$

The Nernst equation for this cell is:

$$E_{cell} = E^\circ_{cell} - \frac{RT}{nF}\ln\frac{[Zn^{2+}]}{[Cu^{2+}]}$$

$$E_{cell} = 1.10\text{ V} - \frac{(8.314\text{ J mol}^{-1}\text{K}^{-1})(298\text{ K})}{2(96{,}500\text{ C mol}^{-1})}\ln\frac{[1.0]}{[0.010]}$$

$$= 1.10\text{ V} - 0.01284(4.605)$$

$$= 1.04\text{ V}$$

21.12

$Cu(s) \rightarrow Cu^{2+}(aq) + 2e^-$ oxidation

$Ag^+(aq) + e^- \rightarrow Ag(s)$ reduction

$E^\circ_{cell} = E^\circ_{Ag^+} - E^\circ_{Cu}$

$E^\circ_{cell} = +0.80\text{ V} - (+0.34\text{ V}) = +0.46\text{ V}$

$$E_{cell} = E^{\circ}_{cell} - \frac{RT}{nF}\ln\frac{[Cu^{2+}]}{[Ag^{+}]^2}$$

$$\ln\frac{[Cu^{2+}]}{[Ag^{+}]^2} = \frac{(E^{\circ}_{cell} - E_{cell})}{\left(\frac{RT}{nF}\right)}$$

$$= \frac{(0.46\ V - 0.57\ V)}{0.01284}$$

$$= -8.5670$$

$$\frac{[Cu^{2+}]}{[Ag^{+}]^2} = e^{-8.5670} = 1.9 \times 10^{-4}$$

Since the $[Ag^+] = 1.00$ M, $[Cu^{2+}] = 1.9 \times 10^{-4}$ M

Substituting the second value into the same expression gives $[Cu^{2+}] = 6.9 \times 10^{-13}$ M.

21.13 We are told that, in this galvanic cell, the chromium electrode is the anode, meaning that oxidation occurs at the chromium electrode.

Now in general, we have the equation:

$$E^{\circ}_{cell} = E^{\circ}_{reduction} - E^{\circ}_{oxidation}$$

which becomes, in particular for this case:

$$E^{\circ}_{cell} = E^{\circ}_{Ni^{2+}} - E^{\circ}_{Cr}$$

The net cell reaction is given by the sum of the reduction and the oxidation half–reactions, multiplied in each case so as to eliminate electrons from the result:

$3 \times [Ni^{2+}(aq) + 2e^{-} \rightarrow Ni(s)]$	reduction
$2 \times [Cr(s) \rightarrow Cr^{3+}(aq) + 3e^{-}]$	oxidation
$3Ni^{2+}(aq) + 2Cr(s) \rightarrow 2Cr^{3+}(aq) + 3Ni(s)$	net reaction

In this reaction, n = 6, and the Nernst equation becomes:

$$E_{cell} = E^{\circ}_{cell} - \frac{RT}{nF} \ln \frac{[Cr^{3+}]^2}{[Ni^{2+}]^3}$$

$$\ln \frac{[Cr^{3+}]^2}{[Ni^{2+}]^3} = \frac{(E^{\circ}_{cell} - E_{cell})}{\left(\frac{RT}{nF}\right)}$$

$$= \frac{(0.487\text{ V} - 0.552\text{ V})}{0.004279}$$

$$= -15.190$$

$$\frac{[Cr^{3+}]^2}{[Ni^{2+}]^3} = e^{-15.190} = 2.5 \text{ x } 10^{-7}$$

Substituting $[Ni^{2+}] = 1.20$ M, we solve for $[Cr^{3+}]$ and get: $[Cr^{3+}] = 6.6 \times 10^{-4}$ M.

21.14 The cathode is always where reduction occurs. We must consider which species could be candidates for reduction, then choose the species with the highest reduction potential from Table 21.1.

$Cd^{2+}(aq) + 2e^- \rightarrow 2Cd(s)$	$E° = -0.40$ V
$Sn^{2+}(aq) + 2e^- \rightarrow 2Sn(s)$	$E° = -0.14$ V
$2H_2O + 2e^- \rightarrow H_2(g) + 2OH^-(aq)$	$E° = -0.83$ V

Tin(II) has the highest reduction potential, so we would expect it to be reduced in this environment. We expect Sn(s) at the cathode.

21.15 The number of Coulombs is: $4.00\text{ A} \times 200\text{ s} = 800\text{ C}$
The number of moles is:

$$\#\,\text{mol OH}^- = 800\,\text{C} \times \frac{1\,\text{F}}{96{,}500\,\text{C}} \times \frac{1\,\text{mol OH}^-}{1\,\text{F}} = 8.29 \times 10^{-3}\text{ mol OH}^-$$

21.16 The number of moles of Au to be deposited is: 3.00 g Au ÷ 197 g/mol = 0.0152 mol Au. The number of Coulombs (A·s) is:

$$\#\,\text{C} = 0.0152\,\text{mol Au} \times \frac{3\,\text{F}}{1\,\text{mol Au}} \times \frac{96{,}500\,\text{C}}{1\,\text{F}} = 4.40 \times 10^3\text{ C}$$

The number of minutes is:

$$\#\,\text{min} = \frac{4.40 \times 10^3\text{ A} \cdot \text{s}}{10.0\text{ A}} \times \frac{1\,\text{min}}{60\,\text{s}} = 7.33\text{ min}$$

21.17 As in Practice Exercise 16 above, the number of Coulombs is 4.40×10^3 C. This corresponds to a current of:

$$\# A = \frac{4.40 \times 10^3 \text{ A} \cdot \text{s}}{20.0 \text{ min}} \times \frac{1 \text{ min}}{60 \text{ s}} = 3.67 \text{ A}$$

21.18 The number of Coulombs is:
$0.100 \text{ A} \times 1.25 \text{ hr}(3600 \text{ s/hr}) = 450 \text{ C}$

The number of moles of copper ions produced is:

$$\# \text{mol } Cu^{2+} = 450 \text{ C} \times \frac{1 \text{ mol } e^-}{96{,}500 \text{ C}} \times \frac{1 \text{ mol } Cu^{2+}}{2 \text{ mol } e^-} = 0.233 \text{ mol } Cu^{2+}$$

Therefore, the increase in concentration is:
$M = \text{mol/L} = (0.233 \text{ mol } Cu^{2+})/0.125 \text{ L} = +\ 0.0187 \text{ M}$

Review Problems

21.66 a)
anode: $Cd(s) \rightarrow Cd^{2+}(aq) + 2e^-$
cathode: $Au^{3+}(aq) + 3e^- \rightarrow Au(s)$
cell: $3Cd(s) + 2Au^{3+}(aq) \rightarrow 3Cd^{2+}(aq) + 2Au(s)$

b)
anode: $Pb(s) + HSO_4^-(aq) \rightarrow PbSO_4(s) + H^+ + 2e^-$
cathode: $PbO_2(s) + HSO_4^-(aq) + 3H^+(aq) + 2e^- \rightarrow PbSO_4(s) + 2H_2O$
cell: $Pb(s) + PbO_2(s) + 2HSO_4^-(aq) + 2H^+(aq) \rightarrow 2PbSO_4(s) + 2H_2O$

c)
anode: $Cr(s) \rightarrow Cr^{3+}(aq) + 3e^-$
cathode: $Cu^{2+}(aq) + 2e^- \rightarrow Cu(s)$
cell: $2Cr(s) + 3Cu^{2+}(aq) \rightarrow 2Cr^{3+}(aq) + 3Cu(s)$

21.68 a) $Fe(s) | Fe^{2+}(aq) || Cd^{2+}(aq) | Cd(s)$
b) $Pt(s) | Br^-(aq), Br_2(g) || Cl_2(aq), Cl^-(aq) | Pt(s)$
c) $Ag(s) | Ag^+(aq) || Au^{3+}(aq) | Au(s)$

21.70 a) $Sn(s)$
b) $Br^-(aq)$
c) $Zn(s)$
d) $I^-(aq)$

21.72 a) $E°_{cell} = -0.40 \text{ V} - (-0.44) \text{ V} = 0.04 \text{ V}$
b) $E°_{cell} = 1.07 \text{ V} - (1.36 \text{ V}) = -0.29 \text{ V}$
c) $E°_{cell} = 1.42 \text{ V} - (0.80 \text{ V}) = 0.62 \text{ V}$

21.74 The reactions are spontaneous if the overall cell potential is positive.

$$E°_{cell} = E°_{substance\ reduced} - E°_{substance\ oxidized}$$

a) $E°_{cell} = 1.42\text{ V} - (0.54\text{ V}) = 0.88\text{ V}$, ∴ spontaneous
b) $E°_{cell} = -0.44\text{ V} - (0.96\text{ V}) = -1.40\text{ V}$, ∴ not spontaneous
c) $E°_{cell} = -0.74\text{ V} - (-2.76\text{ V}) = 2.02\text{ V}$, ∴ spontaneous

21.76 The given equation is separated into its two half–reactions:

$MnO_4^-(aq) + 8H^+(aq) + 5e^- \rightarrow Mn^{2+}(aq) + 4H_2O$ reduction
$5Fe^{2+}(aq) \rightarrow 5Fe^{3+}(aq) + 5e^-$ oxidation

$E°_{cell} = E°_{reduction} - E°_{oxidation} = 1.51\text{ V} - 0.77\text{ V} = 0.74\text{ V}$

21.78 a) $Zn(s) | Zn^{2+}(aq) || Co^{2+}(aq) | Co(s)$
$E°_{cell} = -0.28\text{ V} - (-0.76\text{ V}) = 0.48\text{ V}$

b) $Mg(s) | Mg^{2+}(aq) || Ni^{2+}(aq) | Ni(s)$
$E°_{cell} = -0.25\text{ V} - (-2.37\text{ V}) = 2.12\text{ V}$

c) $Sn(s) | Sn^{2+}(aq) || Au^{3+}(aq) | Au(s)$
$E°_{cell} = 1.42\text{ V} - (-0.14\text{ V}) = 1.56\text{ V}$

21.80 The half–cell with the more positive $E°_{cell}$ will appear as a reduction, and the other half–reaction is reversed, to appear as an oxidation:

$BrO_3^-(aq) + 6H^+(aq) + 6e^- \rightarrow Br^-(aq) + 3H_2O$ reduction
$3 \times (2I^-(aq) \rightarrow I_2(s) + 2e^-)$ oxidation
$BrO_3^-(aq) + 6I^-(aq) + 6H^+(aq) \rightarrow 3I_2(s) + Br^-(aq) + 3H_2O$ net reaction

$E°_{cell} = E°_{substance\ reduced} - E°_{substance\ oxidized}$ or
$E°_{cell} = E°_{reduction} - E°_{oxidation} = 1.44\text{ V} - (0.54\text{ V}) = 0.90\text{ V}$

21.82 The half-reaction having the more positive standard reduction potential is the one that occurs as a reduction. The other occurs as an oxidation:

$2 \times (2HOCl(aq) + 2H^+(aq) + 2e^- \rightarrow Cl_2(g) + 2H_2O)$ reduction
$3H_2O + S_2O_3^{2-}(aq) \rightarrow 2H_2SO_3(aq) + 2H^+(aq) + 4e^-$ oxidation
$4HOCl(aq) + 4H^+(aq) + 3H_2O + S_2O_3^{2-}(aq) \rightarrow$
$2Cl_2(g) + 4H_2O + 2H_2SO_3(aq) + 2H^+(aq)$ overall

This simplifies to give the following net reaction:
$4HOCl(aq) + 2H^+(aq) + S_2O_3^{2-}(aq) \rightarrow 2Cl_2(g) + H_2O + 2H_2SO_3(aq)$

21.84 The two half–reactions are:

$SO_4^{2-}(aq) + 2e^- + 4H^+(aq) \rightarrow H_2SO_3(aq) + H_2O$ reduction
$2I^-(aq) \rightarrow I_2(s) + 2e^-$ oxidation

$E°_{cell} = E°_{reduction} - E°_{oxidation} = 0.17\text{ V} - (0.54\text{ V}) = -0.37\text{ V}$

Since the overall cell potential is negative, we conclude that the reaction is not spontaneous in the direction written.

21.86 First, separate the overall reaction into its two half–reactions:

$2Br^{-}(aq) \rightarrow Br_2(aq) + 2e^{-}$ oxidation

$I_2(s) + 2e^{-} \rightarrow 2I^{-}(aq)$ reduction

$E°_{cell} = E°_{reduction} - E°_{oxidation} = 0.54\text{ V} - (1.07\text{ V}) = -0.53\text{ V}$

The value of n is 2:

$\Delta G° = -nFE°_{cell} = -(2)(96{,}500\text{ C})(-0.53\text{ J/C}) = 1.0 \times 10^5\text{ J} = 1.0 \times 10^2\text{ kJ}$

21.88 a) $E°_{cell} = E°_{reduction} - E°_{oxidation} = 2.01\text{ V} - (1.47\text{ V}) = 0.54\text{ V}$

b) Since n = 10, $\Delta G° = -nFE°_{cell} = -(10)(96{,}500\text{ C})(0.54\text{ J/C}) = -5.2 \times 10^5\text{ J}$

$\Delta G° = -5.2 \times 10^2\text{ kJ}$

c)

$$E°_{cell} = \frac{RT}{nF}\ln K_c$$

$$0.54\text{ V} = \frac{(8.314\text{ J mol}^{-1}\text{K}^{-1})(298\text{ K})}{10(96{,}500\text{ C mol}^{-1})}\ln K_c$$

$$\ln K_c = 210.3$$

Taking the antilog (e^x) of both sides of this equation:

$K_c = 2.1 \times 10^{91}$

21.90 Sn is oxidized by two electrons and Ag is reduced by two electrons:

$$E°_{cell} = \frac{RT}{nF}\ln K_c$$

$$-0.015\text{ V} = \frac{(8.314\text{ J mol}^{-1}\text{K}^{-1})(298\text{ K})}{2(96{,}500\text{ C mol}^{-1})}\ln K_c$$

$$\ln K_c = -1.168$$

$K_c = e^{(-1.168)} = 0.31$

21.92 This reaction involves the oxidation of Ag by two electrons and the reduction of Ni by two electrons. The concentration of the hydrogen ion is derived from the pH of the solution: $[H^+]$ = antilog (–pH) = antilog (–5) = 1 x 10^{-5} M.

$$E_{cell} = E^\circ_{cell} - \frac{RT}{nF}\ln Q$$

$$E_{cell} = 2.48 - \frac{(8.314\ \text{J mol}^{-1}\text{K}^{-1})(298\ \text{K})}{2(96{,}500\ \text{C mol}^{-1})}\ln\left(\frac{[Ag^+]^2[Ni^{2+}]}{[H^+]^4}\right)$$

$$E_{cell} = 2.48 - 0.0128\ln\left(\frac{[1.0\times10^{-2}]^2[1.0\times10^{-2}]}{[1.0\times10^{-5}]^4}\right)$$

$$E_{cell} = 2.48 - 0.0128\ln(1.0\times10^{14})$$

$$E_{cell} = 2.48 - 0.413$$

$$E_{cell} = 2.07\ \text{V}$$

21.94

$$E_{cell} = E^\circ_{cell} - \frac{RT}{nF}\ln\frac{[Mg^{2+}]}{[Cd^{2+}]}$$

$$E_{cell} = 1.97 - \frac{(8.314\ \text{J mol}^{-1}\text{K}^{-1})(298\ \text{K})}{2(96{,}500\ \text{C mol}^{-1})}\ln\frac{[1.00]}{[Cd^{2+}]}$$

$$1.54\ \text{V} = 1.97\ \text{V} - 0.01284\ln\frac{1}{[Cd^{2+}]}$$

$$\ln(1/[Cd^{2+}]) = 33.489$$

Taking (e^x) of both sides:

$1/[Cd^{2+}] = 3.50 \times 10^{14}$

$[Cd^{2+}] = 2.86 \times 10^{-15}$ M

*21.96 In the iron half–cell, we are initially given:

$$0.0500\ \text{mL} \times 0.100\ \text{mol/L} = 5.00 \times 10^{-3}\ \text{mol } Fe^{2+}(aq)$$

The precipitation of $Fe(OH)_2(s)$ consumes some of the added hydroxide ion, as well as some of the iron ion: $Fe^{2+}(aq) + 2OH^-(aq) \rightarrow Fe(OH)_2(s)$. The number of moles of OH^- that have been added to the iron half–cell is:

$$0.500\ \text{mol/L} \times 0.0500\ \text{L} = 2.50 \times 10^{-2}\ \text{mol } OH^-$$

The stoichiometry of the precipitation reaction requires that the following number of moles of OH^- be consumed on precipitation of 5.00 x 10^{-3} mol of $Fe(OH)_2(s)$:

$$5.00 \times 10^{-3}\ \text{mol } Fe(OH)_2 \times 2\ \text{mol } OH^-/\text{mol } Fe(OH)_2 = 1.00 \times 10^{-2}\ \text{mol } OH^-$$

The number of moles of OH^- that are unprecipitated in the iron half–cell is:

$$2.50 \times 10^{-2}\ \text{mol} - 1.00 \times 10^{-2}\ \text{mol} = 1.50 \times 10^{-2}\ \text{mol } OH^-$$

Since the resulting volume is 50.0 mL + 50.0 mL, the concentration of hydroxide ion in the iron half–cell becomes, upon precipitation of the $Fe(OH)_2$:

$$[OH^-] = 1.50 \times 10^{-2}\ \text{mol}/0.100\ \text{L} = 0.150\ \text{M } OH^-$$

We have assumed that the iron hydroxide that forms in the above precipitation reaction is completely insoluble. This is not accurate, though, because some small amount does dissolve in water according to the following equilibrium:

$$Fe(OH)_2(s) \rightarrow Fe^{2+}(aq) + 2OH^-(aq)$$

This means that the true $[OH^-]$ is slightly higher than 0.150 M as calculated above. Thus we must set up the usual equilibrium table, in order to analyze the extent to which $Fe(OH)_2(s)$ dissolves in 0.150 M OH^- solution:

	$[Fe^{2+}]$	$[OH^-]$
I	–	0.150
C	+x	+2x
E	+x	0.150+2x

The quantity x in the above table is the molar solubility of $Fe(OH)_2$ in the solution that is formed in the iron half–cell.

$$K_{sp} = [Fe^{2+}][OH^-]^2 = (x)(0.150 + 2x)^2$$

The standard cell potential is:

$$E°_{cell} = E°_{reduction} - E°_{oxidation} = 0.3419 - (-0.447) = 0.7889$$

The Nernst equation is:

$$E_{cell} = E^{\circ}_{cell} - \frac{RT}{nF}\ln\frac{[Fe^{2+}]}{[Cu^{2+}]}$$

$$1.175 = 0.7889 - \frac{(8.314\ \text{J mol}^{-1}\text{K}^{-1})(298\ \text{K})}{2(96{,}500\ \text{C mol}^{-1})}\ln\frac{[Fe^{2+}]}{[1.00]}$$

$$1.175 = 0.7889 - 0.01284\ln[Fe^{2+}]$$

$$\ln[Fe^{2+}] = -30.07$$

$$[Fe^{2+}] = 8.72 \times 10^{-14}\ \text{M}$$

This is the concentration of Fe^{2+} in the saturated solution, and it is the value to be used for x in the above expression for K_{sp}.

$$K_{sp} = (x)(0.150 + 2x)^2 = (8.72 \times 10^{-14})[0.150 + (2)(8.72 \times 10^{-14})]^2$$

$$K_{sp} = 1.96 \times 10^{-15}$$

21.98 1 C = 1 A·s

a) 4.00 A × 600 s = 2.40×10^3 C

b) 10.0 A × 20.0 min × 60 s/min = 1.20×10^4 C

c) 1.50 A × 6.00 hr × 3600 s/hr = 3.24×10^4 C

21.100 a) $Fe^{2+}(aq) + 2e^- \rightarrow Fe(s)$ 0.20 mol Fe^{2+} × 2 mol e^-/mol Fe^{2+} = 0.40 mol e^-

b) $Cl^-(aq) \rightarrow ½ Cl_2(g) + e^-$ 0.70 mol Cl^- × 1 mol e^-/mol Cl^- = 0.70 mol e^-

c) $Cr^{3+}(aq) + 3e^- \rightarrow Cr(s)$ 1.50 mol Cr^{3+} × 3 mol e^-/mol Cr^{3+} = 4.50 mol e^-

d) $Mn^{2+}(aq) + 4H_2O \rightarrow MnO_4^-(aq) + 8H^+(aq) + 5e^-$

1.0×10^{-2} mol Mn^{2+} × 5 mol e^-/mol Mn^{2+} = 5.0×10^{-2} mol e^-

21.102 $Ag^+(aq) + e^- \rightarrow Ag(s)$, and $Cr^{3+}(aq) + 3e^- \rightarrow Cr(s)$

This shows that there are three moles of electrons per mole of Cr but only one mole of electrons per mole of Ag. The number of moles of electrons involved in the silver reaction is:

$$\#\,\text{mol e}^- = (12.0\text{ g Ag})\left(\frac{1\text{ mol Ag}}{107.9\text{ g Ag}}\right)\left(\frac{1\text{ mol e}^-}{1\text{ mol Ag}}\right) = 0.111\text{ mol e}^-$$

The amount of Cr is then:

$$\#\,\text{mol Cr}^{3+} = (0.111\text{ mol e}^-)\left(\frac{1\text{ mol Cr}^{3+}}{3\text{ mol e}^-}\right) = 0.0371\text{ mol Cr}^{3+}$$

21.104 $Fe(s) + 2OH^-(aq) \rightarrow Fe(OH)_2(s) + 2e^-$

The number of Coulombs is: 12.0 min × 60 s/min × 8.00 C/s = 5.76×10^3 C. The number of grams of $Fe(OH)_2$ is:

$$\#\,\text{g Fe(OH)}_2 = (5.76\times10^3\text{ C})\left(\frac{1\text{ mol e}^-}{96500\text{ C}}\right)\left(\frac{1\text{ mol Fe(OH)}_2}{2\text{ mol e}^-}\right)\left(\frac{89.86\text{ g Fe(OH)}_2}{1\text{ mol Fe(OH)}_2}\right)$$

$$= 2.68\text{ g Fe(OH)}_2$$

21.106 $Cr^{3+}(aq) + 3e^- \rightarrow Cr(s)$

The number of Coulombs that will be required is:

$$\#\,\text{C} = (75.0\text{ g Cr})\left(\frac{1\text{ mol Cr}}{52.00\text{ g Cr}}\right)\left(\frac{3\text{ mol e}^-}{1\text{ mol Cr}}\right)\left(\frac{96500\text{ C}}{1\text{ mol e}^-}\right) = 4.18\times10^5\text{ C}$$

The time that will be required is:

$$\#\,\text{hr} = (4.18\times10^5\text{ C})\left(\frac{1\text{ s}}{2.25\text{ C}}\right)\left(\frac{1\text{ hr}}{3600\text{ s}}\right) = 51.5\text{ hr}$$

21.108 $Mg^{2+}(aq) + 2e^- \rightarrow Mg(\ell)$

The number of Coulombs that will be required is:

$$\#C = (60.0\text{ g Mg})\left(\frac{1\text{ mol Mg}}{24.31\text{ g Mg}}\right)\left(\frac{2\text{ mol e}^-}{1\text{ mol Mg}}\right)\left(\frac{96500\text{ C}}{1\text{ mol e}^-}\right) = 4.76\times10^5\text{ C}$$

The number of amperes is: 4.76×10^5 C ÷ 7200 s = 66.2 A

21.110 First, let's look at what will happen in this electrolytic cell. Candidates for reduction are:

$Na^+(aq) + e^- \rightarrow Na(s)$ $E° = -2.71$ V

$2H_2O + 2e^- \rightarrow H_2(g) + 2OH^-(aq)$ $E° = -0.83$ V

Since the second reaction has the greater standard reduction potential (higher in the table), it will occur instead of the first reaction. This electrolytic cell receives 2.00 A of current for 20 minutes (1200 s), so the amount of OH^- produced will be:

$$\text{mol OH}^- = (1200\text{ s})\left(\frac{2.00\text{ C}}{1\text{ s}}\right)\left(\frac{1\text{ F}}{96{,}500\text{ C}}\right)\left(\frac{2\text{ mol OH}^-}{2\text{ F}}\right) = 0.0249\text{ mol OH}^-$$

According to the balanced equation below, 1 mol HCl is needed to neutralize 1 mol OH^-. This, along with the given molarity of HCl, allows us to calculate the HCl needed to titrate the solution:

$$HCl + OH^- \rightarrow H_2O + Cl^-$$

$$\text{ml HCl} = (0.0249\text{ mol OH}^-)\left(\frac{1\text{ mol HCl}}{1\text{ mol OH}^-}\right)\left(\frac{1\text{ L acid solution}}{0.620\text{ mol HCl}}\right) = 0.0402\text{ L HCl}$$

or, 40.2 mL HCl.

21.112 The electrolysis of NaCl solution results in the reduction of water, together with the formation of hydroxide ion: $2H_2O + 2e^- \rightarrow H_2(g) + 2OH^-(aq)$. The number of Coulombs is: 2.50 A × 15.0 min × 60 s/min = 2.25×10^3 C. The number of moles of OH^- is:

$$\#\text{mol OH}^- = (2.25\times10^3\text{ C})\left(\frac{1\text{ mol e}^-}{96500\text{ C}}\right)\left(\frac{2\text{ mol OH}^-}{2\text{ mol e}^-}\right) = 0.0233\text{ mol OH}^-$$

The volume of acid solution that will neutralize this much OH^- is:

$$\#\text{mL HCl} = (0.0233\text{ mol OH}^-)\left(\frac{1\text{ mol HCl}}{1\text{ mol OH}^-}\right)\left(\frac{1000\text{ mL HCl}}{0.100\text{ mol HCl}}\right) = 233\text{ mL HCl}$$

21.114 Possible cathode reactions:

$Al^{3+} + 3e^- \rightleftharpoons Al(s)$ E° = –1.66 V
$2H_2O + 2e^- \rightleftharpoons H_2(g) + 2OH^-(aq)$ E° = –0.83 V

Cathode reaction: $2H_2O + 2e^- \rightleftharpoons H_2(g) + 2OH^-(aq)$

Possible anode reactions:

$2H_2O \rightleftharpoons O_2(g) + 4H^+(aq) + 4e^-$ E° = –1.23 V
$2SO_4^{2-} \rightleftharpoons S_2O_8^{2-} + 2e^-$ E° = –2.01 V

Anode reaction: $2H_2O \rightleftharpoons O_2(g) + 4H^+(aq) + 4e^-$

Overall reaction: $2H_2O \rightleftharpoons 2H_2(g) + O_2(g)$

21.116 a) Possible anode reactions:

$2H_2O \rightleftharpoons O_2(g) + 4H^+(aq) + 4e^-$ E° = –1.23 V
$2SO_4^{2-} \rightleftharpoons S_2O_8^{2-} + 2e^-$ E° = –2.01 V

Anode reaction: $2H_2O \rightleftharpoons O_2(g) + 4H^+(aq) + 4e^-$

b) Possible anode reactions:

$2H_2O \rightleftharpoons O_2(g) + 4H^+(aq) + 4e^-$ E° = –1.23 V
$2Br^- \rightleftharpoons Br_2 + 2e^-$ E° = –1.07 V

Anode reaction: $2Br^- \rightleftharpoons Br_2 + 2e^-$

c) Possible anode reactions:

$2H_2O \rightleftharpoons O_2(g) + 4H^+(aq) + 4e^-$ E° = –1.23 V
$2Br^- \rightleftharpoons Br_2 + 2e^-$ E° = –1.07 V
$2SO_4^{2-} \rightleftharpoons S_2O_8^{2-} + 2e^-$ E° = –2.01 V

Anode reaction: $2Br^- \rightleftharpoons Br_2 + 2e^-$

21.118 Possible cathode reactions:

$K^+ + e^- \rightleftharpoons K(s)$ E° = –2.92 V
$Cu^{2+} + 2e^- \rightleftharpoons Cu(s)$ E° = +0.34 V
$2H_2O + 2e^- \rightleftharpoons H_2(g) + 2OH^-(aq)$ E° = –0.83 V

Cathode reaction: $Cu^{2+} + 2e^- \rightleftharpoons Cu(s)$

Possible anode reactions:

$2SO_4^{2-} \rightleftharpoons S_2O_8^{2-} + 2e^-$ E° = –2.01 V
$2Br^- \rightleftharpoons Br_2 + 2e^-$ E° = –1.07 V
$2H_2O \rightleftharpoons O_2(g) + 4H^+(aq) + 4e^-$ E° = –1.23 V

Anode reaction: $2Br^- \rightleftharpoons Br_2 + 2e^-$

Overall reaction: $Cu^{2+} + 2Br^- \rightleftharpoons Br_2 + Cu(s)$

Additional Exercises

*21.120 $\Delta G° = -nFE°_{cell}$, $E°_{cell} = 1.34\ V = 1.34\ J/C$, and n = 2

$\Delta G° = -(2)(96{,}500\ C)(1.34\ J/C) = -2.59 \times 10^5$ J per mol of HgO

The maximum amount of work that can be derived from this cell, on using 1.00 g of HgO, is thus:

$$\#\,J = (1.00\ g\ HgO)\left(\frac{1\ mol\ HgO}{216.6\ g\ HgO}\right)\left(\frac{2.59\times10^5\ J}{1\ mol\ HgO}\right) = 1.20\times10^3\ J$$

Now, since 1 watt = 1 J s^{-1}, then 5×10^{-4} watt = 5×10^{-4} J s^{-1}, and the time required for this process is:

$$\#\,hr = (1.20\times10^3\ J)\left(\frac{1\ s}{5\times10^{-4}\ J}\right)\left(\frac{1\ min}{60\ s}\right)\left(\frac{1\ hr}{60\ min}\right) = 6.67\times10^2\ hr$$

21.122 The cathode is positive in a galvanic cell, so we conclude that reduction of platinum ion takes place: $Pt^{2+}(aq) + 2e^- \rightarrow Pt(s)$ $\qquad E° = ?$

The anode reaction is: $2\,Ag(s) + 2Cl^-(aq) \rightarrow 2AgCl(s) + 2e^-$ $\quad E° = -0.2223$ V

The overall cell potential is calculated using the Nernst equation:

$$E_{cell} = E°_{cell} - \frac{RT}{nF}\ln\frac{1}{[Pt^{2+}][Cl^-]^2}$$

$$0.778\ V = E°_{cell} - \frac{(8.314\ J\ mol^{-1}K^{-1})(298\ K)}{2(96{,}500\ C\ mol^{-1})}\ln\frac{1}{[0.0100][0.100]^2}$$

$E°_{cell} = 0.778\ V + 0.118\ V = 0.896\ V$

$$E°_{Pt^{2+}} - 0.2223\ V = 0.896\ V$$

$$E°_{Pt^{2+}} = 0.896\ V + 0.2223\ V$$

$$= 1.118\ V$$

*21.124 Our strategy will be thus:

Step A:

Pressure H_2 (wet) → partial pressure H_2 → mol H_2 → # e^- used

Step B:
Find total charge used = (current)(time)

Step C:
The charge per electron can be arrived at by:

Charge per e^- = total charge/total $\#e^-$ used = Step A/Step B

Step A:
Pressure H_2 (wet) → partial pressure H_2 → mol H_2 → # e^- used

The total pressure of wet hydrogen is 767 torr, but some of this is provided by water vapor. Consulting the water vapor pressure table in the appendices, we find that at 27 °C, the vapor pressure of water is 26.7 torr. Therefore pressure solely due to hydrogen gas (the partial pressure of hydrogen gas) is:

$P_{H_2} = 767 - 26.7 = 740$ torr
740 torr(1 atm/760 torr) = 0.974 atm

Using the ideal gas law,

PV = nRT
(0.974 atm)(0.288 L) = n(0.0821 L·atm $mol^{-1}·K^{-1}$)(27 + 273K)
n = 0.0114 mol H_2

According to the electrolysis equation $2H^+ + 2e^- \rightleftharpoons H_2(g)$, 2 moles of electrons are required per mol of H_2 gas formed. Therefore,

electrons = 0.0114 mol H_2(2 mol e^-/1 mol H_2)(6.022×10^{23} electrons/mol)
= 1.35×10^{22} electrons

Step B:
Total charge used = (1.22 A)(1800 s) = 2200 C

Step C:
Charge per e^- = total charge/total $\#e^-$ used = 2200 C/1.35×10^{22} electrons
= 1.63×10^{-19} C per electron

(This is a fairly good estimate; the accepted value is 1.60×10^{-19} C.)

21.126 a) $E^\circ_{cell} = 1.507\text{ V} + (-1.451\text{ V}) = 0.056\text{ V}$

b) $$\Delta G^\circ = -nFE^\circ = -\left(30\text{ mol e}^-\right)\left(\frac{96500\text{ C}}{1\text{ mol e}^-}\right)\left(\frac{0.056\text{ J}}{1\text{ C}}\right) = 1.62 \times 10^5\text{ J}$$

c) $$E^\circ_{cell} = \frac{RT}{nF}\ln K_c$$

Since n = 30, we write:

$$0.056 = 8.558 \times 10^{-4} \times \ln K_c$$
$$\ln K_c = 65.436 \text{ and } K_c = 2.6 \times 10^{28}$$

d)

$$E_{cell} = E^{\circ}_{cell} - \frac{RT}{nF}\ln\left(\frac{[Mn^{2+}]^6[ClO_3^-]^5}{[MnO_4^-]^6[Cl^-]^5[H^+]^{18}}\right)$$

$$E_{cell} = 0.056\text{ V} - \frac{(8.314\text{ J mol}^{-1}\text{K}^{-1})(298\text{ K})}{30(96{,}500\text{ C mol}^{-1})}\ln\frac{[Mn^{2+}]^6[ClO_3^-]^5}{[MnO_4^-]^6[Cl^-]^5[H^+]^{18}}$$

$$E_{cell} = 0.056\text{ V} - \left[8.558\times10^{-4}\ln\frac{[Mn^{2+}]^6[ClO_3^-]^5}{[MnO_4^-]^6[Cl^-]^5[H^+]^{18}}\right]$$

e) $$E_{cell} = 0.056\text{ V} - \left[8.558\text{ x }10^{-4}\ln\frac{[0.050]^6[0.110]^5}{[0.20]^6[0.0030]^5[5.62\times10^{-5}]^{18}}\right]$$

$$= -0.103\text{ V}$$

*21.128 The approach is as follows:

area, thickness Cr → volume Cr → mass Cr → moles Cr → $\#e^-$ needed → current

V_{Cr} = (area)(thickness) = $(1.00\text{ m}^2)(5.0 \times 10^{-5}\text{ m}) = 5.0 \times 10^{-5}\text{ m}^3$

$$\#e^- = (5.0\times10^{-5}\text{ m}^3)\left(\frac{100\text{ cm}}{1\text{ m}}\right)^3\left(\frac{7.19\text{ g Cr}}{1\text{ cm}^3}\right)\left(\frac{1\text{ mol Cr}}{52.0\text{ g Cr}}\right)\left(\frac{6\text{ F}}{1\text{ mol Cr}}\right)\left(\frac{96{,}500\text{ C}}{1\text{ F}}\right)$$
$$= 4.00 \times 10^6\text{ C}$$

This is done in 4.50 hr (16,200 s). So the current must be:

Current = charge/time = 4.00×10^6 C/16,200 s = 247 A

*21.130 The balanced half-reactions are as follows:

$$4H_2O + Mn^{2+} \rightarrow MnO_4^- + 8H^+ + 5e^-$$

$$6e^- + 14H^+ + Cr_2O_7^{2-} \rightarrow 2Cr^{3+} + 7H_2O$$

Multiplying the top equation by 6 and the bottom by 5, and combining, we obtain:

$$24H_2O + 6Mn^{2+} \rightarrow 6MnO_4^- + 48H^+ + 30e^-$$

$$30e^- + 70H^+ + 5Cr_2O_7^{2-} \rightarrow 10Cr^{3+} + 35H_2O$$

Combining the two gives:

$$30e^- + 22H^+ + 5Cr_2O_7^{2-} + 6Mn^{2+} \rightarrow 6MnO_4^- + 10Cr^{3+} + 11H_2O + 30e^-$$

This gives n = 30.

Under standard (1.0 M) conditions,

$$Mn^{2+} \rightarrow MnO_4^- \qquad E° = -1.49\ V$$
$$2Cr^{3+} \rightarrow Cr_2O_7^{2-} \qquad E° = -1.33\ V$$

However in this case, the second reaction is reversed, giving:

$$Cr_2O_7^{2-} \rightarrow 2Cr^{3+} \qquad E° = +1.33\ V$$

Therefore $E°_{cell}$ = 1.33 V + (–1.49 V) = –0.16 V.

Based on concentrations given, and the balanced chemical equation above,

$$E_{cell} = E^{\circ}_{cell} - \frac{RT}{nF}\ln\left(\frac{[MnO_4^-]^6[Cr^{3+}]^{10}}{[Mn^{2+}]^5[Cr_2O_7^{2-}]^6[H^+]^{22}}\right)$$

$$E_{cell} = -0.16\ V - \frac{(8.314\ J\ mol^{-1}K^{-1})(298\ K)}{30(96{,}500\ C\ mol^{-1})}\ln\left(\frac{[0.0010]^6[0.0010]^{10}}{[0.10]^5[0.010]^6[10^{-6}]^{22}}\right)$$

$$E_{cell} = -0.16\ V - \frac{(8.314\ J\ mol^{-1}K^{-1})(298\ K)}{30(96{,}500\ C\ mol^{-1})}\ln\left(\frac{[10^{-3}]^6[10^{-3}]^{10}}{[10^{-1}]^5[10^{-2}]^6[10^{-6}]^{22}}\right)$$

$$E_{cell} = -0.16\ V - \left[8.558\times10^{-4}\ \ln\left(\frac{[10^{-18}][10^{-30}]}{[10^{-5}][10^{-12}][10^{-132}]}\right)\right]$$

$$E_{cell} = -0.16\ V - \left[8.558\times10^{-4}\ \ln(10^{101})\right]$$

$$E_{cell} = -0.16\ V - \left[8.558\times10^{-4}(233)\right]$$

$$E_{cell} = -0.16\ V - [0.1994]$$

$$E_{cell} = -0.36\ V$$

(Standard temperature is assumed. H^+ concentration is obtained from the expression pH = –log $[H^+]$.)

$$\Delta G = -nFE_{cell} = -(30\ mol\ e^-)\left(\frac{96500\ C}{1\ mol\ e^-}\right)\left(\frac{-0.36\ J}{1\ C}\right) = 1.04\times10^6\ J$$

This is 1.04×10^3 kJ.

Since ΔG is *positive*, the reaction will proceed in the reverse direction.

21.132 $H_2(g) + \frac{1}{2}O_2(g) \rightarrow H_2O(g)$

The theoretical maximum amount of work is given by the change in free energy:

$\Delta G = \Delta H - T\Delta S$

We can find ΔH, ΔS, and T as follows:

$\Delta H = [\Delta H^\circ_{f\,products}] - [\Delta H^\circ_{f\,reactants}] = [-241.8\text{ kJ}] - [0 + 0] = -241.8\text{ kJ}$

$\Delta S = [\Delta S^\circ_{products}] - [\Delta S^\circ_{reactants}] = [188.7\text{ J}] - [130.6\text{ J} + (\frac{1}{2})205.0\text{ J}] = -44.4\text{ J}$
$= -0.0444\text{ kJ}$

$T = 110^\circ C + 273 = 373\text{ K}$

Inserting these values into the equation for ΔG, we obtain:

$\Delta G = (-241.8\text{ kJ}) - 373(-0.0444\text{ kJ}) = -224.8\text{ kJ}$

At 70% efficiency, this is $(-224.8)(0.70) = -157.4$ kJ
(The negative sign simply tells us that work is done by the system.)

1 kW = 1 kJ/s

$$\#\,g\,H_2/sec = \left(\frac{1\,kJ}{1\,sec}\right)\left(\frac{1\,mol\,H_2}{157.4\,kJ}\right)\left(\frac{2.016\,g\,H_2}{1\,mol\,H_2}\right) = 0.01281\,g\,H_2/sec$$

$$\#\,g\,O_2/sec = \left(\frac{1\,kJ}{1\,sec}\right)\left(\frac{0.5\,mol\,O_2}{157.4\,kJ}\right)\left(\frac{32.00\,g\,O_2}{1\,mol\,O_2}\right) = 0.1065\,g\,0_2/sec$$

*21.134

$$E_{cell} = E^\circ_{cell} - \frac{RT}{nF}\ln[Cl^-]$$

$$-0.0435\,V = 0.2223\,V - \frac{(8.314\,J\,mol^{-1}K^{-1})(298\,K)}{1(96{,}500\,C\,mol^{-1})}\ln[Cl^-]$$

$$-0.2655\,V = \frac{(8.314\,J\,mol^{-1}K^{-1})(298\,K)}{1(96{,}500\,C\,mol^{-1})}\ln[Cl^-]$$

$$-10.34 = \ln[Cl^-]$$

$$3.23\times10^{-5} = [Cl^-]$$

Chapter 22

Practice Exercises

22.1 $^{226}_{88}Ra \rightarrow ^{222}_{86}Rn + ^{4}_{2}He + ^{0}_{0}\gamma$

22.2 $^{90}_{38}Sr \rightarrow ^{90}_{39}Y + ^{0}_{-1}e$

22.3 Half life for Rn-222 = 4 days(24 h/day)(3600 s/hr) = 345,500 s
$k = \ln 2/t_{½} = (0.6931)/(345,500\ s) = 2.01 \times 10^{-6}\ s^{-1}$

Activity = 6 pCi = 6×10^{-12} Ci (3.7×10^{10} dps/1Ci)
= 0.222 Bq
= 0.222 disintegrations per second
Activity = disintegrations/sec = kN
$0.222 = (2.01 \times 10^{-6}\ s^{-1})N$
$N = 1.10 \times 10^{5}$ atoms Rn-222

22.4 We make use of the Inverse Square Law:

$$\frac{I_1}{I_2} = \frac{d_2^{\ 2}}{d_1^{\ 2}}$$

$$\frac{1.4\ \text{units}}{I_2} = \frac{(1.2\ \text{m})^2}{(10\ \text{m})^2} = 100\ \text{units (assuming 1 significant figure)}$$

Review Problems

22.43 Solve the Einstein equation for Δm:
$\Delta m = \Delta E/c^2$
$1\ kJ = 1.00 \times 10^3\ J = 1.00 \times 10^3\ kg\ m^2\ s^{-2}$
$\Delta m = 1.00 \times 10^3\ kg\ m^2\ s^{-2} \div (3.00 \times 10^8\ m/s)^2 = 1.11 \times 10^{-14}\ kg = 1.11 \times 10^{-11}\ g$

22.45 The joule is equal to one kg m^2/s^2, and this is employed directly in the Einstein equation: $\Delta m = \Delta E/c^2$, where ΔE is the enthalpy of formation of liquid water, which is available in the appendices.

$H_2(g) + ½\ O_2(g) \rightarrow H_2O(\ell)$, $\Delta H°_f = -285.9$ kJ/mol
$\Delta m = (-285.9 \times 10^3\ kg\ m^2/s^2) \div (3.00 \times 10^8\ m/s^2) = -3.18 \times 10^{-12}\ kg$
$-3.18 \times 10^{-12}\ kg \times 1000\ g/kg \times 10^9\ ng/g = -3.18\ ng$
The negative value for the mass implies that mass is lost in the reaction.

22.47 The mass of the deuterium nucleus is the mass of the proton (1.00727252 u) plus that of a neutron (1.008665 u), or 2.015938 u. The difference between this calculated value and the observed value is equal to Δm:

$\Delta m = (2.015938 - 2.0135) = 2.4 \times 10^{-3}$ u
$\Delta E = \Delta mc^2 = (2.4 \times 10^{-3}\ u)(1.6606 \times 10^{-27}\ kg/u)(3.00 \times 10^8\ m/s)^2$
$\Delta E = 3.6 \times 10^{-13}\ kg\ m^2/s^2 = 3.6 \times 10^{-13}$ J

Since there are two nucleons per deuterium nucleus, we have:
$\Delta E = 3.6 \times 10^{-13}$ J/2 nucleons = 1.8×10^{-13} J per nucleon

22.49 a) $^{211}_{83}Bi$ b) $^{177}_{72}Hf$ c) $^{216}_{84}Po$ d) $^{19}_{9}F$

22.51 a) $^{242}_{94}Pu \rightarrow {}^{4}_{2}He + {}^{238}_{92}U$
b) $^{28}_{12}Mg \rightarrow {}^{0}_{-1}e + {}^{28}_{13}Al$
c) $^{26}_{14}Si \rightarrow {}^{0}_{1}e + {}^{26}_{13}Al$
d) $^{37}_{18}Ar + {}^{0}_{-1}e \rightarrow {}^{37}_{17}Cl$

22.53 a) $^{261}_{102}No$ b) $^{211}_{82}Pb$ c) $^{141}_{61}Pm$ d) $^{179}_{74}W$

22.55 $^{87}_{36}Kr \rightarrow {}^{86}_{36}Kr + {}^{1}_{0}n$

22.57 The more likely process is positron emission, because this produces a product having a higher neutron–to–proton ratio: $^{38}_{19}K \rightarrow {}^{0}_{1}e + {}^{38}_{18}Ar$.

22.59 Six half–life periods correspond to the fraction 1/64 of the initial material. That is, one sixty–fourth of the initial material is left after 6 half lives: 3.00 mg x 1/64 = 0.0469 mg remaining.

22.61 $^{53}_{24}Cr^*$; $^{51}_{23}V + {}^{2}_{1}H \rightarrow {}^{53}_{24}Cr^* \rightarrow {}^{1}_{1}p + {}^{52}_{23}V$

22.63 $^{80}_{35}Br$

22.65 $^{55}_{26}Fe$; $^{55}_{25}Mn + {}^{1}_{1}p \rightarrow {}^{1}_{0}n + {}^{55}_{26}Fe$

22.67 $^{70}_{30}Zn + {}^{208}_{82}Pb \rightarrow {}^{278}_{112}Uub \rightarrow {}^{1}_{0}n + {}^{277}_{112}Uub$

22.69 Radiation $\alpha \dfrac{1}{d^2}$

$$\frac{I_1}{I_2} = \frac{d_2^{\,2}}{d_1^{\,2}}$$

$$d_2 = d_1\sqrt{\frac{I_1}{I_2}} = 2.0m\sqrt{\frac{2.8}{0.28}} = 6.3\ m$$

22.71 This calculation makes use of the Inverse Square Law:

$$\frac{I_1}{I_2} = \frac{d_2^2}{d_1^2}$$

$$\frac{8.4 \text{ rem}}{0.50 \text{ rem}} = \frac{d_2^2}{(1.60 \text{ m})^2}$$

$d_2 = 6.6\text{m}$

22.73 Half life for Am-241 = 1.70×10^5 days(24 h/day)(3600 s/hr) = 1.47×10^{10} s
$k = \ln 2/t_{1/2} = (0.6931)/(1.47 \times 10^{10} \text{ s}) = 4.71 \times 10^{-11} \text{ s}^{-1}$

N = 0.20×10^{-3} g Am(1 mol Am/243 g Am)(6.022×10^{23} atoms Am/1 mol)
= 5.0×10^{17} atoms Am

Activity = kN = (4.71×10^{-11} s^{-1})(5.0×10^{17} atoms Am)
= 2.4×10^7 Bq
= 2.4×10^7 Bq (1 Ci/3.7×10^{10} Bq) = 6.5×10^{-4} Ci = 6.5×10^2 μCi

22.75 N = 1.00 mg I-131(1 g/1000 mg)(1 mol/131 g)(6.022×10^{23} atoms/1 mol)
= 4.60×10^{18} atoms I-131

Activity = kN
4.6×10^{12} Bq = k(4.60×10^{18})
$k = 1.0 \times 10^{-6}$

$t_{1/2} = \ln 2/k = 0.6931/1.0 \times 10^{-6} = 6.9 \times 10^5$ s (this is about 8 days)

22.77 Recall that for a first order process $k = 0.693/t_{1/2}$. Therefore,

$k = 0.693/30 \text{ yr} = 2.30 \times 10^{-2}/\text{yr}$

$$\ln\frac{[A]_o}{[A]_t} = kt$$

$[A]_t = [A]_o \exp(-kt)$

$= \exp[-(2.30 \times 10^{-2}/\text{yr})(150 \text{ yr}) = 0.0317$

So 3.2 % of the original sample remains. The chemical product is $BaCl_2$.

22.79 This calculation makes use of the first order rate equation, where knowing $[A]_t$, we need to calculate $[A]_0$: $\ln\frac{[A]_0}{[A]_t} = kt$.

$k = 0.693/t_{1/2} = 0.693/8.07 \text{ d} = 8.59 \times 10^{-2} \text{ d}^{-1}$

$$\ln\frac{[A]_0}{(25.6\times10^{-5}\ ^{Ci}\!/\!_{g})}=(8.59\times10^{-2}\ d^{-1})(28.0\ d)$$

Taking the antiln of both sides of the above equation gives:

$$\frac{[A]_0}{(25.6\times10^{-5}\ ^{Ci}\!/\!_{g})}=e^{2.41}=11.1$$

Solving for the value of $[A]_0$ gives: $[A]_0 = 2.84 \times 10^{-3}$ Ci/g

22.81 In order to solve this problem, it must be assumed that all of the argon–40 that is found in the rock must have come from the potassium–40, i.e., that the rock contains no other source of argon–40. If the above assumption is valid, then any argon–40 that is found in the rock represents an equivalent amount of potassium–40, since the stoichiometry is 1:1. Since equal molar amounts of potassium–40 and argon–40 have been found, this indicates that the amount of potassium–40 that remains is exactly half the amount that was present originally. In other words, the potassium–40 has undergone one half–life of decay by the time of the analysis. The rock is thus seen to be 1.3×10^9 years old.

22.83 Using equation 22.2 we may determine how long it has been since the tree died.

$$\frac{^{14}C}{^{12}C}=1.2\times10^{-12}\ e^{-t/8270}$$

Taking the natural log we determine:

$$\ln\left(\frac{4.8\times10^{-14}}{1.2\times10^{-12}}\right)=-t/8270$$

$$t=-8270\times\ln\left(\frac{4.8\times10^{-14}}{1.2\times10^{-12}}\right)=2.7\times10^{4}\ \text{yr}$$

According to the calculations, the tree died 27,000 years ago. This is when the volcanic eruption would have occured.

22.85 $^{235}_{92}U + ^{1}_{0}n \rightarrow ^{94}_{38}Sr + ^{140}_{54}Xe + 2\,^{1}_{0}n$

Additional Exercises

22.87 a) $^{30}_{13}Al \rightarrow ^{0}_{-1}e + ^{30}_{14}Si$

b) $^{252}_{99}Es \rightarrow ^{4}_{2}He + ^{248}_{97}Bk$

c) $^{93}_{42}Mo + ^{0}_{-1}e \rightarrow ^{93}_{41}Nb$

d) $^{28}_{15}P \rightarrow ^{0}_{1}e + ^{28}_{14}Si$

*22.89 Uranium–235 has 92 protons and 143 neutrons. The mass of one nucleus is:

m: $(92 \times 1.007276470\ u) + (143 \times 1.008664904\ u) = 236.908517\ u$

$\Delta m = (236.908517\ u - 235.0439\ u) = 1.8646\ u$

$\Delta E = \Delta mc^2 = (1.8646\ u)(1.6605 \times 10^{-27}\ kg/u)(3.00 \times 10^8\ m/s)^2$

$\Delta E = 2.787 \times 10^{-10}\ kg\ m^2/s^2 = 2.787 \times 10^{-10}\ J$

Since there are 235 neucleons per uranium–235 nucleus, we have:

$\Delta E = 2.786 \times 10^{-10}\ J/235\ nucleons = 1.186 \times 10^{-12}\ J$ per nucleon

*22.91 None. The parent mass may be exactly the same as the daughter mass since the positron has a mass of zero.

22.93 $^{214}_{83}Bi \rightarrow\ ^{4}_{2}He +\ ^{210}_{81}Tl$

$^{210}_{81}Tl \rightarrow\ ^{0}_{-1}e +\ ^{210}_{82}Pb$

$^{210}_{82}Pb \rightarrow\ ^{0}_{-1}e +\ ^{210}_{83}Bi$

$^{210}_{83}Bi \rightarrow\ ^{0}_{-1}e +\ ^{210}_{84}Po$

$^{210}_{84}Po \rightarrow\ ^{4}_{2}He +\ ^{206}_{82}Pb$

Element E is $^{206}_{82}Pb$

22.95 Positron decay

22.97 Recall that for a first order process $k = 0.693/t_{½}$.

$k = 0.693/5730\ yr = 1.21 \times 10^{-4}/yr$

$$\ln\frac{[A]_o}{[A]_t} = kt$$

$$t = \frac{1}{k}\ln\frac{[A]_o}{[A]_t} = \frac{1}{1.21\times10^{-4}\ /yr}\ln\frac{8}{1}$$

$= 17,000$ years

22.99 As time passes, equal parts of CH_3HgI and $CH_3{}^*HgI$ will be produced. Since half the atoms are labeled, half the products will contain radioactive Hg.

*22.101 This problem is similar to a dilution problem, i.e., $C_1V_1 = C_2V_2$. We will use cpm as the concentration unit. First, we need to account for the density difference.

$$\text{Methanol: concentration} = \left(\frac{580\text{ cpm}}{\text{g}}\right)\left(\frac{0.792\text{ g}}{\text{mL}}\right) = 460\frac{\text{cpm}}{\text{mL}}$$

$$\text{Coolant: concentration} = \left(\frac{29\text{ cpm}}{\text{g}}\right)\left(\frac{0.884\text{ g}}{\text{mL}}\right) = 26\frac{\text{cpm}}{\text{mL}}$$

Now we want the volume of the cooling system. Solve the following:

$$\frac{\left(\frac{459\text{ cpm}}{\text{mL}}\right)(10.0\text{ mL})}{\frac{26\text{ cpm}}{\text{mL}}}$$

$= 180\text{ mL}$

22.103

$$\text{Necessary activity} = 20\text{ g}\left(\frac{86\ \mu\text{Ci}}{1\text{ g}}\right)\left(\frac{1\text{ Ci}}{1\times10^{6}\ \mu\text{Ci}}\right)\left(\frac{3.7\times10^{10}\text{ Bq}}{1\text{ Ci}}\right)$$

$$= 6.4\times10^{7}\text{ Bq}$$

Activity = kN (k for I-131 = 1.0×10^{-6}, see Problem 22.75)

6.4×10^{7} Bq = (1.0×10^{-6})N

N = 6.4×10^{13} atoms I-131(1 mol/6.022×10^{23} atoms)(131 g/1 mol)

= 1.4×10^{-8} g

= 14 ng I-131

Chapter 23

Practice Exercises

23.1 $\Delta H = \Delta H_f° = 601.7$ kJ
$\Delta S° = S°(Mg_{(s)}) + 1/2S°(O_{2(g)}) - S°(MgO_{(s)})$
$= 32.5 J/K + ½(205 J/K) - 26.9 J/K$
$= 108.1$ J/K
$= 0.108$ kJ/K

$\Delta G_T° = \Delta H° - T\Delta S°$
$= 601.7$ kJ $- T(0.108$ kJ/K$)$

Decomposition occurs when $\Delta G<0$. Solve for $\Delta G° = 0$

$$T = \frac{601.7 \text{ kJ}}{0.108 \text{kJ/K}}$$
$= 5570$ K

23.2 The net charge on the complex ion must first be determined. Two $S_2O_3^{2-}$ ions contribute a charge of 4–; the metal contributes a charge of 1+. The sum of these is 3–. The formula of the complex ion is therefore $[Ag(S_2O_3)_2]^{3-}$. The ammonium salt of this ion would have the formula $(NH_4)_3[Ag(S_2O_3)_2]$.

23.3 The salt must include the six hydrated water molecules. We know that Al exists as a 3+ ion and that chloride has a charge of 1–. The hydrate would have the formula $AlCl_3{\cdot}6H_2O$. The complex ion most likely has the formula $[Al(H_2O)_6]^{3+}$.

23.4 a) potassium hexacyanoferrate(III)
b) dichlorobis(ethylenediamine)chromium(III) sulfate

23.5 a) $[SnCl_6]^{2-}$
b) $(NH_4)_2[Fe(CN)_4(H_2O)_2]$

23.6 a) Since there are three ligands and $C_2O_4^{2-}$ is a bidentate ligand, the coordination number is six.
b) The coordination number is six. There are two bidentate ligands and two unidentate ligands.
c) The coordination number is six. Both $C_2O_4^{2-}$ and ethylenediamine are bidentate ligands. Since there are three bidentate ligands, the coordination number must be six.
d) EDTA is a hexadentate ligand so the coordination number is six.

Review Problems

23.89 First we need $\Delta H°$ for this reaction:

$$\Delta H° = 2\Delta H_f°(Hg_{(g)}) + \Delta H_f°(O_{2(g)}) - 2\Delta H_f°(HgO_{(s)})$$
$$= 2(61.3 \text{ kJ}) + 0 - 2(-90.8 \text{ kJ}) = 304 \text{ kJ}$$

Similarly, $\Delta S° = 2(175 \text{ J/K}) + 205 \text{ J/K} - (70.3 \text{ J/K}) = 414 \text{ J/K}$

Decomposition occurs when $\Delta G = 0$

$\Delta G° = \Delta H° - T\Delta S°$

$$T = \frac{\Delta H°}{\Delta S°} = \frac{304 \text{ kJ}}{0.414 \text{ kJ/K}} = 734 \text{ K}$$

23.91 a) Bi_2O_5 b) PbS c) PbI_2

23.93 a) SnO b) SnS

23.95 a) $CaCl_2$ b) BeF_2

23.97 a) HgS b) Ag_2S

23.99 a) HgS b) Ag_2S

23.101 The net charge is −3, and the formula is $[Fe(CN)_6]^{3-}$.

23.103 $[CoCl_2(en)_2]^+$

23.105 a) $C_2O_4^{2-}$ (oxalato)
b) S^{2-} (sulfide, or thio)
c) Cl^- (chloro)
d) $(CH_3)_2NH$ (dimethylamine)

23.107 a) hexaamminenickel(II) ion
b) triamminetrichlorochromate(II) ion
c) hexanitrocobaltate(III) ion
d) diamminetetracyanomanganate(II) ion
e) trioxalatoferrate(III) ion or trisoxalatoferrate(III) ion

23.109 (a) $[Fe(CN)_2(H_2O)_4]^+$
(b) $Ni(C_2O_4)(NH_3)_4$
(c) $[Fe(CN)_4(H_2O)_2]^-$
(d) $K_3[Mn(SCN)_6]$
(e) $[CuCl_4]^{2-}$

23.111 The coordination number is six, and the oxidation number of the iron atom is +2.

23.113

The curved lines represent $-CH_2-CO-O-$ groups.

23.115 Since both are the *cis* isomer, they are identical. One can be superimposed on the other after simple rotation.

23.117

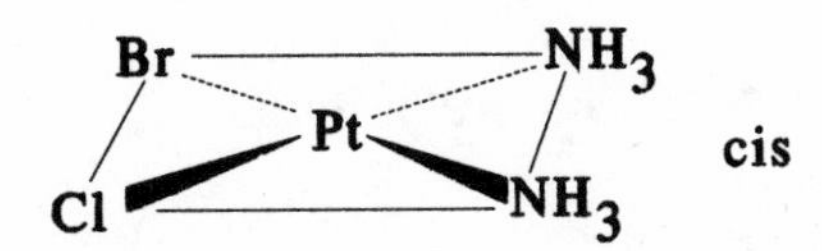

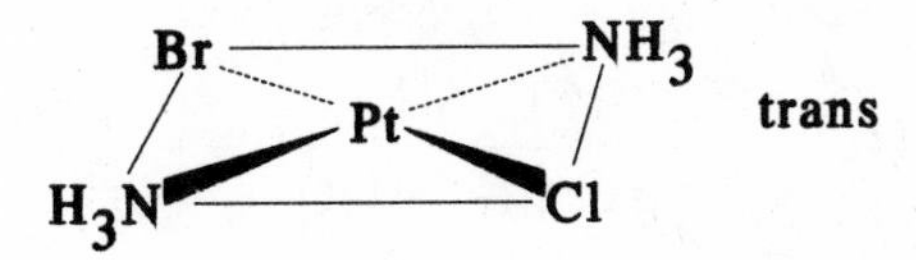

23.119

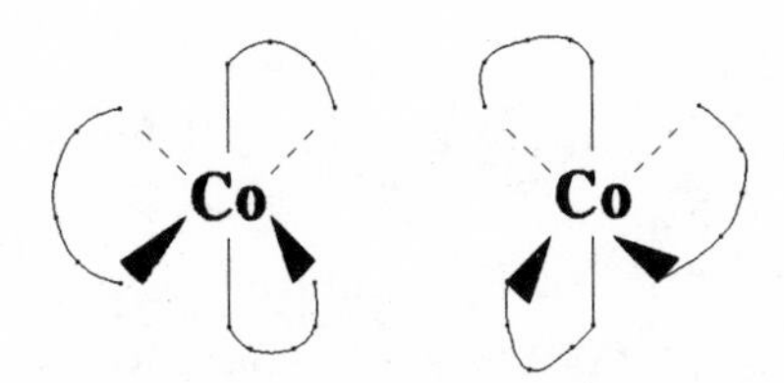

23.121 a) $Cr(H_2O)_6{}^{3+}$ b) $Cr(en)_3{}^{3+}$

23.123 $[Cr(CN)_6]^{3-}$

23.125 a) The value of Δ increases down a group. Therefore, we choose: $[RuCl(NH_3)_5]^{3+}$

b) The value of Δ increases with oxidation state of the metal. Therefore, we choose: $[Ru(NH_3)_6]^{3+}$

23.127 This is the one with the strongest field ligand, since Co^{2+} is a d^7 ion: $[CoA_6]^{3+}$

23.129 This is a weak field complex of Co^{2+}, and it should be a high–spin d^7 case. It cannot be diamagnetic; even if it were low spin, we would still have one unpaired electron.

Additional Exercises

*23.131 The value of the equilibrium constant when $\Delta G = 0$ is 1. Thus, the equilibrium concentrations (pressures) are 1 atm for both species. This makes sense since we are thermally decomposing the solid so the vapor pressure of the sample should equal the atmospheric pressure.

23.133 The compound is chiral:

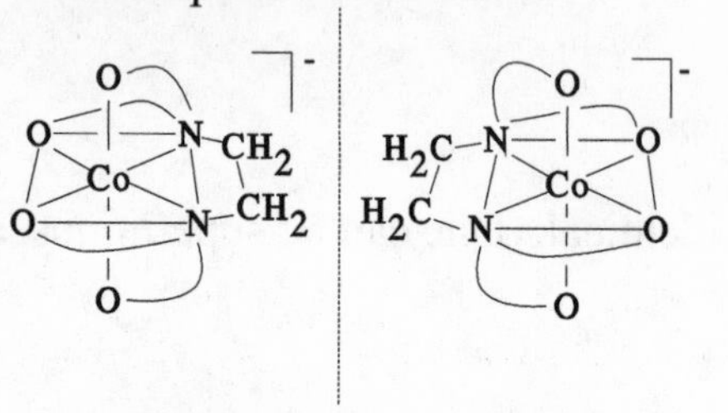

The mirror images are not superimposable.

*23.135 a) The number of moles of chloride that have been precipitated is:

$$\#\,\text{mol AgCl} = (0.538\text{ g AgCl})\left(\frac{1\text{ mol AgCl}}{143.32\text{ g AgCl}}\right) = 3.75\times10^{-3}\text{ mol AgCl}$$

The number of moles of Cr that were originally present is:

$$\#\,\text{mol Cr} = (0.500\text{ g CrCl}_3\cdot6\text{H}_2\text{O})\left(\frac{1\text{ mol CrCl}_3\cdot6\text{H}_2\text{O}}{266.4\text{ g CrCl}_3\cdot6\text{H}_2\text{O}}\right)$$

$$\times\left(\frac{1\text{ mol Cr}}{1\text{ mol CrCl}_3\cdot6\text{H}_2\text{O}}\right) = 1.88\times10^{-3}\text{ mol Cr}$$

The ratio of moles of Cl^- per mole of Cr is therefore: 3.75/1.88 = 1.99.

This means that there were 2 mol of Cl^- that were free to precipitate. The other mole of chloride ion must have been bound as a ligand to the Cr. In other words, the complex ion was $[Cr(Cl)(H_2O)_5]^{2+}$.

b) $[Cr(Cl)(H_2O)_5]Cl_2\cdot H_2O$

c)

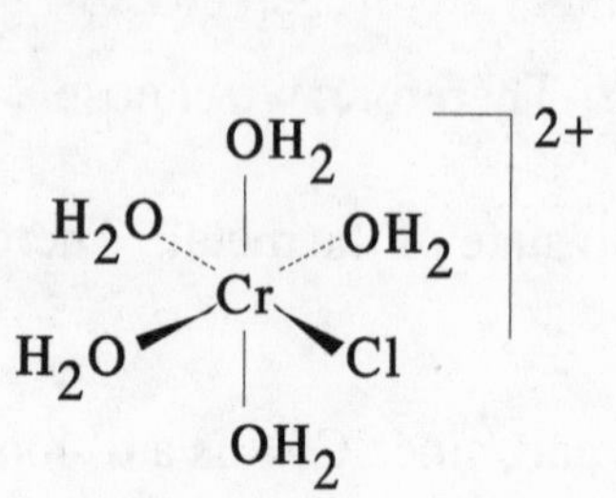

d) There is only one isomer.

Chapter 24

Review Problems

24.89 a) $AlP(s) + 3H_2O \rightarrow Al(OH)_3(s) + PH_3(g)$
b) $Mg_2C(s) + 4H_2O \rightarrow CH_4(g) + 2Mg(OH)_2(s)$
c) $FeS(s) + 2HCl(aq) \rightarrow FeCl_2(aq) + H_2S(aq)$
d) $MgSe(s) + H_2SO_4(aq) \rightarrow H_2Se(aq) + MgSO_4(aq)$

24.91 a) $2CO + O_2 \rightarrow 2CO_2$
b) $2C_2H_6 + 7O_2 \rightarrow 4CO_2 + 6H_2O$
c) $P_4O_6 + 2O_2 \rightarrow P_4O_{10}$
d) $4NH_3 + 5O_2 \rightarrow 4NO + 6H_2O$

24.93 $\Delta G = 0 + 2(0) - 2(51.9 \text{ kJ/mol}) = -103.8 \text{ kJ}$

$$K_p = \frac{(P_{N_2})(P_{O_2})^2}{(P_{NO_2})^2}$$

$$\Delta G = -RT\ln K_p$$

$$K_p = \exp\left(-\frac{\Delta G}{RT}\right) = \exp\left(\frac{103.8 \times 10^3 \text{ J/mol}}{(8.314 \text{ J/molK})(298 \text{ K})}\right) = 1.57 \times 10^{18}$$

Because K_p is so large, we expect the equilibrium to lie to the right. Since we observe that NO_2 is stable, we must conclude that the reaction kinetics are too slow to achieve equilibrium.

24.95 $\Delta H = 9.7\text{kJ} - 2(33.8 \text{ kJ}) = -57.9 \text{ kJ}$

The reaction is exothermic so heat can be considered to be a product of the reaction. If we increase the temperature, effectively adding heat (a product), Le Châtelier's Principle states that the equilibrium will shift to the left and N_2O_4 will dissociate.

24.97 In each case, describe the structure without any hydrogen atoms, i.e., describe the anion structure:
a) IO_3^- is trigonal pyramidal
b) ClO_2^- is bent
c) IO_6^- is octahedral
d) ClO_4^- is tetrahedral

24.99 a) trigonal pyramidal

:Cl—P—Cl:
:Cl:

b) T-shaped

:F—Br—F:
:F:

c) octahedral

d) planar triangular

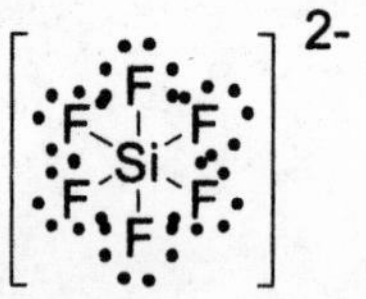

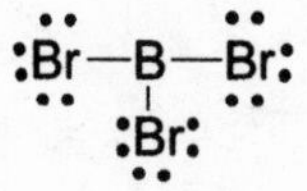

e) square planar

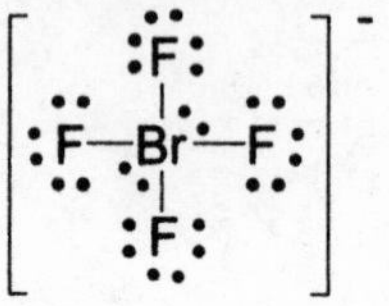

Additional Exercises

24.101 $N_2 + 3H_2 \rightarrow 2NH_3$

$4NH_3 + 5O_2 \rightarrow 4NO + 6H_2O$

$2NO + O_2 \rightarrow 2NO_2$

$3NO_2 + H_2O \rightarrow 2HNO_3 + NO$

24.103

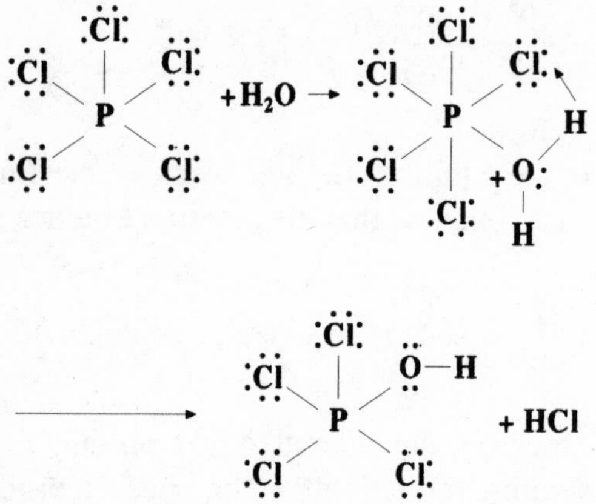

Chapter 25

Practice Exercises

25.1 a) 3–methylhexane
b) 4–ethyl–2,3–dimethylheptane
c) 5–ethyl–2,4,6–trimethyloctane

25.2 a)

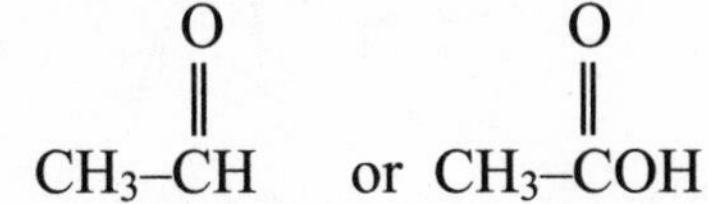

$CH_3\text{–}\overset{\overset{\displaystyle O}{\|}}{C}H$ or $CH_3\text{–}\overset{\overset{\displaystyle O}{\|}}{C}OH$

(depending on the strength of the oxidizing agent)

b) $CH_3CH_2\overset{\overset{\displaystyle O}{\|}}{C}CH_2CH_3$

Review Problems

25.82 a)

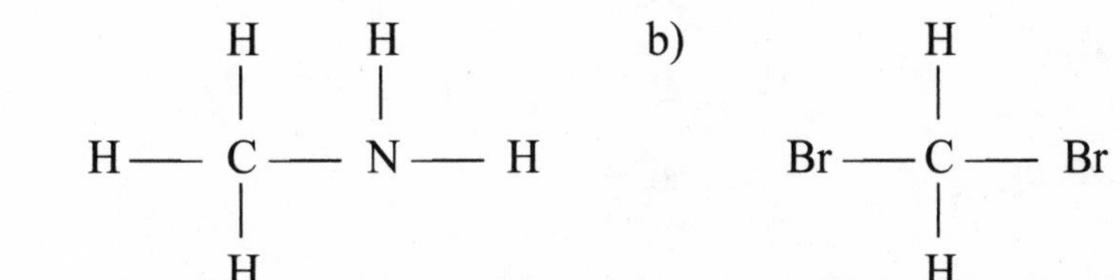

b)

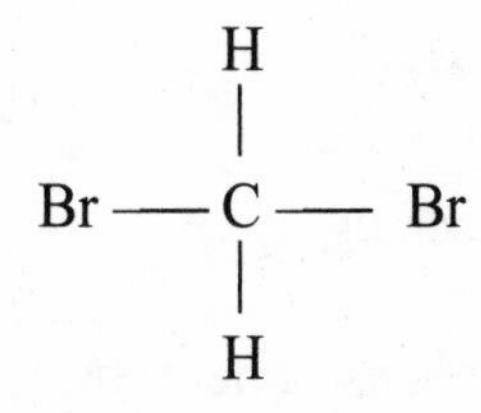

c)

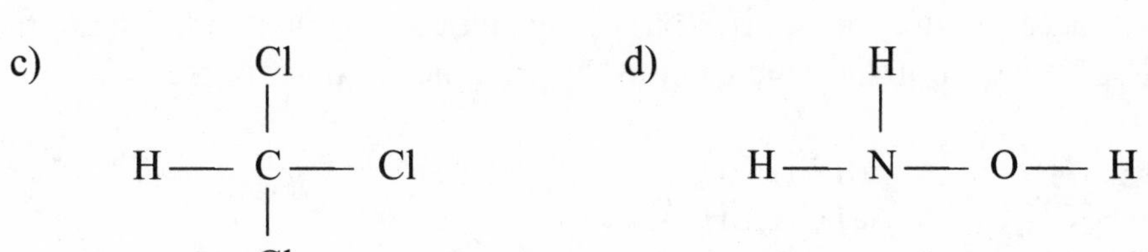

d)

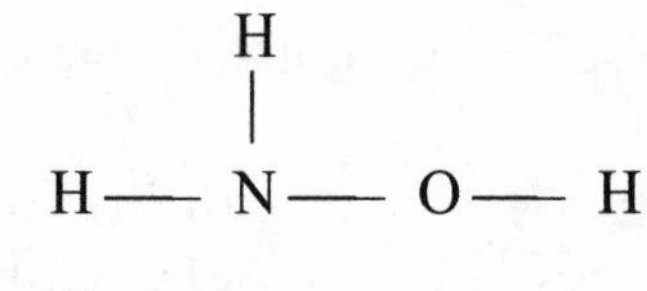

e) $H—C\equiv C—H$

f)

```
     H     H
     |     |
H—   N—    N—  H
```

25.84 a) alkene
b) alcohol
c) ester
d) carboxylic acid
e) amine
f) alcohol

25.86 The saturated compounds are b, e, and f.

25.88 a) amine b) amine c) amide d) amine, ketone

25.90 a) These are identical, being oriented differently only.
b) These are identical, being drawn differently only.

c) These are unrelated, being alcohols with different numbers of carbon atoms.
d) These are isomers, since they have the same empirical formula, but different structures.
e) These are identical, being oriented differently only.
f) These are identical, being drawn differently only.
g) These are isomers, since they have the same empirical formula, but different structures.

25.92 a) pentane
b) 2–methylpentane
c) 2,4–dimethylhexane

25.94 a) no isomers
b)

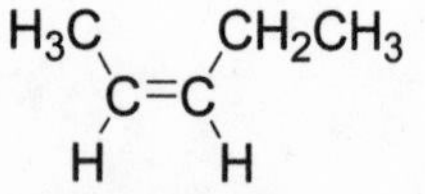

H_3C CH_2CH_3
C=C
H H

cis

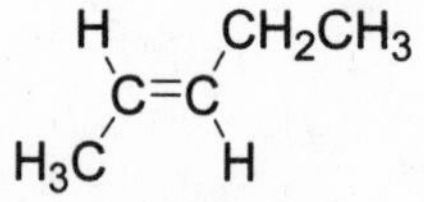

H CH_2CH_3
C=C
H_3C H

trans

c)

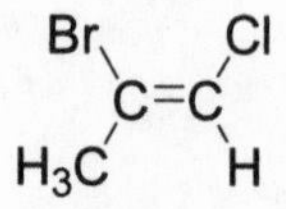

Br Cl
C=C
H_3C H

"Z" isomer

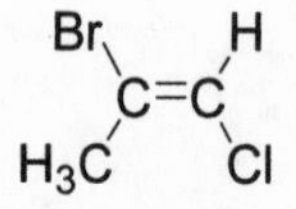

Br H
C=C
H_3C Cl

"E" isomer

Two isomers exist here, but in this case, the terms "cis" and "trans" are not useful because there are 4 different groups on the double bond. The terms "E" and "Z" are used in Organic Chemistry…a topic for that course.

25.96 a) CH_3CH_3
b) $ClCH_2CH_2Cl$
c) $BrCH_2CH_2Br$
d) CH_3CH_2Cl
e) CH_3CH_2Br
f) CH_3CH_2OH

25.98 a) $CH_3CH_2CH_2CH_3$
b) $CH_3\,CH—CHCH_3$ (Cl on each CH: Cl Cl)
c) $CH_3\,CH—CHCH_3$ (Br Br)
d) $CH_3\,CH—CHCH_3$ (H Cl)
e) $CH_3\,CH—CHCH_3$ (H Br)

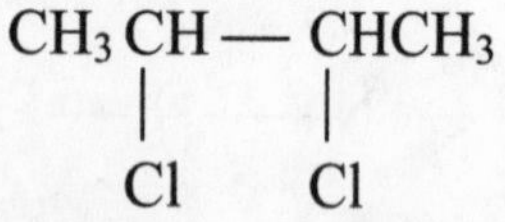

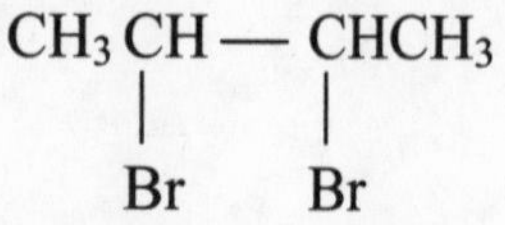

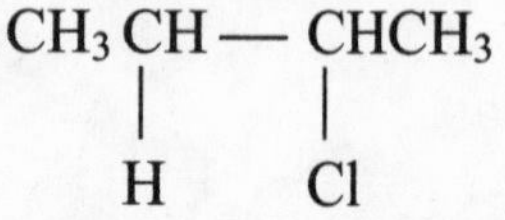

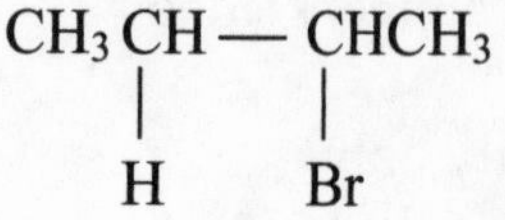

f) $CH_3CH—CHCH_3$ (with H on the first CH and OH on the second CH)

$$\begin{array}{ll} CH_3\,CH & —\ CHCH_3 \\ \ \ \ \ \ \ \ | & \ \ \ \ | \\ \ \ \ \ \ \ \ H & \ \ \ OH \end{array}$$

25.100 This sort of reaction would disrupt the π-delocalization of the benzene ring. The subsequent loss of resonance energy would not be favorable.

25.102
CH_3OH IUPAC name = methanol; common name = methyl alcohol
CH_3CH_2OH IUPAC name = ethanol; common name = ethyl alcohol
$CH_3CH_2CH_2OH$ IUPAC name = 1–propanol; common name = propyl alcohol
$CH_3\underset{\displaystyle OH}{\underset{|}{C}}HCH_3$ IUPAC name = 2–propanol; common name = isopropyl alcohol

25.104
$CH_3CH_2CH_2–O–CH_3$ methyl propyl ether
$CH_3CH_2–O–CH_2CH_3$ diethyl ether
$(CH_3)_2CH–O–CH_3$ methyl 2–propyl ether

*25.106

a) (cyclopentene structure)

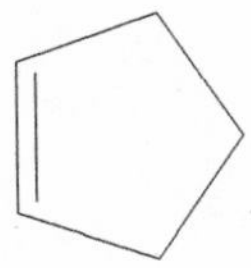

b) (benzene ring)– CH=CH2

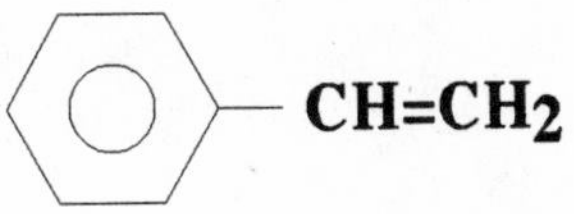

c) (benzene ring)– CH=CH2

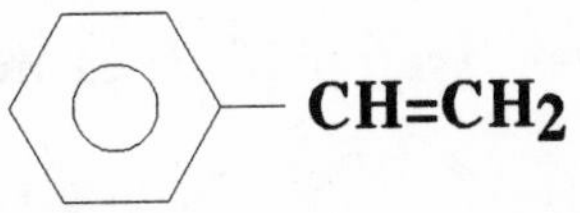

25.108

a) (cyclopentanone structure: O double-bonded to cyclopentane ring)

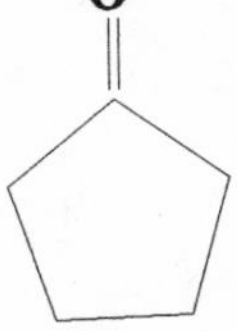

b) (benzene ring)– C(=O)-CH3

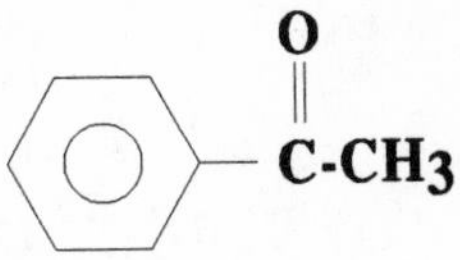

c) (benzene ring)-CH2-C(=O)-H

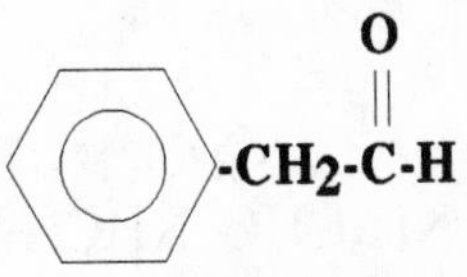

*25.110 The elimination of water can result in a C=C double bond in two locations:

$CH_2{=}CHCH_2CH_3$ — 1–butene

$CH_3CH{=}CHCH_3$ — 2–butene

25.112 The aldehyde is more easily oxidized. The product is:

$$CH_3CH_2\overset{O}{\overset{\|}{C}}OH$$

25.114 a) $CH_3CH_2CO_2H$
b) $CH_3CH_2CO_2H\ +\ CH_3OH$
c) $Na^+ + CH_3CH_2CH_2CO_2^- \ +\ H_2O$

*25.116 $CH_3CO_2H\ +\ CH_3CH_2NHCH_2CH_3$

25.118

$$\begin{array}{l} CH_2\text{–}O\text{–}\overset{O}{\overset{\|}{C}}\text{–}(CH_2)_7CH{=}CH(CH_2)_7CH_3 \\ | \\ CH\text{–}O\text{–}\overset{O}{\overset{\|}{C}}\text{–}(CH_2)_7CH{=}CHCH_2CH{=}CH(CH_2)_4CH_3 \\ | \\ CH_2\text{–}O\text{–}\overset{O}{\overset{\|}{C}}\text{–}(CH_2)_{14}CH_3 \end{array}$$

25.120

$$\begin{array}{l} CH_2\text{–}O\text{–}\overset{O}{\overset{\|}{C}}\text{–}(CH_2)_{16}CH_3 \\ | \\ CH_2\text{–}O\text{–}\overset{O}{\overset{\|}{C}}\text{–}(CH_2)_{12}CH_3 \\ | \\ CH_2\text{–}O\text{–}\overset{O}{\overset{\|}{C}}\text{–}(CH_2)_{16}CH_3 \end{array}$$

25.122 Hydrophobic sites are composed of fatty acid units. Hydrophilic sites are composed of charged units.

25.124

$$^+NH_3CH_2\overset{O}{\overset{\|}{C}}NHCH_2\overset{O}{\overset{\|}{C}}O^-$$

25.126

$$^+NH_3\underset{\underset{\displaystyle CH_2C_6H_5}{|}}{CH}\overset{O}{\overset{\|}{C}}\text{–}NHCH_2\overset{O}{\overset{\|}{C}}O^- \qquad\qquad ^+NH_3CH_2\overset{O}{\overset{\|}{C}}\text{–}NH\text{–}\underset{\underset{\displaystyle CH_2C_6H_5}{|}}{CH}\overset{O}{\overset{\|}{C}}O^-$$

Additional Exercises

*25.128 a) $(CH_3)_2CHCH_2OH$ and $(CH_3)_3COH$

b) $(CH_3)_3COH$

c)

```
        CH3                         CH3
        |                           |
  CH3— C— CH3               CH3— CH— CH2
        +                                +
```

d) The first one is more stable, since it is the one that leads to the observed product. It is a tertiary carbocation.

*25.130 The dimethylamine has a higher boiling point, in spite of the lower formula mass due to its ability to form hydrogen bonds between the nitrogen of one molecule and the hydrogen of another.

*25.132 There are six possibilities, GAP, GPA, AGP, APG, PGA, and PAG.

*25.134 The original number of moles of hydroxide are:

$0.1016\ M \times 0.05000\ L = 5.080 \times 10^{-3}\ mol\ OH^-$

The moles of hydroxide not neutralized are:

$0.1182\ M \times 0.02378\ L = 0.002811\ mol\ OH^-$

Therefore, the moles of hydroxide that were neutralized by the acid were:

$5.080 \times 10^{-3}\ mol - 2.811 \times 10^{-3}\ mol = 2.269 \times 10^{-3}\ mol\ OH^-$

This is also equal to the number of moles of the organic acid.

The formula mass is therefore:

$0.2081\ g/2.269 \times 10^{-3}\ mol = 91.71\ g/mol$

This is equal to the molecular mass *only if the unknown is a monoprotic acid.*

NOTES

NOTES

NOTES

NOTES

NOTES

NOTES

NOTES

NOTES

NOTES

NOTES

NOTES